Frontiers of X-Ray Astronomy

X-ray astronomy has undergone a revolution in recent years. With the launch of two new orbiting observatories, Chandra and XMM-Newton, astronomers are now able to obtain spectra and images at a higher resolution than ever before. New observations have had a major impact on topics ranging from protostars to cosmology. The contributions in this work, by leading authorities in the field, originate from a Royal Society Discussion Meeting that was held to review the most recent results from the current generation of X-ray telescopes, and set them in context. This book is a valuable reference for research astronomers and graduate students wishing to understand the latest developments in this exciting field.

ANDREW FABIAN is a Royal Society Research Professor at the Institute of Astronomy, University of Cambridge.

KENNETH POUNDS is Professor of Space Physics at the University of Leicester.

ROGER BLANDFORD is the Richard Chace Tolman Professor of Theoretical Astrophysics at Caltech, California.

FRONTIERS OF X-RAY ASTRONOMY

Edited by

A. C. FABIAN
University of Cambridge

K. A. POUNDS
University of Leicester

R. D. BLANDFORD
Caltech

CAMBRIDGE UNIVERSITY PRESS
Cambridge, New York, Melbourne, Madrid, Cape Town,
Singapore, São Paulo, Delhi, Mexico City

Cambridge University Press
The Edinburgh Building, Cambridge CB2 8RU, UK

Published in the United States of America by Cambridge University Press, New York

www.cambridge.org
Information on this title: www.cambridge.org/9780521827591

First published 2004

A catalogue record for this publication is available from the British Library

Library of Congress Cataloguing in Publication Data

Frontiers of X-ray astronomy / edited by A.C. Fabian, K.A. Pounds, R.D. Blandford.
p. cm.
Includes bibliographical references and index.
ISBN 0 521 82759 0 – ISBN 0 521 53487 9 (paperback)
1. X-ray astronomy. I. Fabian, A. C., 1948– II. Pounds, K. (Ken) III. Blandford, Roger D.

QB472.F76 2003
522′.6863–dc21 2003053213

ISBN 978-0-521-82759-1 Hardback
ISBN 978-0-521-53487-1 Paperback

Contents

Contributors

K. Pounds, S. M. Kahn, E. Behar, A. Kinkhabwala, D. W. Savin, M. Gudel, M. Cropper, G. Ramsay, C. Hellier, K. Mukai, C. Mauche, D. Pandel, C. Done, C. R. Canizares, M. Ward, S. W. Allen, R. Mushotzky, A. C. Fabian, G. Matt, W. N. Brandt, D. M. Alexander, F. E. Bauer, A. E. Hornschemeier, G. Hasinger, R. D. Blandford.

Preface

X-ray astronomy reached its fortieth anniversary in 2002. In the four decades since the discovery of Scorpius X-1 and the cosmic-X-ray background radiation, X-ray observations of increasing sensitivity and resolution have led the development of the new field of high-energy astrophysics, which has had a major impact on the whole of astronomy and much of physics, and is now, arguably, the most dynamic field of astronomical research.

With the launch of two world-class X-ray observatory missions, Chandra and XMM-Newton, X-ray astronomy has now entered its 'third age', bringing data of unsurpassed quality and richness. The first detailed studies of the physical environment and nature of a wide range of objects are now being made with the high-resolution X-ray spectra provided by dispersive grating spectrometers on both satellites. In addition, CCD cameras are producing a wealth of X-ray images in distinct X-ray 'colours', yielding direct information on plasma temperatures and chemical abundance.

Particularly powerful are the sub-arc-second X-ray images provided by the high-resolution Chandra telescope mirrors, revealing previously unseen structure in objects as distinct as supernova remnants and jets in active galaxies. The complementary power of XMM's EPIC cameras, with the highest 'photon capture' yet, is particularly beneficial in the study of faint extended sources. Together, Chandra and XMM-Newton are yielding new insights on the physical processes by which powerful X-ray emission is produced in objects as diverse as comets, active stars, accreting binary systems, supernovae, active galactic nuclei and galaxy clusters. Furthermore, very deep observations are targeting the final fraction of the X-ray background radiation, and yielding the first X-ray data on cosmologically interesting sources, such as those being found at visible and infrared wavelengths in the Hubble deep fields.

This collection of articles is based on the talks given at a Royal Society Discussion Meeting, in February 2002, by key researchers in the field, and represents a unique review of the current status and exciting potential of a major branch of astronomy.

1

Forty years on from Aerobee 150: a personal perspective

BY KEN POUNDS

University of Leicester

Introduction

Before the historic discovery (Giacconi *et al.* 1962) of a bright X-ray source in the constellation of Scorpius, in June 1962, the expectation of astronomers was that observations in the ultraviolet (UV) and gamma-ray bands offered the best promise for exploiting the exciting new potential of space research. In fact the forecast for X-ray studies was limited to the study of active stars, with fluxes scaled from that of the solar corona, the only known X-ray source at that time. Optimistic flux predictions ranged up to a thousand times the Sun's X-ray luminosity, but seemed beyond the reach of detection with then-current technology. As a reflection of the contemporary thinking, the recently formed US National Aeronautics and Space Agency (NASA) were planning a series of Orbiting Astronomical Observatories, with the first missions devoted to UV astronomy. Despite those limited expectations a proposal from University College London (UCL) and University of Leicester groups, for simultaneous X-ray observations of the primary UV targets, was made in 1961, and eventually the instrument was flown on OAO-3 (Copernicus) 11 years later. In the USA, Riccardo Giacconi and Bruno Rossi had, still earlier, published the design of a grazing-incidence X-ray telescope with nested mirrors (Giacconi & Rossi 1960). Rossi, then at MIT, made a characteristically visionary statement around that time in declaring that 'nature so often leaves the most daring imagination of man far behind'.

The Aerobee 150 sounding-rocket flight from the White Sands Missile Range in June 1962, which found in Sco X-1 a cosmic-X-ray source a million times more luminous than the Sun (and actually brighter than the non-flaring corona at a few keV), began a transformation that has led, over the intervening 40 years, to the most vibrant area of space science, and the foundations for a revolution in high-energy

Frontiers of X-Ray Astronomy, ed. A.C. Fabian, K.A. Pounds and R.D. Blandford. Published by Cambridge University Press.

astrophysics. Looking back, it is interesting to recall that the discovery of the first cosmic-X-ray source did not become widely known for half a year, until the paper by Giacconi and his colleagues at American Science and Engineering (AS&E) was published in *Physics Review Letters* (Giacconi *et al.* 1962), a time-scale in sharp contrast with the immediacy of communicating results on a global scale today via the Internet.

The 1960s: solar physics and X-ray astronomy with sounding rockets

Progress in the remainder of the 1960s was rapid, with the emerging discipline of X-ray astronomy developed largely by physicists with a background in cosmic-ray studies or solar physics. Further sounding-rocket observations followed, by the US Naval Laboratory (NRL) group (responsible for still earlier but unsuccessful flights (Friedman 1959), confirming Sco X-1 and finding a further source in Taurus (Bowyer *et al.* 1964a)), the AS&E group (Gursky *et al.* 1963), and a team at Lockheed (Fisher & Meyerott 1964). As momentum in this new field built up, the NRL group identified extended X-ray emission from the Crab Nebula supernova remnant in a classic use of the Moon as an occulting disc (Bowyer *et al.* 1964b), and an accurate position for Sco X-1 led to its optical identification with a 13th-magnitude blue star (Sandage *et al.* 1966).

In the UK there was a rapid development of interest and activity, building on research already underway in studies of the solar X-ray emission, led by groups at the Culham Laboratory and Leicester University, and made possible by the availability of the competitive Skylark sounding rocket. Skylark, which evolved during the 1960s to be an excellent platform for space astronomy, with the ability to point at the Sun, the Moon or a star, was meanwhile being used to obtain the first good-quality X-ray images of the solar corona (Fig. 1.1; Russell & Pounds 1966) and high-resolution spectra (Evans *et al.* 1967). The first Skylark flights to search for cosmic-X-ray sources from the Southern Hemisphere were launched in 1967 (Cooke *et al.* 1967).

Solar X-ray studies remained at a much higher profile than cosmic-X-ray astronomy throughout the 1960s, with NASA's series of Orbiting Solar Observatory spacecraft leading the way. The first international space science satellite, Ariel 1 (Fig. 1.2), including a UCL/Leicester proportional-counter spectrometer to measure the X-ray emission from the solar corona, was successfully placed in orbit on a Thor Delta from Cape Canaveral on 26 April 1962. That was just two months before the historic Aerobee 150 rocket flight and brought about a first (unplanned) link with future colleagues at AS&E.

All went well with Ariel 1 for several weeks post launch and it yielded X-ray spectra of both the quiet and the flaring corona (Bowen *et al.* 1964). Then, on 9 July

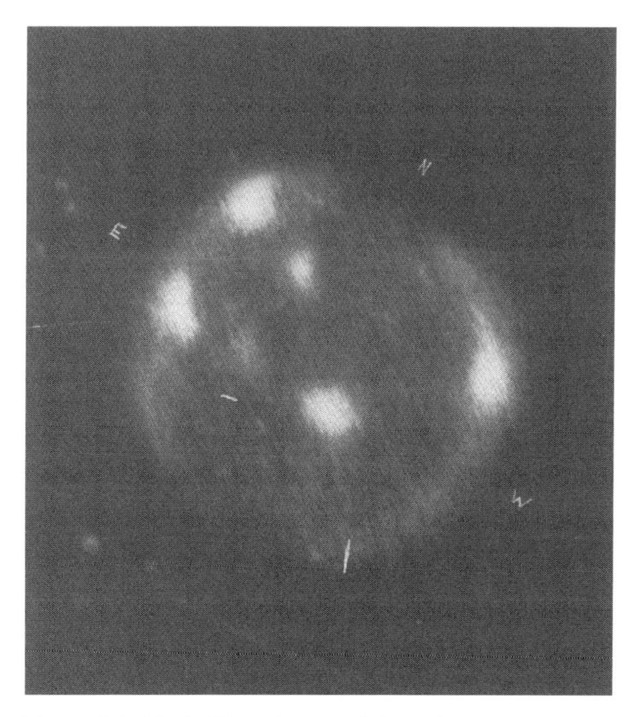

Figure 1.1. Early X-ray image of the solar corona.

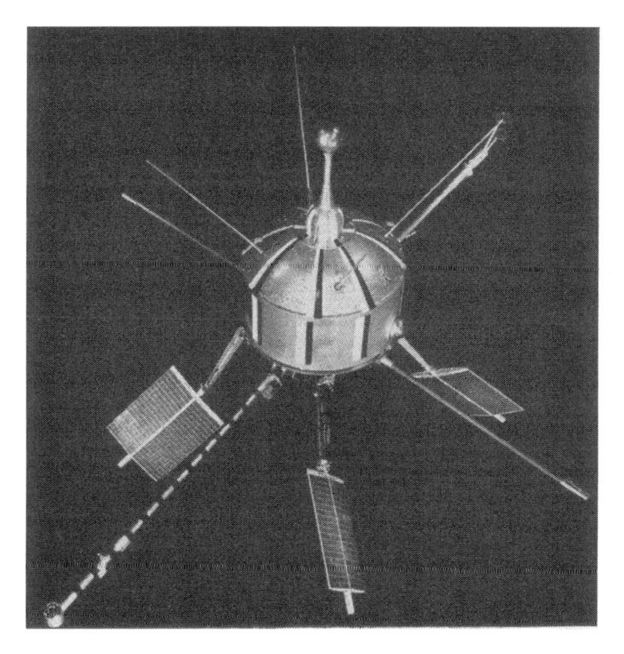

Figure 1.2. The Ariel-1 satellite.

1962, the US Air Force chose to detonate a nuclear bomb in the atmosphere 400 km above Johnston Island in the Pacific. The explosion put radioactive debris, which was to take many months to decay, into the atmosphere. The 'artificial radiation belts' induced spectacular count rates in several of the Ariel 5 instruments, and were a bonus for that part of the mission. Sadly, my proportional counters used methane quench gas, which was broken down by the extreme count rates to form a polymerized deposit on the anode wires, leading to a rapid loss of 'gain' and sensitivity. (Only 30 years later did I learn from Herb Gursky of Riccardo Giacconi's complicity in the destruction of my first in-orbit instrument, as senior AS&E staff were present at Johnston Island to support their weapons-testing contract!)

Looking back now at those early days in space astronomy, the pace of development was remarkable. On the big stage, of course, the Apollo programme dominated attention. However, space science was also hectic, with many satellites being launched in the USA and the USSR. The UCL/Leicester solar studies developed apace, with an evolution of the Ariel 1 spectrometer flown successfully on OSO-D in 1967 (Culhane *et al.* 1969), and on Europe's first orbiting satellite, ESRO-2 (1968). A continuous sequence of X-ray images of the corona was provided by an imaging instrument orbited on OSO-F from 1969 (Parkinson & Pounds 1971) and was published routinely in *Solar-Geophysical Data* to 1975. Within the UK national programme the frequency of Skylark launches from Woomera peaked at 20 in 1965, with a remarkable 198 flights between 1957 and 1978 (Massey & Robins 1986).

Skylark (Fig. 1.3) provided the means for the first X-ray surveys of the sky in the Southern Hemisphere from 1967 (Cooke & Pounds 1971). As the global number

Figure 1.3. Skylark cosmic-X-ray payload.

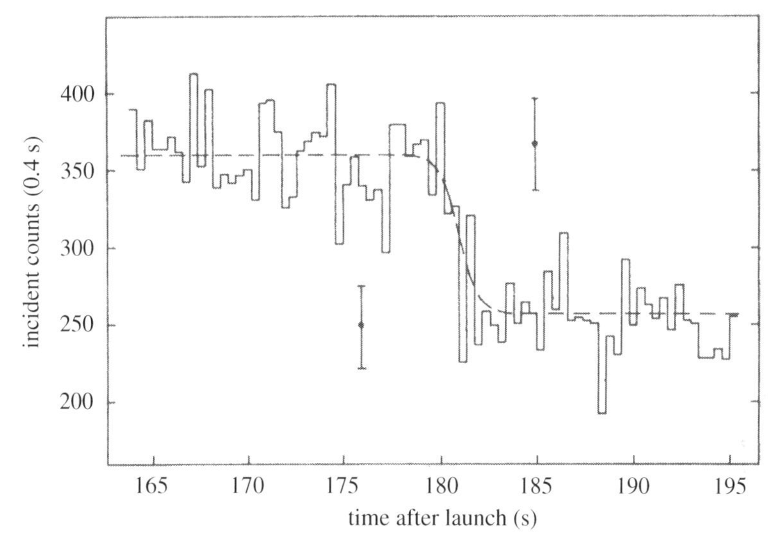

Figure 1.4. X-ray flux from GX3+1 cut off by lunar occultation.

of X-ray sources grew well into double figures, a continuing challenge was their optical identification. In some cases, the typical few-tenths-of-a-degree accuracy of X-ray source location was sufficient for a reliable association with an outstanding candidate, e.g. the bright quasar 3C 273, but ingenious techniques were developed to obtain much more precise positions for other sources (Oda *et al.* 1965).

Occasionally there were disappointments, as in the technically successful Leicester flight of a variable-length modulation collimator intended to better locate and identify the bright source Cen X-3, only for it to have disappeared! An explanation came later when Uhuru observations showed Cen X-3 to be an eclipsing binary in which the X-ray source is occulted by its companion star for a quarter of each two-day binary period.

Another heroic source-identification attempt involved two Skylark launches, in September and October 1972, to locate the source GX3+1 with arc-second precision, in order to identify an optical counterpart within the crowded sky close to the centre of our Milky Way galaxy. The idea was to observe the *c.* 2 arcmin SAS-3 error box of GX3+1 as it was being occulted by the Moon.

Given a predicted in-flight rate of *c.* 0.5 arcsec s^{-1} for the Moon's disc to travel across the star field, the launch window was less than one minute wide. The first firing, of SL 1002, went perfectly, the Sun-pointing rocket being held in the predetermined roll direction by locking X-ray detectors onto Sco X-1, placing the Moon in the field of view of the main X-ray detectors 96 s post launch.

The occultation of GX3+1 was successfully recorded to ± 0.5 s, or ± 0.3 arcsec (Fig. 1.4). A second Skylark flight one lunar month later was also successful, though the reduced sensitivity allowed by the use of an unstabilized (spinning)

Table 1.1. *Standard deviations of the limb positions*

experiment	timing and rocket position	lunar ephemeris	combined error
Leicester	0.2″	0.3″	0.4″
UCL/MSSL	3.4″	0.3″	3.4″

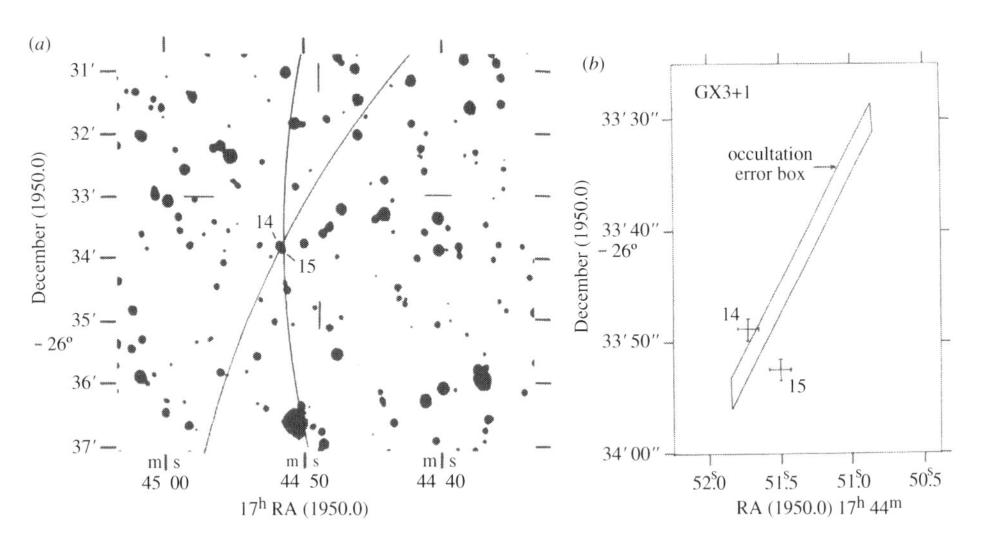

Figure 1.5. (*a*) Lunar arcs at occultation and (*b*) the GX3+1 error box.

rocket yielded a lower positional accuracy. Overall, the X-ray position of GX3+1 was determined (Janes *et al.* 1972) with a precision (Fig. 1.5 and Table 1.1) that was unchallenged until the launch of Chandra. Sadly, due to the large obscuration in the Galactic Bulge region, no optical counterpart was found in the GX3+1 error box to a limit of 21st magnitude.

Those combined Skylark launches were typical of the technical ingenuity and uncertain scientific returns which made the first decade of X-ray astronomy with sounding rockets both an exciting and an educational experience.

The second decade: the era of small dedicated satellites

X-ray astronomy enjoyed a major advance during the 1970s, with the first small orbiting satellites dedicated to observations of cosmic-X-ray sources. Uhuru led the way in December 1970, placing an array of proportional counters (effective area 840 cm^2) into a circular equatorial orbit. Within a few months the extended observations made possible from orbit had shown that many X-ray sources were

Figure 1.6. The Ariel 5 satellite.

variable. This led to the historic discovery that many of the most luminous galactic sources were in binary star systems. The equally dramatic discovery of extended X-ray emission from galaxy clusters followed, while the number of known X-ray sources multiplied. The 3U catalogue (Giacconi *et al.* 1974), listing 161 sources, was an important milestone in the development of X-ray astronomy, and the major scientific impact of Uhuru is well recorded in Giacconi & Gursky (1974).

Other Uhuru-class satellites followed, with Ariel 5 (UK), SAS-3 (USA) and Hakucho (Japan) dedicated to X-ray observations and OSO-7 (USA) and ANS (Netherlands) being solar and UV astronomy missions with secondary X-ray instrumentation.

For astronomers in the UK, Ariel 5 (Fig. 1.6) brought an ideal opportunity to play a part in the rapid advances taking place. Like Uhuru, Ariel 5 was launched on a Scout rocket into a circular near Earth orbit from a disused oil platform off the coast of Kenya. It carried six experiments, including a Sky Survey Instrument (SSI), similar to that on Uhuru, an All Sky Monitor from Goddard Space Fight Center (GSFC) and three X-ray spectrometers viewing along the satellite spin axis. The Ariel 5 orbit was a good choice, not only in minimizing background due to cosmic rays and trapped radiation, but in allowing regular data dumps from the small on-board data recorder. With a direct microwave, cable and satellite link to the UK (Fig. 1.7), we received six orbits of 'quick look' data within an hour of ground-station contact. The remaining 'bulk' data were received within 24 hours, an immediacy that contributed substantially to the excitement of the mission operations, while also ensuring a rapid response to new discoveries.

One such discovery was particularly well timed, with the SSI detecting a previously unseen source in the constellation Monoceros just two days before the start of

Figure 1.7. Ariel 5 data link to UK.

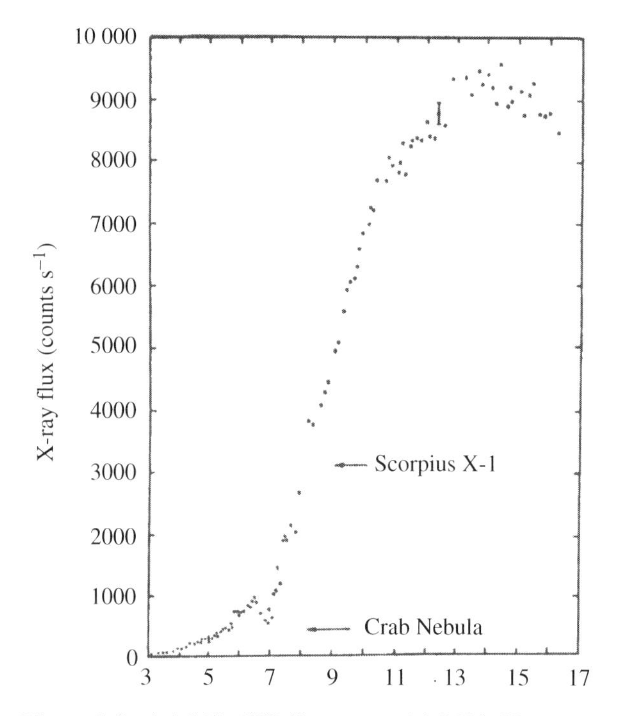

Figure 1.8. Ariel 5's SSI discovery of A0620-00.

the first European Astronomy Society (EAS) meeting in August 1975 in Leicester, where new X-ray results from the on-going satellite missions were high on the agenda. Variable X-ray sources were by then commonplace, but what set Mon X-1 apart was its strength. By day two of the EAS meeting it was brighter than the Crab Nebula, while two days later it outshone Sco X-1 (Fig. 1.8) to become, for a few weeks, the brightest cosmic-X-ray source seen (Elvis *et al.* 1975), a record still held today. After peaking at a flux level three times that of Sco X-1, the new source (by then renamed A0620-00) was being monitored by Ariel 5, SAS-3 and other space- and ground-based telescopes around the world (Fig. 1.9(*a*)) as it gradually faded from view (Kaluzienski *et al.* 1975).

Optical and radio counterparts were quickly identified and spectroscopy of the binary companion (Fig. 1.9(*b*)) later revealed a mass estimate for the compact X-ray-emitting component in A0620-00 to be a strong black-hole candidate (McClintock & Remillard 1986).

Soft X-ray transients became relatively common as the Ariel 5 mission continued until the satellite's attitude-control gas ran out, ending observations in 1980. A number of other discoveries had by then marked Ariel 5 as a highly successful mission. One was the detection of X-ray line emission from the Perseus Cluster galaxies (Fig. 1.10), showing that the luminous diffuse radiation was of thermal origin at *c*. 10^8 K (Mitchell *et al.* 1976).

Another important and enduring result from the Ariel 5 SSI was to establish powerful X-ray emission (alongside the bright optical nucleus and broad permitted lines) as a characteristic property of Seyfert galaxies (Fig. 1.11). The challenge of correctly identifying many previously unidentified sources, individually located to only a few tenths of a square degree, was possible only because Seyfert nuclei are also unusually bright in the optical band; even so, the initial identifications were made on a statistical basis, but held up well as new data emerged to establish Seyfert galaxies and active galactic nuclei (AGN) in general as the dominant class of extragalactic X-ray source (Elvis *et al.* 1978). In recalling the scientific contributions of Ariel 5, in an introduction to this book, based on the Royal Society Discussion Meeting, it is interesting to note that the (only) previous meeting in the same series, held 24 years ago, was largely devoted to results from Ariel 5 (Massey *et al.* 1979).

Beyond 1980: X-ray astronomy becomes a global enterprise

Notwithstanding the contributions of Ariel 5, Hakucho and ANS (the last especially for the discovery of X-ray burst sources), the US programme continued to set the pace in X-ray astronomy with two large spacecraft, HEAO-1 (a sky-survey mission from NRL and GSFC) and HEAO-2 (the Einstein Observatory), being launched in 1978. The Einstein Observatory, again led by Giacconi's team, then at the Harvard–Smithsonian Center for Astrophysics, marked another milestone in the development of X-ray astronomy, being the first imaging telescope devoted to the study of cosmic-X-ray sources, and bringing X-ray astronomy close to optical and radio astronomies as a major branch of observational astrophysics. Among many advances brought by

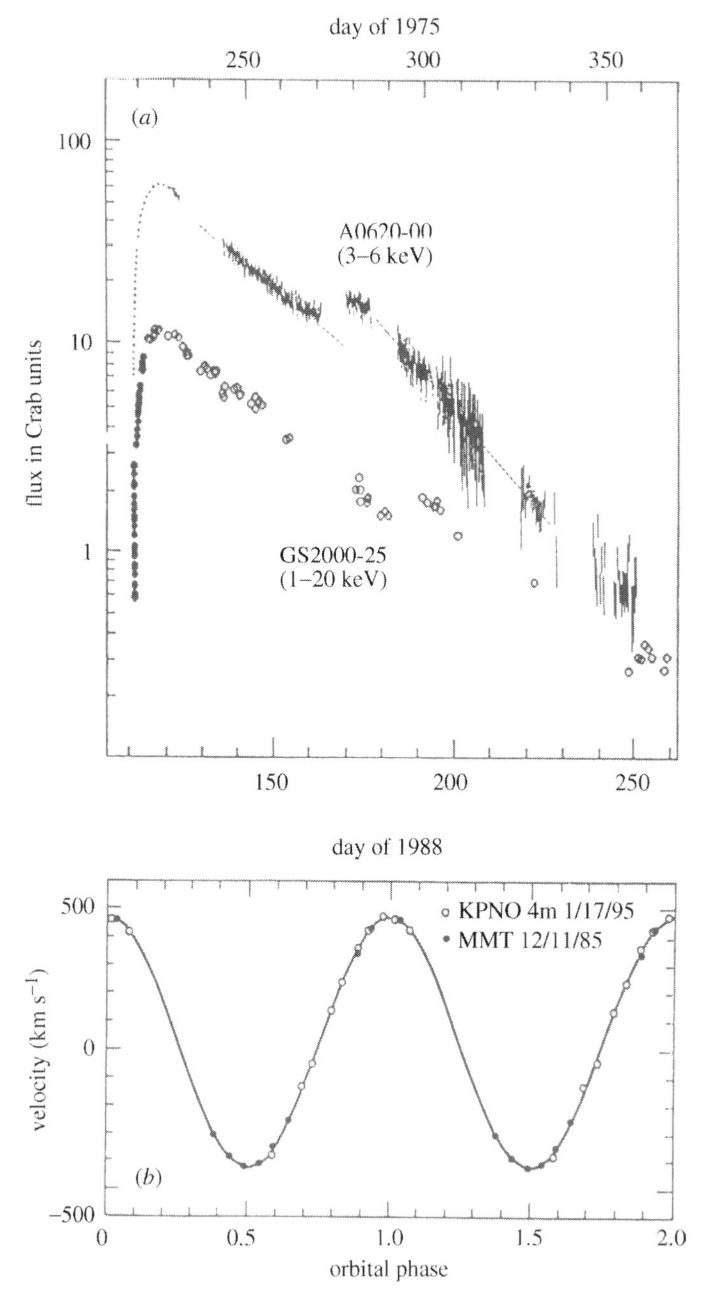

Figure 1.9. (*a*) Contemporary X-ray light curves of A0620-00 (in 1975) and GS2000-25 (in 1988). (*b*) Optical radial velocity curve of the companion to A0620-00.

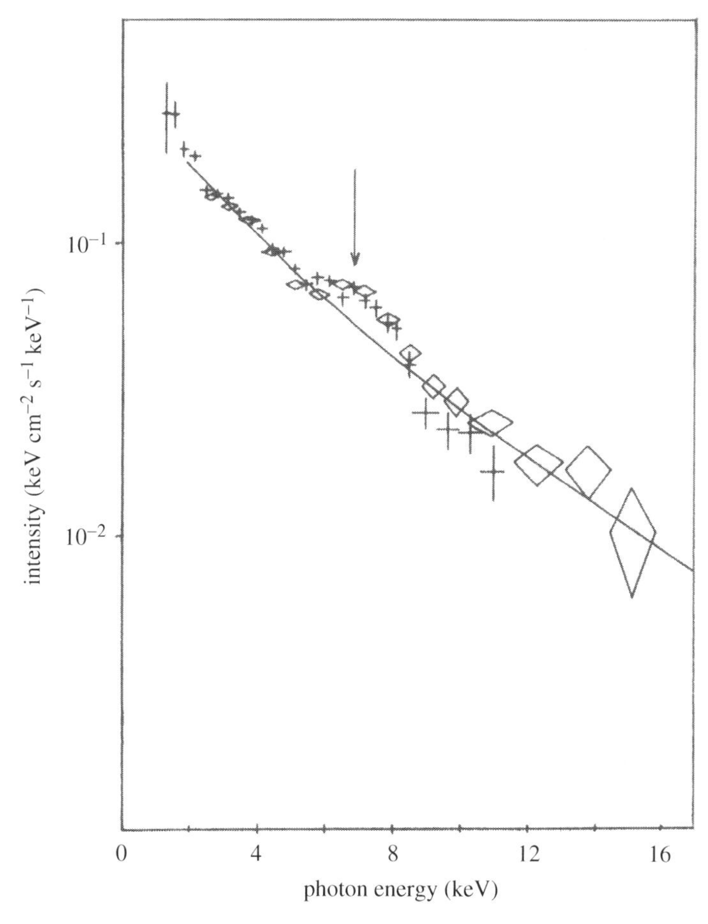

Figure 1.10. Fe K emission from the Perseus Cluster.

the powerful HEAO missions, the imaging properties of the Einstein Observatory stand out (for an early summary see Giacconi (1980)), mapping the distribution of hot gas in galaxy clusters, and producing detailed X-ray images of many supernova remnants and star fields. Stellar X-ray astronomy essentially began with the Einstein Observatory, while the faint sources reached by long observations of an imaging telescope led to a significant advance in 'resolving' the cosmic-X-ray background (Giacconi *et al.* 1979).

Following the end of the HEAO missions in 1981–2, a lengthy gap was to develop before the next NASA mission in X-ray astronomy was put into orbit. With hindsight, that delay has been attributed to NASA placing too much emphasis on the next 'big mission', namely the Advanced X-ray Astrophysics Facility (AXAF). For instance, Freeman Dyson, speaking at the meeting on 'Observations in near-Earth orbit and beyond' (Dyson 1990), said "the Great Observatory missions have kept

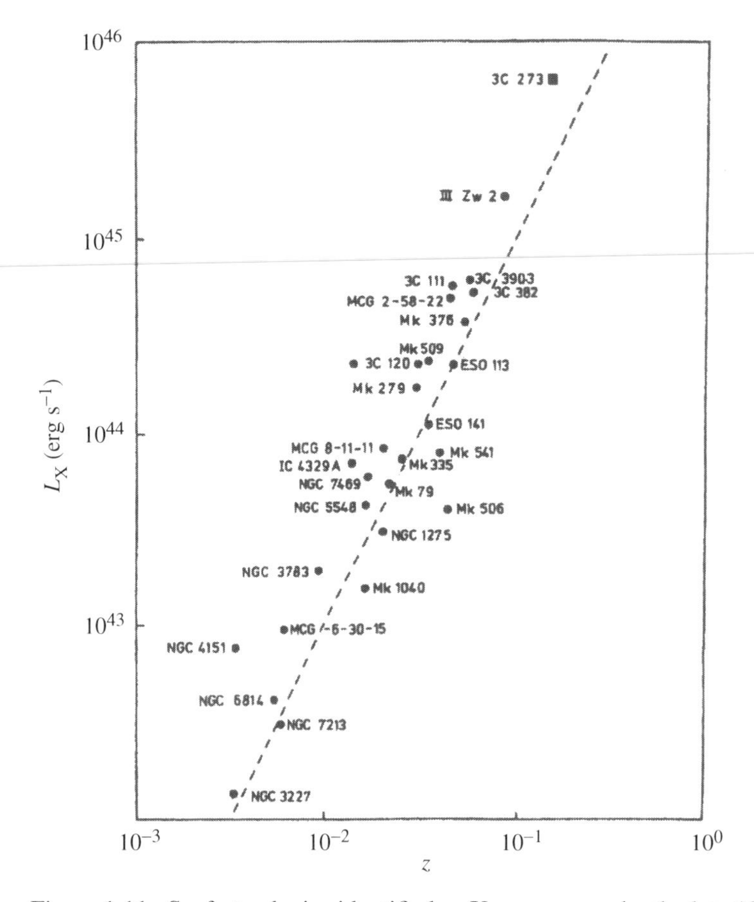

Figure 1.11. Seyfert galaxies identified as X-ray sources by the late 1970s.

us waiting too long. They have had a bad effect on astronomy in discouraging the building of smaller instruments that cost less and could fly sooner".

Fortunately for the continued development of X-ray astronomy, ambition and momentum were building up in Europe and Japan as the NASA programme was being focused on the development of AXAF as a 'Great Observatory' mission fully comparable with the Hubble Telescope, SIRTF, GRO and the upcoming 8-m ground-based optical telescopes (Fig. 1.12). The recently formed European Space Agency had agreed to proceed with an X-ray mission, as first promised when COS-A missed out in the earlier competition with the gamma-ray satellite COS-B (Pounds 1999). Initially conceived as a lunar occultation mission (HELOS) to provide accurate locations and optical identification of hard-X-ray sources not accessible to imaging telescopes, the redesigned satellite EXOSAT carried a sensitive combination of soft- and medium-energy X-ray detectors into a highly eccentric orbit.

In the event, EXOSAT's unusual orbit (Fig. 1.13) led directly to many of its most important scientific contributions over a three-year lifetime (1983–6), reviewed in

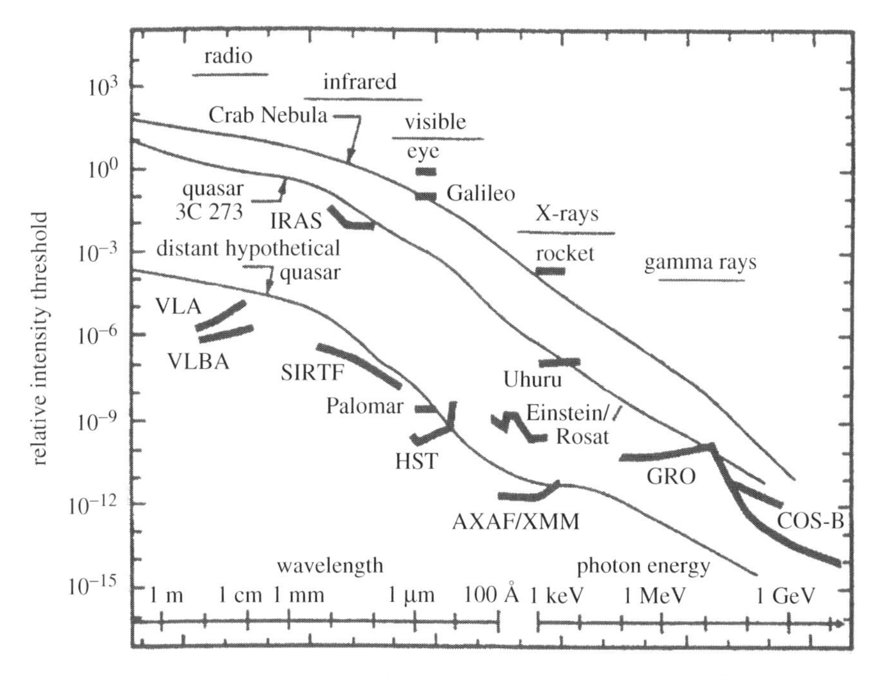

Figure 1.12. The comparable sensitivity for the observation of AGN was a key factor in the case made for AXAF.

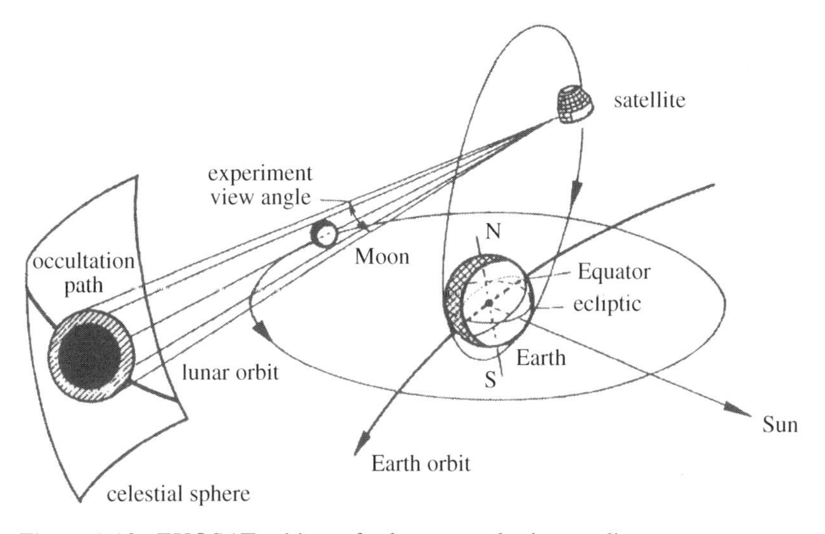

Figure 1.13. EXOSAT orbit set for lunar occultation studies.

Pallavicini & White (1988). In addition to providing a vital observatory facility for the international community over that period, EXOSAT also gave a considerable boost to X-ray astronomy across Europe. The resulting momentum undoubtedly helped greatly in the case that quickly followed (Bleeker *et al.* 1984) for an even

Figure 1.14. The GINGA satellite.

more ambitious ESA mission in X-ray astronomy (realized some 14 years later with the launch of XMM).

Although interagency planning was still very limited at the time, a well-connected community ensured that other international missions during the period 1980–2000 were strongly complementary. So, in 1987, the Institute for Space and Astronautical Science in Japan launched GINGA (Fig. 1.14) to give X-ray astronomers a uniquely powerful means of obtaining broadband spectra and high-resolution time-series data on a wide variety of bright sources for the next four years. The UK enjoyed a substantial role in the GINGA mission, from the joint development of the Large Area Counters (LAC) detectors to the scientific exploitation of the data. A special issue of *Publications of the Astronomical Society of Japan* **41** (1989) gives a measure of the major scientific impact of GINGA.

Continuing the sequence of international X-ray astronomy missions with new observing capabilities, the German ROSAT satellite was launched in 1990, carrying a high-resolution imaging telescope to undertake the first deep all-sky survey with imaging optics. An early success of ROSAT was in finally detecting X-rays from

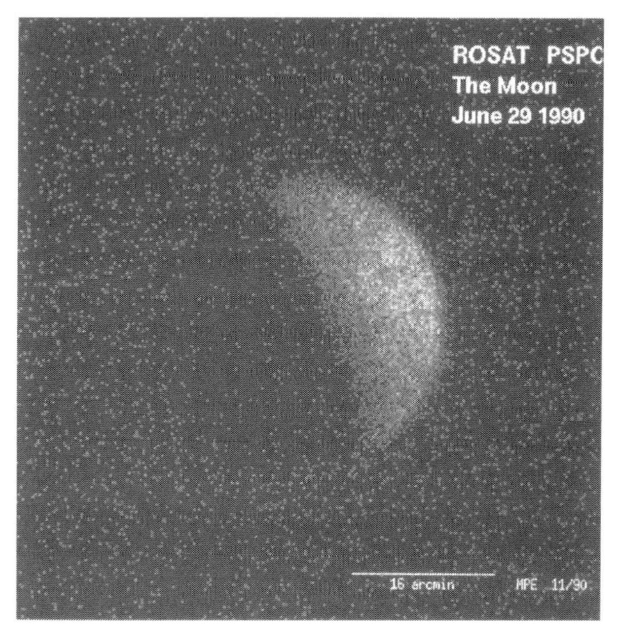

Figure 1.15. ROSAT X-ray image of the Moon.

the Moon (Fig. 1.15), 28 years after the initial, unsuccessful, attempt on AS&E's Aerobee 150 rocket flight. In its all-sky survey, ROSAT increased the number of catalogued cosmic-X-ray sources by over an order of magnitude, to some 65 000, with a similar number being added from pointing observations over the remainder of ROSAT's highly successful term, setting the agenda for the observatory missions to come.

As noted earlier, a feature of the X-ray satellites over the 1981–99 period, each modest in scale and cost compared with Chandra and XMM-Newton, was the highly complementary strengths in the kind of data they obtained. The scientific advances that correspondingly followed can be illustrated well with reference to the study of Seyfert galaxies, a long-standing personal interest (Fig. 1.16). Thus, the combination of low- and medium-energy detectors on EXOSAT showed that a 'soft excess' in the continuum emission below *c.* 1 keV was present in at least 50% of the sources observed, suggesting thermal emission from the putative accretion disc (Turner & Pounds 1989). Then the broadband sensitivity of GINGA was able to demonstrate that 'X-ray reflection' (from the disc material?) was also common in Seyfert spectra (Pounds *et al.* 1990; Nandra & Pounds 1994), opening an important new means of studying matter in the innermost regions of the active nucleus. This diagnostic potential was later to be underlined with the improved spectral resolution of the Advanced Satellite for Cosmology and Astrophysics (ASCA), charge-coupled devices yielding a broad profile for the Fe K fluorescence line indicating reprocessing

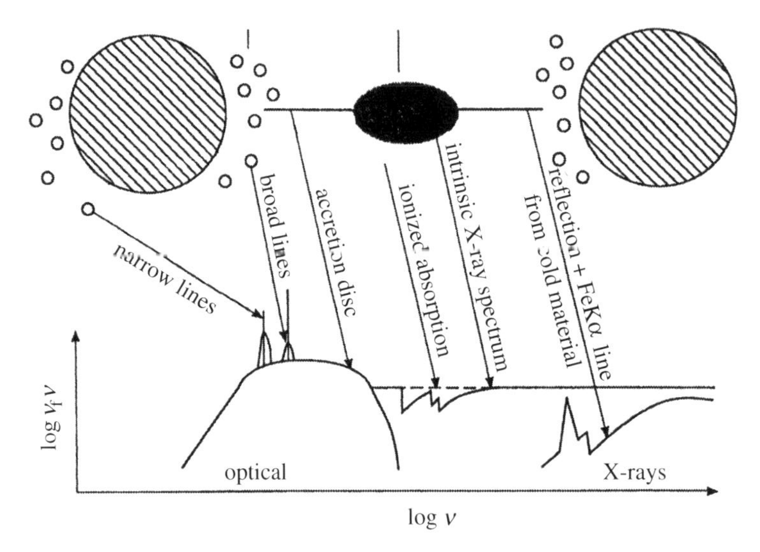

Figure 1.16. Schematic of Seyfert nucleus and X-ray emission components.

by matter moving relativistically in a regime of strong gravity (Tanaka *et al.* 1995; Nandra *et al.* 1997). Meanwhile, ROSAT's improved sensitivity and spectral resolution below *c.* 2 keV showed further unexpected spectral features indicating a surprisingly large column density of ionized gas (the 'warm absorber') in the line of sight to the nucleus of many Seyfert galaxies (Turner *et al.* 1993).

Completing the series of X-ray satellites that so richly filled the gap between Einstein and AXAF were the aforementioned Japanese X-ray satellite, ASCA (launched in 1993), the US Rossi X-Ray Timing Explorer (RXTE) satellite (1995) and the Italian–Dutch *Beppo*SAX mission (1996), each providing critical new capabilities. The rich scientific returns from the improved spectral resolution of the ASCA instruments are well documented in Makino & Mitsuda (1997). The large-area proportional-counter array on the RXTE has provided the best-ever timing data on a range of relatively bright sources, while a major achievement of *Beppo*SAX was the detection of X-ray afterglows from gamma-ray-burst sources, establishing these mysterious high-energy flashes as among the most luminous objects in the Universe. The RXTE and *Beppo*SAX missions are reviewed in detail in Scarsi *et al.* (1998).

Now and into the future

At present, X-ray astronomy is enjoying the simultaneous availability of two long-awaited Great Observatories. Since July 1999, the sub-arc-second imaging telescope on the AXAF, a mission pursued energetically for more than 20 years as the successor to the Einstein Observatory, has been in orbit and is producing outstanding images and high-resolution X-ray spectra. Now named Chandra, this powerful mission was joined in orbit in December 1999 by the ESA's XMM-Newton Observatory. This book is appropriately dominated by the exciting results from these two observatory spacecraft, which take X-ray astronomy and high-energy astrophysics into a new era.

For the immediate future we can look forward to Chandra and XMM-Newton yielding major new science for several years, and being joined in orbit in 2005 by a third 'world-class mission', ASTRO-E2. This latest Japanese X-ray satellite will add powerful complementary capabilities in high-resolution spectroscopy and high-energy sensitivity, advances that are eagerly awaited to build on the discoveries coming from Chandra and XMM-Newton.

Further ahead, the planning of NASA and ESA is once more dominated by the next big step, with Constellation-X and XEUS, respectively, scheduled for development over the coming decade. The exciting advances now being made by Chandra and XMM-Newton demonstrate that further order-of-magnitude increases in overall observational capability are essential to maintain progress in X-ray astronomy. Those next big steps must remain a high priority (or they will not occur). However, history suggests the words of Freeman Dyson and others should be kept in mind in urging the parallel need for smaller missions with unique 'niche' capabilities that can come along more quickly. Such a 'niche mission' may be the LOBSTER X-ray telescope (Fraser *et al.* 2001), now being considered by ESA for flight on the International Space Station, an instrument whose unusual optics will provide a uniquely wide field of view, promising a rich yield of data on variable and transient sources such as the 'orphan' X-ray afterglows from outside the direct beam of (hence unseen) gamma-ray bursts.

References

Bleeker, J. *et al.* 1984 *Physica Scr.* **T7**, 224–234.

Bowen, P. J., Norman, K., Pounds, K. A., Sanford, P. & Willmore, A. P. 1964 *Proc. R. Soc. Lond.* A **281**, 538–552.

Bowyer, S., Byram, E. T., Chubb, T. A. & Friedman, H. 1964a *Nature* **201**, 1307.

Bowyer, S., Byram, E. T., Chubb, T. A. & Friedman, H. 1964b *Science* **146**, 912.

Cooke, B. A. & Pounds, K. A. 1971 *Nature* **229**, 144–147.

Cooke, B. A., Pounds, K. A., Stewardson, E. A. & Adams, D. J. 1967 *Astrophys. J.* **150**, 189–191.

Culhane, J. L. *et al.* 1969 *Mon. Not. R. Astron. Soc.* **145**, 435–455.

Dyson, F. J. 1990 In *Observatories in Earth Orbit and Beyond* (ed. Y. Kondo). Astrophysics and Space Science Series, vol. 166, pp. 413–416. Holland: Kluwer.

Elvis, M., Page, C. G., Pounds, K. A., Ricketts, M. J. & Turner, M. J. L. 1975 *Nature* **257**, 656–657.

Elvis, M. *et al.* 1978 *Mon. Not. R. Astron. Soc.* **183**, 129–157.

Evans, K. D., Pounds, K. A. & Culhane, J. L. 1967 *Nature* **214**, 41–42.
Fisher, P. C. & Meyerott, A. J. 1964 *Astrophys. J.* **139**, 123–142.
Fraser, G. W. *et al.* 2001 *Proc. SPIE* **4497**, 115–126.
Friedman, H. 1959 *Proc. Inst. Radio Engrs* **47**, 278.
Giacconi, R. (ed.) 1980 *X-ray Astronomy with the Einstein Satellite*. Astrophysics and Space Science Series, vol. 87, pp. 1–323. Dordrecht: Reidel.
Giacconi, R. & Gursky, H. (eds) 1974 *X-ray Astronomy*. Astrophysics and Space Science Series, vol. 43, pp. 1–443. Dordrecht: Reidel.
Giacconi, R. & Rossi, B. 1960 *J. Geophys. Res.* **65**, 773.
Giacconi, R., Gursky, H., Paolini, F. & Rossi, B. 1962 *Phys. Rev. Lett.* **9**, 439.
Giacconi, R. *et al.* 1974 *Astrophys. J. Suppl.* **237**, 37.
Giacconi, R. *et al.* 1979 *Astrophys. J.* **234**, L1–L7.
Gursky, H., Giacconi, R., Paolini, F. & Rossi, B. 1963 *Phys. Rev. Lett.* **11**, 530.
Janes, A. F., Ricketts, M., Willmore, A. P. & Morrison, L. 1972 *Nature* **235**, 152–156.
Kaluzienski, L. *et al.* 1975 *Astrophys. J.* **201**, L121–124.
Makino, F. & Mitsuda, K. (eds) 1997 *X-ray Imaging and Spectroscopy of Cosmic Hot Plasmas*, pp. 1–640. Tokyo: Universal Academy Press.
Massey, H. & Robins, M. O. (eds) 1986 *History of British Space Science*. Cambridge: Cambridge University Press.
Massey, H., Boyd, R. L. F. & Willmore, A. P. 1979 *Proc. R. Soc. Lond.* A **366**, 279–489.
McClintock, J. E. & Remillard, R. A. 1986 *Astrophys. J.* **308**, 110–122.
Mitchell, R. J., Culhane, J. L., Davison, P. J. N. & Ives, J. C. 1976 *Mon. Not. R. Astron. Soc.* **175**, 29P–34P.
Nandra, K. & Pounds, K. A. 1994 *Mon. Not. R. Astron. Soc.* **268**, 405–429.
Nandra, K. *et al.* 1997 *Astrophys. J.* **477**, 602.
Oda, M. *et al.* 1965 *Nature* **554**, 207.
Pallavicini, R. & White, N. E. (eds) 1988 *Mem. Soc. Astron. Ital.* **59**, 1–290. (Special issue.)
Parkinson, J. H. & Pounds, K. A. 1971 *Solar Phys.* **17**, 146–159.
Pounds, K. A. 1999 In *The History of the European Space Agency* (ed. R. A. Harris). European Space Agency Special Publication no. SP-436, pp. 151–168.
Pounds, K. A., Nandra, K., Stewart, G. C., George, I. & Fabian, A. C. 1990 *Nature* **344**, 132–133.
Russell, P. C. & Pounds, K. A. 1966 *Nature* **209**, 490–491.
Sandage, A. *et al.* 1966 *Astrophys. J.* **146**, 314.
Scarsi, L., Bradt, H., Giommi, P. & Fiore, F. (eds) 1998 *The Active X-ray Sky: Results from* Beppo*SAX and RXTE*, pp. 1–738. Amsterdam: North-Holland.
Tanaka, Y. *et al.* 1995 *Nature* **375**, 659.
Turner, T. J. & Pounds, K. A. 1989 *Mon. Not. R. Astron. Soc.* **240**, 833–880.
Turner, T. J., Nandra, K., George, I. M., Fabian, A. C. & Pounds, K. A. 1993 *Astrophys. J.* **419**, 127–135.

2

X-ray spectroscopy of astrophysical plasmas

BY STEVEN M. KAHN, EHUD BEHAR,
ALI KINKHABWALA AND DANIEL W. SAVIN

Columbia University

Introduction

The science of X-ray astronomy was born in 1962 with the serendipitous discovery of the very bright X-ray binary source, Scorpius X-1 (Giacconi *et al.* 1962). In the ensuing 40 years, this field has progressed to the point where it is now one of the 'standard' disciplines of observational astrophysics (see Chapter 1 and the other chapters in this book). X-ray observatories have revealed a diverse collection of sources ranging from the nearest stars to the most distant galaxies in the Universe.

However, until very recently, X-ray spectroscopy of astrophysical sources has been largely unavailable, due to instrumental limitations. Non-dispersive detectors, chosen for their high quantum efficiency and large collecting area in most early missions, provided only crude spectral resolution, insufficient for the unambiguous identification of discrete features. As in other wavebands, spectroscopy is crucial to the derivation of quantitative constraints on physical conditions in the sources under investigation.

This need has now finally been met, due to the launches of two major space-observatory facilities. The Chandra X-Ray Observatory, launched by NASA in July 1999, carries two separate transmission-grating spectrometers (Brinkman *et al.* 2000; Canizares *et al.* 2000), collectively providing high-resolution spectra over the broad wavelength band 1–200 Å (0.05–10 keV). The XMM-Newton Observatory, launched by ESA in December 1999, incorporates reflection-grating spectrometers behind two of the three grazing incidence telescopes carried on board (den Herder *et al.* 2001). These provide high throughput and high resolution in the range 5–38 Å (0.3–2.5 keV). Together, these instruments are providing magnificent spectra of nearly all classes of astrophysical X-ray sources.

Frontiers of X-Ray Astronomy, ed. A.C. Fabian, K.A. Pounds and R.D. Blandford. Published by Cambridge University Press.

X-ray spectroscopy is very important for astronomy, because X-ray-emitting gas is often the key component of the system. For many objects (e.g. elliptical galaxies, clusters of galaxies), the 'virial temperature' $kT \sim GMm_p/R$ lies in the range 10^6–10^8 K, where most of the emission comes out at X-ray energies. In others (e.g. supernova remnants, binary sources), shocks heat gas into the same temperature regime.

In addition, the conventional X-ray band (0.1–10 keV) is unusually rich in discrete spectral features: the K-shell transitions of carbon through iron and the L-shell transitions of silicon through iron fall in this range. In contrast to other wavebands, many charge states of the same element, as well as entire series of lines are all visible in a single X-ray spectrum. This makes the interpretation of the spectrum fairly unambiguous. For example, one can derive relative elemental abundances without invoking a large number of assumptions about the thermal state of the gas.

Finally, because of the high radiative decay rates of X-ray transitions, astrophysical emitting plasmas are generally not in local thermodynamic equilibrium (LTE), except at very high densities (e.g. in the photospheres of neutron stars and white dwarfs). This means that the details of the observed spectra are characterized by the explicit balance between the various microphysical processes that feed and deplete the relevant quantum levels. While that can occasionally lead to complications in the interpretation of the data, it also implies that they are quite sensitive to physical conditions in the source. Hence, X-ray spectra have high diagnostic utility.

Types of equilibria

As indicated above, astrophysical X-ray-emitting plasmas are rarely in LTE. Nevertheless, other types of equilibrium, or steady-state conditions, can apply. The detailed characteristics of the X-ray spectra depend crucially on the nature of these equilibria. Two cases are especially important.

Collisional equilibrium. Here, the excitations and ionizations are dominated by electron–ion collisions. The electrons are hot, with characteristic temperatures comparable with the energies of the spectral lines observed. The emergent spectrum is nearly a unique function of the electron temperature distribution and the elemental abundances. These conditions apply in stellar coronae, in the intracluster media of clusters of galaxies, in elliptical galaxies, and in the shocked gas in older supernova remnants.

Photoionization equilibrium. In this case, the presence of an intense continuum radiation field has a significant effect on the ionization and thermal structure of the surrounding gas. The electrons are generally too cool to excite prominent X-ray lines. Instead, excited levels are populated by direct radiative recombination, by radiative cascades following recombination onto higher levels, and by direct photoexcitation from the continuum. These conditions can apply in the circumsource media of accretion-powered sources, such as X-ray binaries and active galactic nuclei.

The principal spectroscopic differences between collisionally ionized and photoionized plasmas are due to the very different electron temperatures that accompany a given charge state in the two cases. For collisional ionization, the electron temperature is comparable with the ionization potential, while for photoionization, the photon field does most of the work, so the electron temperature can be much lower. For example, the characteristic temperature for helium-like oxygen (O^{6+}) in a collisional plasma is $kT_e \sim 100$ eV (Mazzotta *et al.* 1998), whereas the same ion is found in photoionized plasmas at $kT_e \sim 4$ eV (Kinkhabwala *et al.* 2002).

Perhaps the most useful spectroscopic diagnostics for distinguishing collisional ionization from photoionization are radiative recombination continua (RRC) (see Liedahl (1999)). Radiative recombination generates continua associated with decay from a Maxwellian electron distribution into individual discrete levels. The emission starts at the ionization energy threshold χ for a particular level and falls off smoothly toward higher energies, with a characteristic width *c.* kT_e. In a collisional plasma, $kT_e \sim \chi$, so the RRC are broad and have low contrast. On the other hand, in a photoionized plasma, $kT_e \ll \chi$. The RRC are strong and fall off steeply with increasing energy, or decreasing wavelength.

This effect is illustrated in Fig. 2.1, where we have plotted the emergent spectrum for pure recombination in hydrogen-like oxygen forming helium-like oxygen, as a function of the electron temperature. The RRC is the central prominent feature in the plot, with a threshold near 16.8 Å. As can be seen, as the electron temperature increases, this feature grows broader, decreasing its contrast with respect to the accompanying discrete line emission. For a collisionally ionized plasma, the RRC is so broad that it is virtually invisible.

In Fig. 2.2, we show the spectrum of the bright Seyfert 2 galaxy NGC 1068, as obtained with the reflection-grating spectrometer on XMM-Newton (Kinkhabwala *et al.* 2002a). As can be seen, the spectrum is rich in emission lines, especially hydrogen-like and helium-like lines of low-Z metals (carbon, nitrogen, oxygen and neon). The features labelled RRC in Fig. 2.2 are the radiative recombination continua, which arise when each of these ions is formed. They are narrow, indicating a low electron temperature of a few eV, characteristic of a photoionized plasma. In NGC 1068, the soft-X-ray spectrum is produced in an ionization cone, which is irradiated by an intense X-ray continuum emanating from a central obscured nucleus.

Cosmic plasmas are not necessarily in ionization equilibrium. The ionization and recombination rates, and thus the equilibration times, scale with the electron

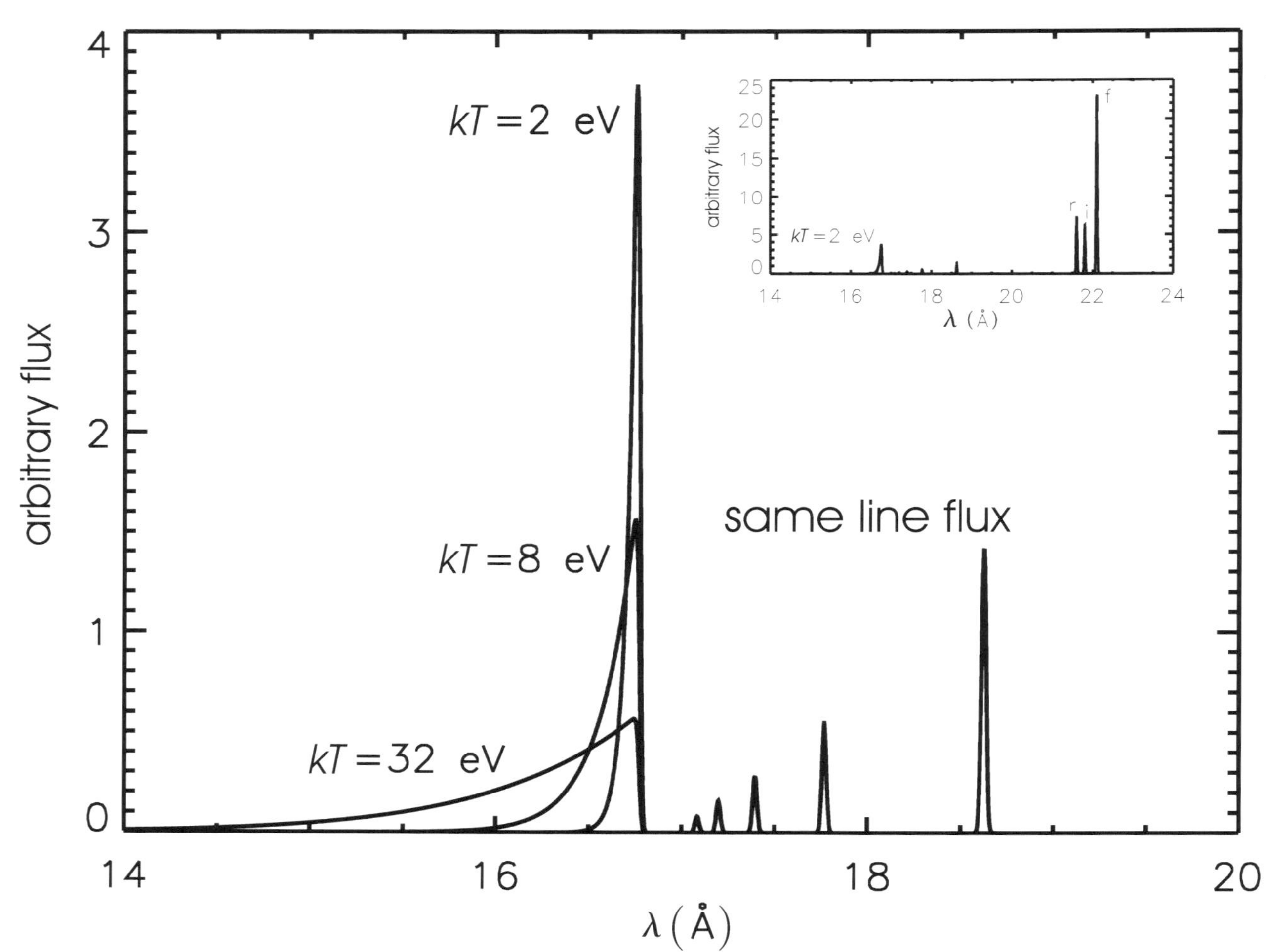

Figure 2.1. The shape of the radiative recombination continua for recombination forming helium-like oxygen as a function of temperature, assuming the same recombination rate for each temperature.

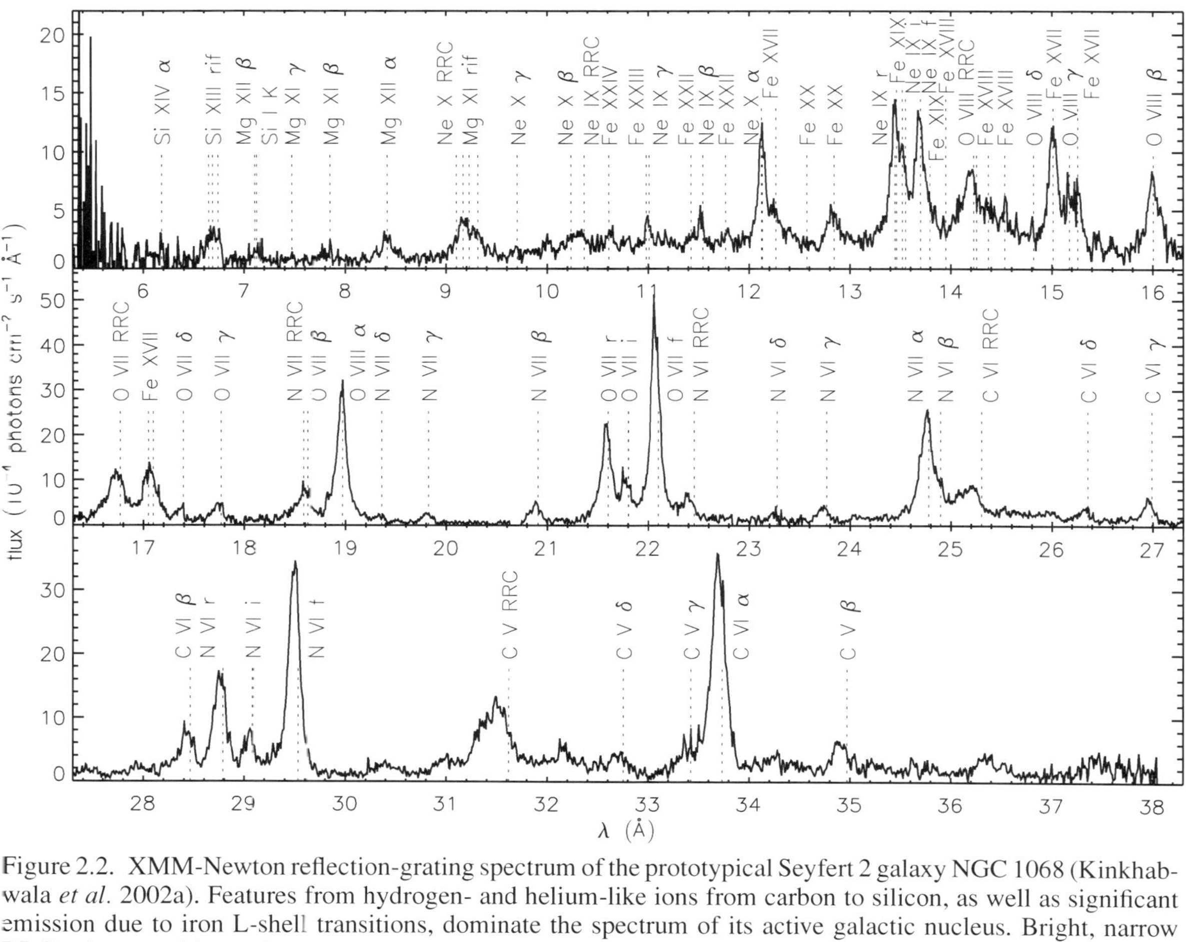

Figure 2.2. XMM-Newton reflection-grating spectrum of the prototypical Seyfert 2 galaxy NGC 1068 (Kinkhabwala *et al.* 2002a). Features from hydrogen- and helium-like ions from carbon to silicon, as well as significant emission due to iron L-shell transitions, dominate the spectrum of its active galactic nucleus. Bright, narrow RRC point unambiguously to the predominance of recombination in a photoionized plasma. Strong higher-order Rydberg transitions (np $\rightarrow$ 1s) are present, implying that photoexcitation contributes to the spectrum as well.

density. The low-density environments in many sources result in rather long equilibration times. The ionization evolution of the plasma depends on the ionization-age parameter $n_e t$, where n_e is the electron density and t is the time that passed since the non-equilibrium conditions were created, either by abrupt heating (hydrodynamical shocks), or by a sudden change in the photoionizing flux. Diagnostics of the ionization age combined with an independent measurement of the electron density can provide valuable information about the age of the X-ray source. The low-density, high-temperature, shocked gas in young supernova remnants (SNR) can take thousands of years to reach ionization equilibrium, a time which often exceeds the age of the remnant. Consequently, non-equilibrium ionization (NEI) conditions are common in these X-ray sources. The most distinct signatures of NEI are emisison lines that arise due to excitations from inner atomic shells by means of collisions with particles that are more energetic than the typical particle energies at which the ion forms in equilibrium.

However, in practice the spectral signatures of NEI are not easily detectable, because it is often difficult to disentangle the ionization state of the plasma from its temperature structure. Different temperature structures can mimic the NEI effects. For example, it has been found (Behar *et al.* 2001b, van der Heyden *et al.* 2002) that exactly the same spectral pattern of helium-like oxygen lines forms under high-temperature (*c.* 600 eV) NEI conditions, as under lower-temperature (*c.* 200 eV) collisional equilibrium conditions. In X-ray photoionized plasmas, probing NEI is even more difficult, because high-energy photons are ubiquitous to the photoionizing source, both in ionization equilibrium, and also in NEI.

Perhaps the least ambiguous evidence for NEI lies in the iron K-shell emission line complex. In the X-ray spectrum of the SNR N103B obtained with the EPIC camera on board XMM-Newton, Fe-K lines have been measured at an unusually low energy of 6.5 keV (van der Heyden *et al.* 2002). The Fe-K region of the N103B spectrum is shown in Fig. 2.3. The 1s–2p lines of He-like Fe^{24+} are expected at 6.7 keV. In N103B, however, these lines occur at lower energies, because they are emitted by lower charge states of Fe, namely Fe^{18+}–Fe^{20+}. At low-temperature equilibrium, where these ions are regularly formed ($kT_e \sim 0.6-1$ keV), the electron energies are insufficient for 1s–2p (inner-shell) excitations. Conversely, at high-temperature equilibrium, only an extremely small fraction of Fe is in these L-shell charge states. Consequently, the observed 1s–2p lines can be produced *only* by L-shell ions at temperatures that are much higher (*c.* 3.5 keV) than their typical temperatures of formation, or in other words only under NEI conditions. Based on this unambiguous detection of NEI measured with the Fe-K feature at 6.5 keV, the Fe-K component of N103B was diagnosed to be very recently shocked (*c.* 100 y) and still ionizing.

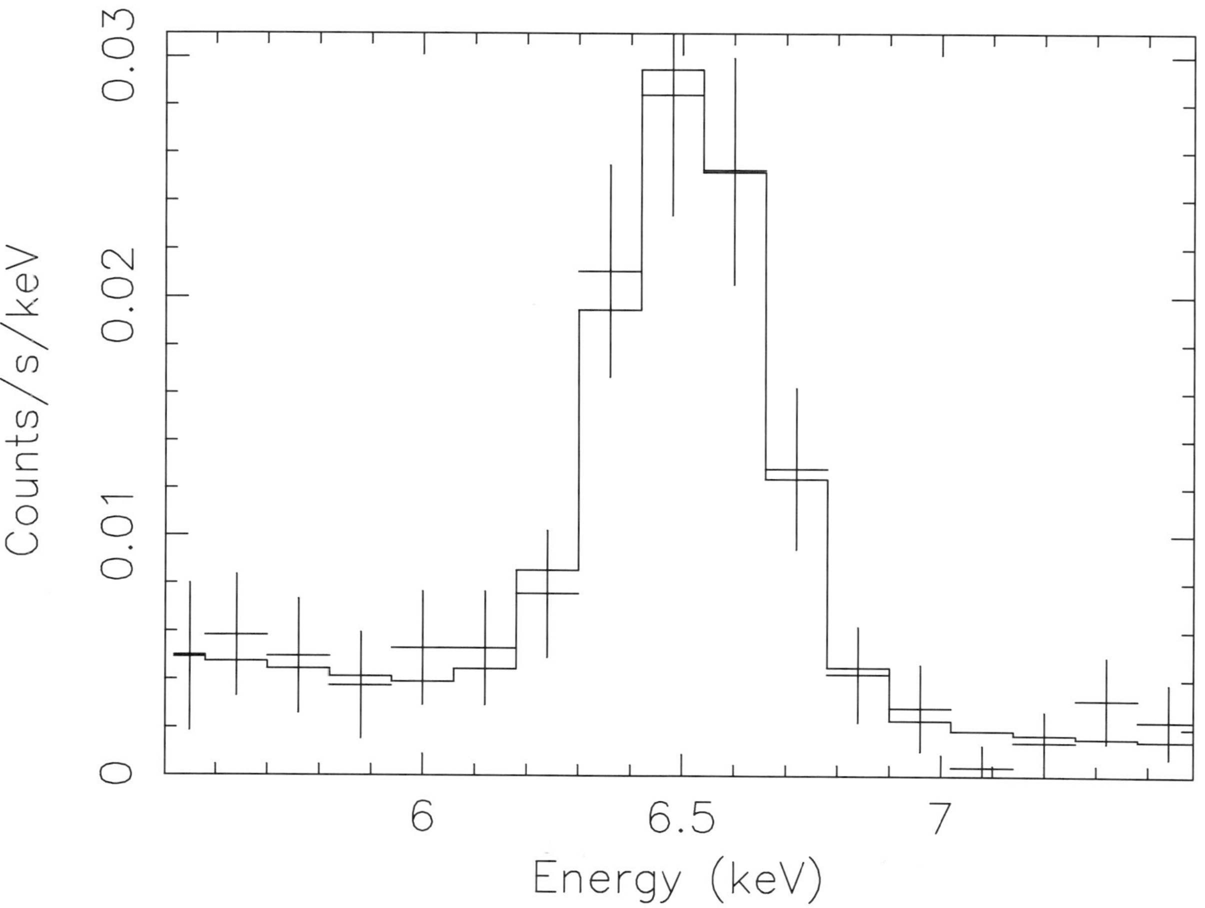

Figure 2.3. XMM-Newton EPIC-pn spectrum of SNR N103B in the range 5.5–7.5 keV (van der Heyden *et al.* 2002), showing the Fe-K emission feature. The solid line is a $kT_e = 3.5$ keV NEI model fit. The line centroid at *c.* 6.5 keV clearly indicates that the gas is not in ionization equilibrium.

Hydrogen-like ions

At the characteristic temperatures of X-ray-emitting plasmas, the low-Z abundant elements (such as carbon, nitrogen and oxygen) are often found in their hydrogen-like charge states. The most prominent emission lines from hydrogen-like ions are the Lyman-series transitions:

$$\begin{array}{l} \text{Lyman } \alpha_{1,2}\text{: } 2\text{p}\,^2P_{3/2,1/2} \rightarrow 1\text{s} \\ \text{Lyman } \beta_{1,2}\text{: } 3\text{p}\,^2P_{3/2,1/2} \rightarrow 1\text{s} \\ \text{Lyman } \gamma_{1,2}\text{: } 4\text{p}\,^2P_{3/2,1/2} \rightarrow 1\text{s} \\ \qquad\qquad\vdots \end{array}$$

These lines are usually quite bright, and are therefore good for abundance and velocity determinations. Examples are shown in Fig. 2.4, which displays the XMM-Newton reflection-grating spectrum of the SNR 1E0102-72.3 in the Small Magellanic Cloud (Rasmussen *et al.* 2001). This young core-collapse remnant is likely to be an oxygen-rich type-1b SNR akin to CasA (Blair *et al.* 2000), so the spectrum is dominated by lines of elements produced by α-burning reactions. The Lyman-series lines (α through γ) of hydrogen-like carbon, oxygen, neon and magnesium are clearly visible in the spectrum, as marked in the figure.

Despite their prominence in astrophysical X-ray spectra, Lyman-series transitions have rather limited utility as density and temperature diagnostics. Lines in this series are all produced through electric dipole transitions, so the radiative decay rates are high, and the collisional couplings are negligible. In addition, because of the n^{-2} dependence of the hydrogen-like energy levels (where n is the principal quantum number), the upper levels for the different transitions in the series are close in energy, so the Boltzmann factor in the excitation rates varies only slightly from transition to transition in the temperature range where the hydrogen-like ion is the dominant species.

At the very low temperatures characteristic of photoionized plasmas, Lyman-series lines are formed by radiative cascades following either photoexcitation, or radiative recombination. The line ratios produced by these processes depend strongly on the balance between those two mechanisms, but in any case differ from the ratios expected for collisional excitation in collisional plasmas. As both photoexcitation and recombination are more efficient than collisional excitation in populating high-lying levels, the high series lines are generally stronger in photoionized plasmas than they are in collisional plasmas.

For photoionized plasmas, the high series emission lines provide a sensitive probe for the column density through the absorbing (and re-emitting) medium. At increasingly higher column densities, the low series lines saturate, thus reducing the efficiency of photoexcitation for producing these lines, while photoexcitation of the

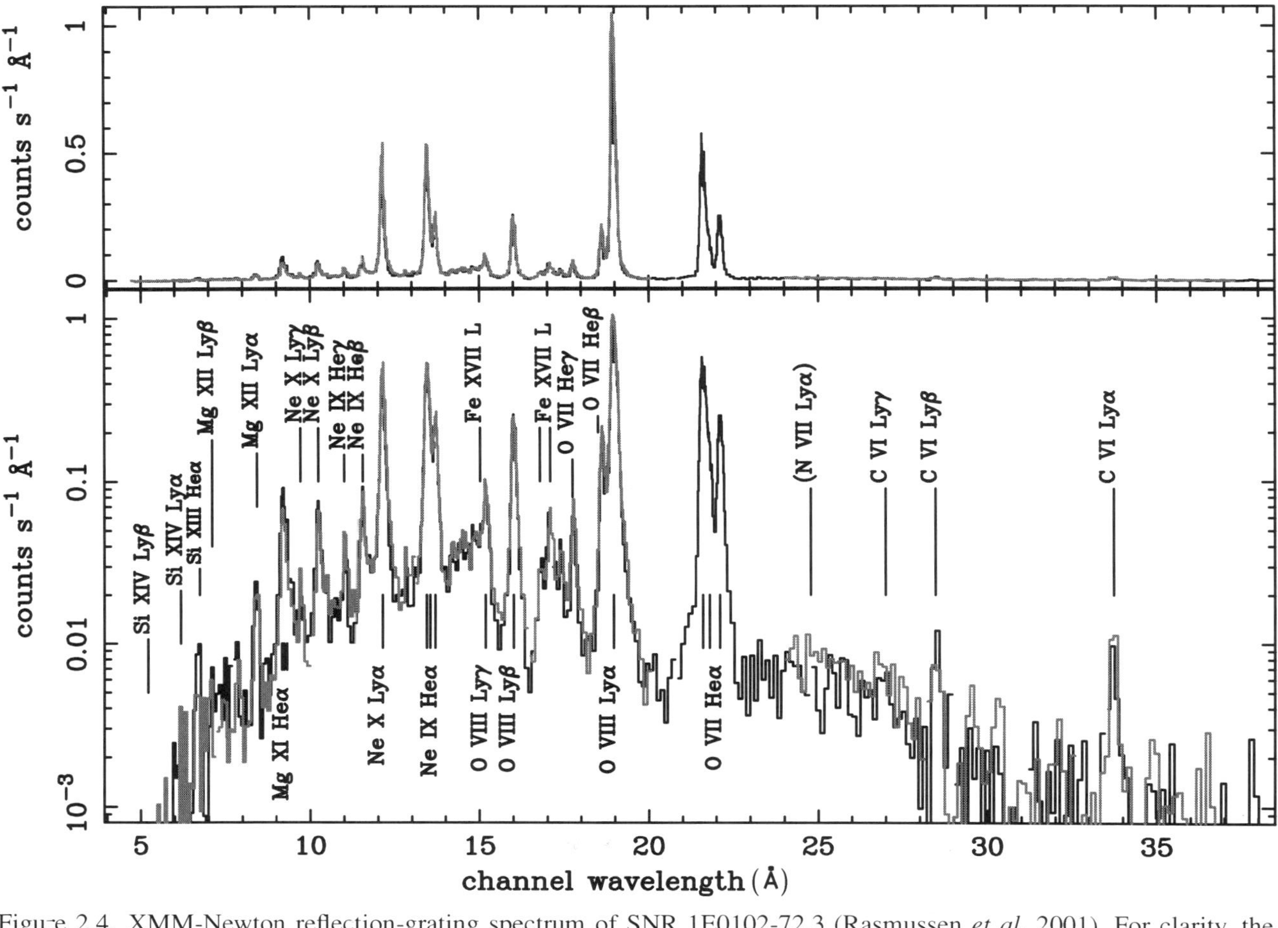

Figure 2.4. XMM-Newton reflection-grating spectrum of SNR 1E0102-72.3 (Rasmussen *et al.* 2001). For clarity, the spectrum is shown in both linear (top) and logarithmic units (bottom). Hydrogen- and helium-like emission lines from carbon to silicon are present with significant emission from iron L-shell transitions as well.

high series lines (lower f-values) persists. At very high column densities, the high series lines also saturate and the spectrum approaches that expected for a purely recombining plasma. The column density at which each line saturates depends directly on the absorption line profile, which in most astrophysical sources is dominated by (macro)turbulence, with characteristic velocities of at least 100 km s^{-1}. The ability to determine column density from emission spectra is particularly useful for X-ray observations of Seyfert 2 galaxies, where the ionizing source is obscured and thus the direct absorption spectrum is unavailable. Perhaps the best example to date of the use of this method is presented in Kinkhabwala *et al.* (2002) for the Seyfert 2 galaxy NGC 1068.

The sensitivity of the Lyman series line ratios to column density is illustrated in Fig. 2.5 for C^{5+}. At the point where Lyα saturates (c. 10^{16} cm^{-2}), the ratios of other lines to Lyα rise, eventually reaching values of up to a factor of a few higher than their respective low column-density values. As the column density increases further, the other members of the series saturate progressively, while the RRC intensity (represented by the diamonds in Fig. 2.5) keeps rising. Although the details of the curves in Fig. 2.5 depend on the turbulent velocity (taken here to be 800 km s^{-1}), the intensity ratio trends depicted are generic to emission by photoionized gas. The ratios measured for NGC 1068 (represented by the asterisks) are clearly in the high ionic column-density regime, at about 10^{18} cm^{-2}. The values for the same ratios, but for collisional equilibrium conditions ($kT_e = 100$ eV for C^{5+}), are considerably lower, as indicated by the horizontal bars in the figure. In short, high intensity ratios of high Lyman series lines to Lyα represent unambiguous evidence for a photoionized source. For more details on the underlying model, see Kinkhabwala *et al.* (2003) and Behar *et al.* (2002).

Helium-like ions

The helium-like K-shell lines are among the most important in X-ray spectra of cosmic sources. Since the helium-like ground state is a tight 'closed' shell, this is the dominant ion species for each element over a wide range in temperature, particularly in collisionally ionized plasmas. In addition, as we explain below, these lines exhibit strong sensitivity to electron density, temperature, the ultraviolet radiation field, and ionization conditions in the emitting plasma.

The most important K-shell helium-like transitions are as follows:

W: $1s2p\,^1P_1 \rightarrow 1s^2\,^1S_0$
X: $1s2p\,^3P_2 \rightarrow 1s^2\,^1S_0$
Y: $1s2p\,^3P_1 \rightarrow 1s^2\,^1S_0$
Z: $1s2s\,^3S_1 \rightarrow 1s^2\,^1S_0$

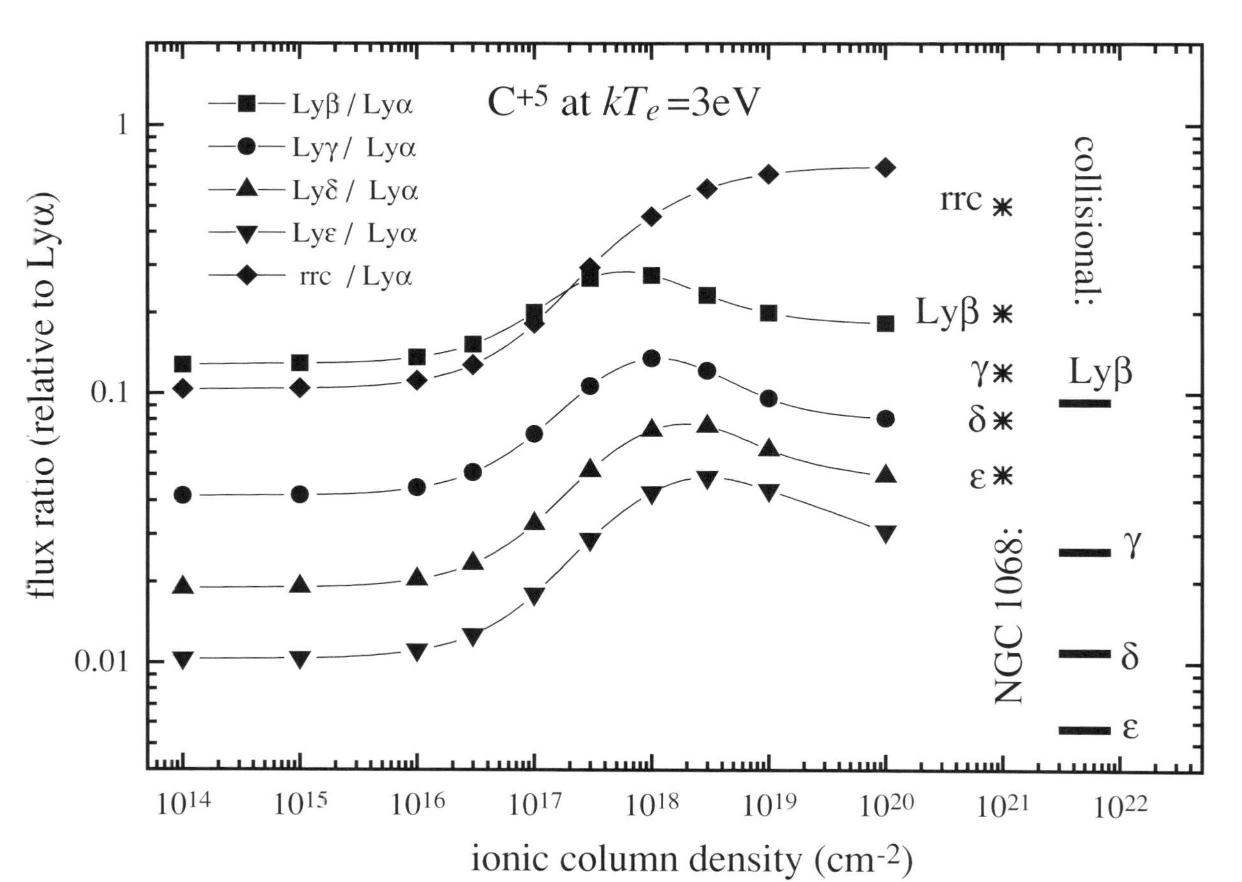

Figure 2.5. Intensity ratios of the Lyman series lines (β–ϵ) and RRC (diamonds) of C^{5+} to Lyα (Behar *et al.* 2002). The values measured for NGC 1068 with the reflection grating spectrometer are plotted as asterisks. The values for collisional C^{5+} gas at a temperature of 100 eV are indicted by horizontal bars. The ratios level out due to gradual saturation of the Lyman series lines. The ratios for collisional gas are very low, even lower than those in the low column-density, photoionized scenario.

W is an electric dipole transition, also called the *resonance* transition, and is sometimes designated by r. X and Y are the so-called *intercombination* lines. These are usually blended (especially for the lower-Z elements), and are collectively designated by i. Z is the *forbidden* line, often designated by f. It is a relativistic magnetic-dipole transition, with a very low radiative-decay rate.

The temperature sensitivity of these lines arises as follows (Gabriel & Jordan 1969; Pradhan 1982; Porquet & Dubau 2000): since W is an electric dipole transition, the collision strength for collisional excitation of this line includes important contributions from higher-order terms in the partial wave expansion, and thus continues to increase with energy above threshold. By contrast, X and Z are both electric-dipole forbidden. The dominant term in the excitation collision strength for these transitions involves electron exchange. Therefore, their excitation collision strengths drop off strongly with energy above threshold, whereas Y, which involves a $J = 1$ upper level, mixes with the ${}^1P_1$ upper level of X, and, as a result, remains relatively constant. Thus, the line ratio $G = (\mathrm{X} + \mathrm{Y} + \mathrm{Z})/\mathrm{W}$ is a decreasing function of electron temperature.

The density sensitivity comes from the fact that the ${}^3S_1$ level can be collisionally excited to the ${}^3P$ levels. At high enough electron density, the process successfully competes with radiative decay of the forbidden line. Therefore, the ratio

$$R = \mathrm{Z}/(\mathrm{X} + \mathrm{Y})$$

drops off above a critical density, n_c. The critical density depends strongly on Z. For C^{4+}, $n_\mathrm{c} \sim 10^9\ \mathrm{cm}^{-3}$, while, for Si^{12+}, $n_\mathrm{c} \sim 10^{13}\ \mathrm{cm}^{-3}$.

These dependencies are illustrated in Fig. 2.6, which shows the helium-like spectra of oxygen, nitrogen and carbon for two stellar coronal sources, Procyon and Capella, as measured with the Chandra low-energy transmission-grating spectrometer (Ness *et al.* 2001). The corona of Procyon is both cooler and of higher density than that of Capella. As can be seen, the forbidden line of carbon is comparatively suppressed due to the higher density in the Procyon spectrum. However, the density is not high enough to alter the oxygen lines. Here the forbidden and intercombination line intensities are comparatively greater for Procyon, because of the lower temperature.

The helium-like line ratios can also be affected by the presence of a significant ultraviolet radiation field (Gabriel & Jordan 1969). In particular, the ${}^3S_1$ level can be photoexcited to ${}^3P$ levels, prior to radiative decay, if there is sufficient ultraviolet intensity at the energy of the relevant transitions. That leads to suppression of the forbidden line and enhancement of the intercombination lines, mimicking the effects of high electron density. This process is an important contributor in the spectra of early-type stars, and some accretion-powered sources (Kahn *et al.* 2001).

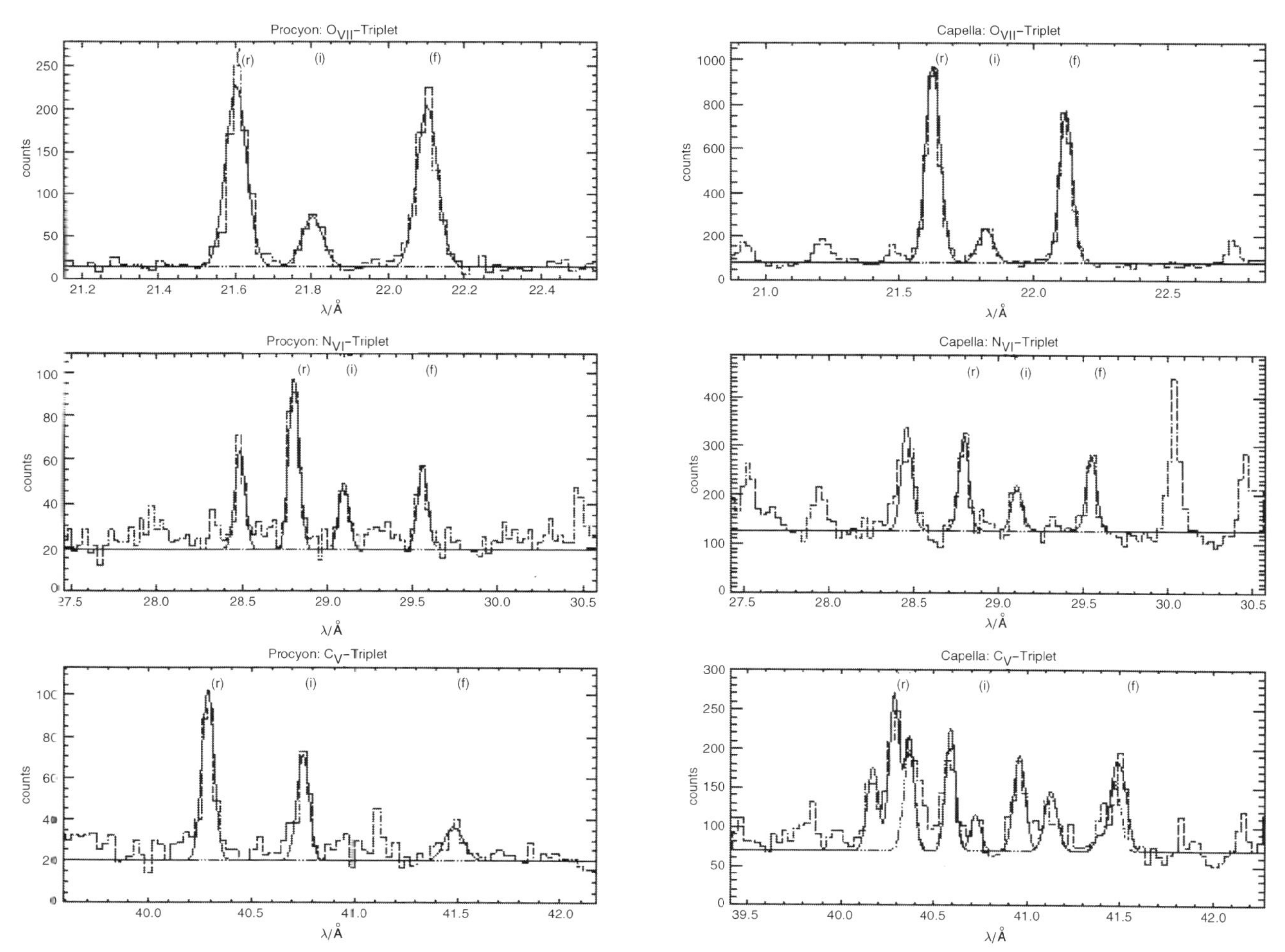

Figure 2.6. Helium-like triplets for O, N and C from the coronal stars Procyon and Capella (Ness *et al*. 2001): Procyon O_{VII} triplet; Capella O_{VII} triplet; Procyon N_{VI} triplet; Capella N_{VI} triplet; Procyon C_V triplet; Capella C_V triplet.

In photoionized plasmas, as in the hydrogen-like case, the excited levels for helium-like ions are fed by photoexcitation, direct radiative recombination, and by radiative cascades following recombination onto higher levels. At high column densities, the forbidden line is most intense, since most of the cascades from high-n and high-l (high-J) levels land on the lowest-lying 1s2s ($J = 1$) level, which produces the forbidden line. This can be seen in the spectrum of NGC 1068 shown in Fig. 2.2, for both the helium-like oxygen lines near 22 Å and the helium-like nitrogen lines near 29 Å. However, when photoexcitation from the continuum is important, the resonance line will also be enhanced, partly offsetting this effect. As for the hydrogen-like case, the various contributions can be disentangled by looking at the higher series helium-like transitions (Kinkhabwala *et al.* 2002, 2003). For high-density photoionized plasmas, the forbidden line is suppressed, and the intercombination lines may actually dominate the spectrum (Cottam *et al.* 2001).

Iron L-shell transitions

Since iron is the most abundant high-Z element, its L-shell spectrum plays a crucial role in astrophysical X-ray spectroscopy. As a result of their higher ionization potentials, the iron L-shell ions contribute significant line emission even when the lower-Z elements are fully stripped. For collisionally ionized plasmas, this complex samples a wide range in temperature (0.2–2 keV). In addition, the L-shell spectrum is very 'rich', and there is significant diagnostic sensitivity.

The brightest iron L-shell lines are of the form:

$$2s^2 2p^{k-1} 3d \rightarrow 2s^2 2p^k$$
$$2s^2 2p^{k-1} 3s \rightarrow 2s^2 2p^k$$
$$2s 2p^k 3p \rightarrow 2s^2 2p^k$$

The 2p–3d lines generally have the highest oscillator strength because these transitions tend to have the highest overlap integrals. The line positions are a strong function of charge state. Thus, the ionization structure is easily discernible, which provides a simple, abundance-independent constraint on the temperature distribution.

This is illustrated in Fig. 2.7, which shows the iron L-spectrum of Capella, as observed with the Chandra high-energy transmission-grating spectrometer. Plotted below the measured data are the calculated contributions from each of the individual charge states, ranging from sodium-like iron (Fe^{15+}) to beryllium-like iron (Fe^{21+}). Note the relatively clean separation between the line complexes from each of these ions, allowing for easy decomposition of the spectrum.

Density sensitivity arises from the fact that the intermediate iron L charge states (e.g. those that are nitrogen-like and carbon-like) possess a number of low-lying

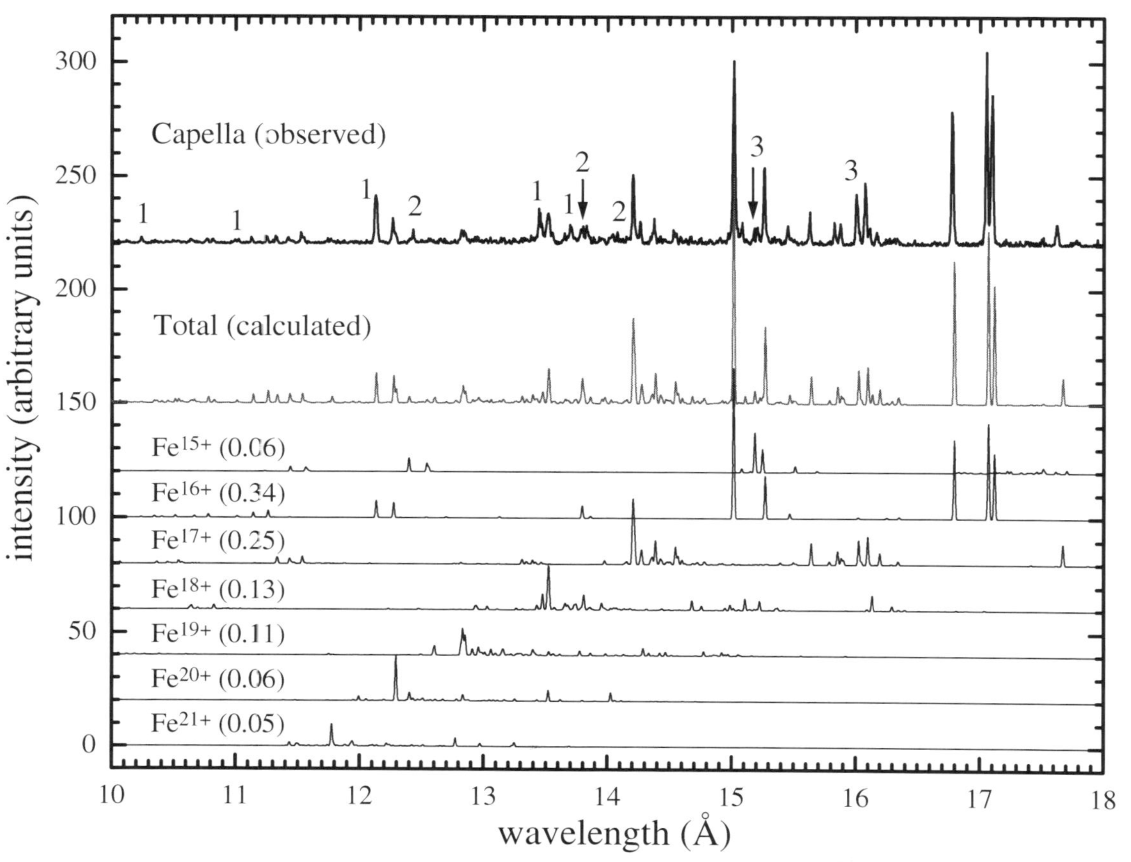

Figure 2.7. Observed Capella spectrum compared with the total calculated iron (only) spectrum (Behar *et al.* 2001). The seven separate plots at the bottom of the figure show the individual ion spectra of Fe^{15+} to Fe^{21+} (with fractional contributions to the total in parentheses). The spectra are calculated assuming plasma conditions of $kT_e = 600$ eV, which provides a very good agreement with the data. The non-iron lines are indicated by the numbers 1–3, corresponding, respectively, to lines of neon, nickel and oxygen.

metastable levels associated with $n = 2 \rightarrow n' = 2$ excitations. These can be populated collisionally, leading to new 'seed' states for $2 \rightarrow 3$ excitations followed by $3 \rightarrow 2$ radiative decays. Such density diagnostics 'turn on' at electron densities of *c.* 10^{13} cm^{-3}.

This is illustrated in Fig. 2.8, where relative line intensities of C-like Fe^{+20} are plotted for increasing electron densities. As the density exceeds a few times 10^{12} cm^{-3}, more relatively intense lines arise as low-lying (seed) levels become populated and available for further excitation.

The iron K-shell complex

The iron K complex is relatively isolated in the spectrum at energies *c.* 6–7 keV, where even non-dispersive detectors have moderate spectral resolution (see Fig. 2.9 for a synthetic iron K-spectrum). Thus, iron K lines were the first discrete atomic features unambiguously detected for cosmic-X-ray sources (see, for example, Sanford *et al.* (1975), Pravdo *et al.* (1976), Mitchell *et al.* (1976)).

An important contributor to iron K emission, especially for accretion-powered sources, is fluorescence from cold material in the vicinity of a bright X-ray continuum. Fluorescence involves a radiative decay following inner-shell photoionization, i.e. a transition of the form $1s2s^2 2p^k nl \rightarrow 1s^2 2s^2 2p^{k-1} nl$. The excited level, in this case, can also decay via autoionization by ejecting one or more of the outer electrons in the valence shell. This latter process dominates for low-Z elements. However, since radiative decay rates scale like Z^4 and autoionization decay rates scale like Z^0 (see e.g. Cowan (1981)), the fluorescence yield becomes appreciable for a high-Z element like iron. The near-neutral iron K fluorescence line falls at 6.4 keV, easily distinguishable from the helium-like lines near 6.7 keV and the Lyman α line at 6.97 keV.

In more diffuse cosmic sources, we often find iron K-shell lines produced by collisional excitation (CE) of helium-like iron and by dielectronic recombination (DR) of helium-like iron forming lithium-like iron. Ratios of lines produced by DR and CE of helium-like iron can be used as temperature diagnostics that are largely independent of the ionization balance in the emitting gas (Dubau & Volonte 1980; Bely-Dubau *et al.* 1982).

DR is a two-step recombination process that begins when an ambient electron excites a bound electron of an ion, and is simultaneously captured. The core excitation process involves an $n - n'$ transition where n is the principal quantum number of the initially bound electron. The incident electron is captured into a higher-lying Rydberg level (and is often referred to as the spectator electron). For capture to occur, energy conservation requires that $E_k = \Delta E - E_b$, where E_k is the kinetic energy of the incident electron, ΔE is the energy for the core excitation, and E_b is the binding energy released when the incident electron is captured onto the excited

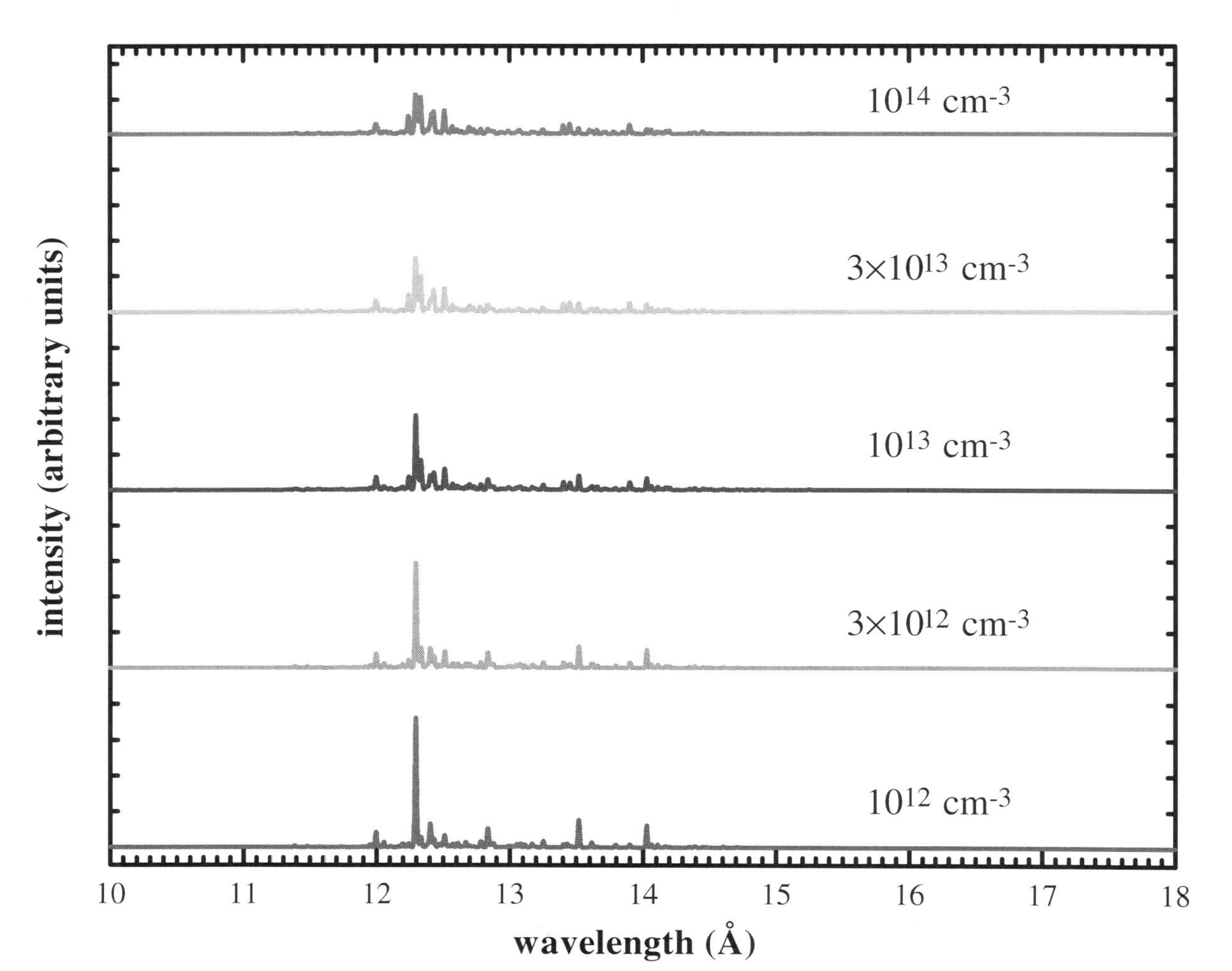

Figure 2.8. Synthetic spectrum of C-like Fe^{20+} as a function of electron density.

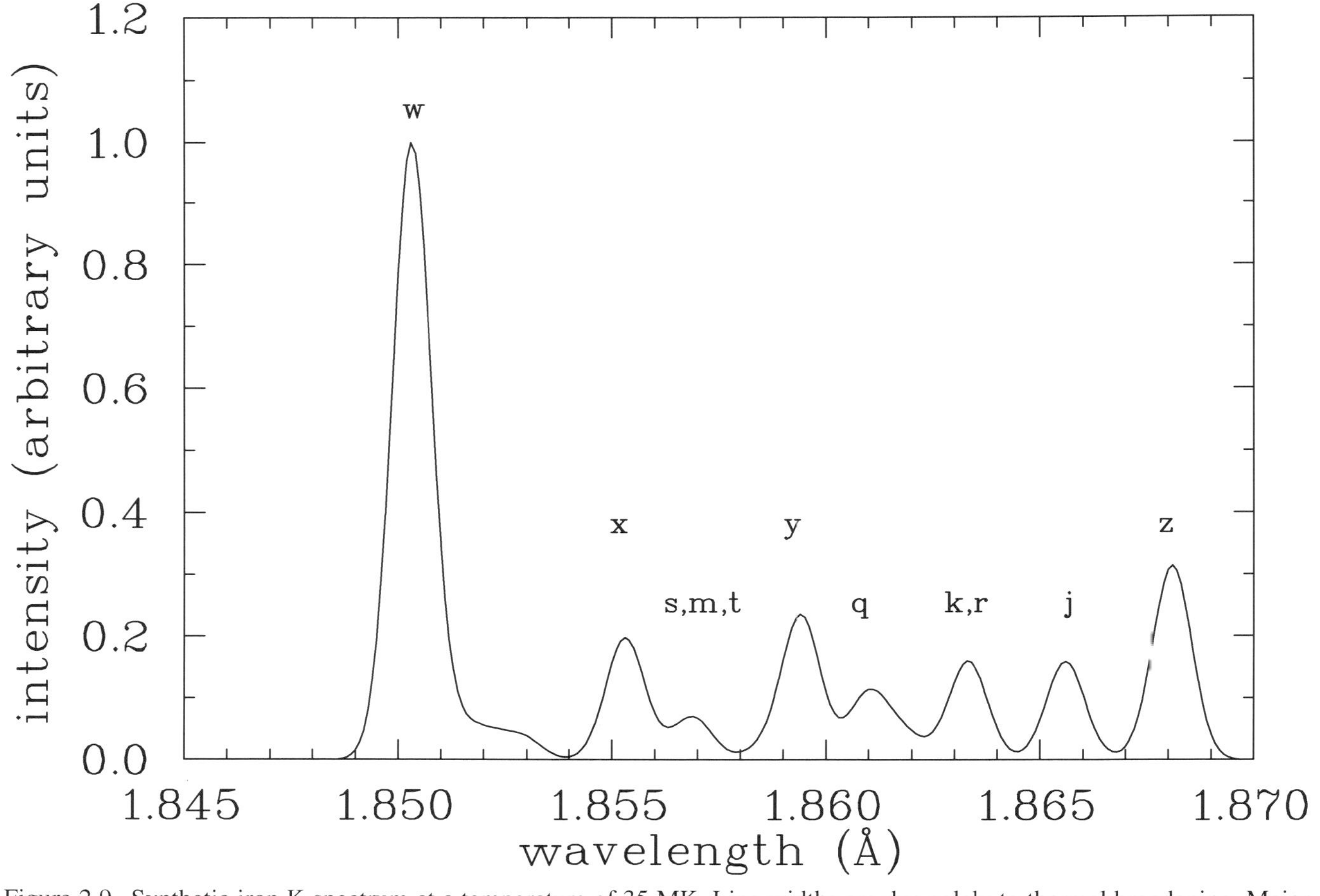

Figure 2.9. Synthetic iron K spectrum at a temperature of 35 MK. Line widths are due solely to thermal broadening. Major features are labelled using the convention of Gabriel (1972). The relevant atomic data are from Bely-Dubau *et al*. (1979a,b, 1982). Wavelengths have been shifted to agree with the laboratory measurements of Beiersdorfer *et al*. (1993).

ion. The total energy of this doubly-excited, intermediate state lies in the continuum of the recombined ion and the system may autoionize. DR is complete when the recombined ion relaxes radiatively and emits a photon which reduces the total energy of the recombined system to below its ionization threshold.

Typically the radiative stabilization results from an n' transition. For capture into very high levels, the spectator electron does not affect the excited ionic core and the energy (wavelength) of this transition will be indistinguishable from n' transitions produced by CE of the target ion. However, when the captured electron falls into a level comparable with that of the excited core electron, it can screen the core and reduce the energy of the n' transition. This results in 'satellite' lines in the recombined ion which are slightly longer in wavelength than the n' CE-generated lines in the target ion. Because DR samples the portion of the free electron energy distribution at energies below ΔE and CE samples the portion at energies of ΔE and above, ratios of DR-generated to CE-generated lines can therefore be used as a temperature diagnostic.

At astrophysical densities, all atoms are in the ground state. Most of the satellite lines cannot be produced by collisional excitation of lithium-like iron (e.g. $1s2p^2 \rightarrow 1s^2 2p$). They come purely from DR on helium-like atoms. However, other lines involve upper level configurations which are single electron excitations of the lithium-like ground-state (e.g. $1s2s2p \rightarrow 1s^2 2s$). These can be produced by both collisional excitation of lithium-like ions and DR on helium-like ions. Hence, the line ratios for these various transitions provide an independent measure of the charge balance. Analysis of the iron K helium-like spectrum thus provides independent constraints on the electron temperature and the level of ionization, and is ideal for investigating departures from ionization equilibrium.

Summary

X-ray spectroscopy provides a unique and powerful new avenue of investigation for the study of a wide range of astrophysical sources. Discrete X-ray line diagnostics can be used to provide unambiguous measurements of temperatures, densities, excitation conditions, ionization balance and elemental abundances. Many of these techniques have already been employed in the analysis and interpretation of data acquired by the grating experiments on Chandra and XMM-Newton, especially at soft-X-ray energies. This field is still in its infancy. There are undoubtedly many surprises yet to come.

References

Behar, E., Cottam, J. & Kahn, S. M. 2001a *Astrophys. J.* **548**, 966.
Behar, E. *et al.* 2001b *Astron. Astrophys.* **365**, L241.

Behar. E. *et al.* (2002) In *Mass Outflows in Active Galactic Nuclei: New Perspectives*, (ed. D. M. Krenshaw, S. B. Kraemer, & I. M. George), p. 43. San Francisco, CA: ASP.
Beiersdorfer, P. *et al.* 1993 *Astrophys. J.* **409**, 846.
Bely-Dubau, F., Gabriel, A. H. & Volonté, S. 1979a *Mon. Not. R. Astron. Soc.* **186**, 405.
Bely-Dubau, F., Gabriel, A. H. & Volonté, S. 1979b *Mon. Not. R. Astron. Soc.* **189**, 801.
Bely-Dubau, F., Dubau, J., Faucher, P. & Gabriel, A. H. 1982 *Mon. Not. R. Astron. Soc.* **198**, 239.
Blair, W. P. *et al.* 2000 *Astrophys. J.* **537**, 667.
Brinkman, A. C. *et al.* 2000 *Astrophys. J.* **530**, L111.
Canizares, C. R. *et al.* 2000 *Astrophys. J.* **539**, L41.
Cottam, J. C., Kahn, S. M., Brinkman, A. C., den Herder, J. W. & Erd, C. 2001 *Astron. Astrophys.* **365**, L277.
Cowan, R. D. 1981 *The Theory of Atomic Stucture and Spectra*. Berkeley, CA: University of California Press.
den Herder, J. W. *et al.* 2001 *Astron. Astrophys.* **365**, L7.
Dubau, J. & Volonte, S. 1980 *Rep. Prog. Phys.* **43**, 199.
Gabriel, A. H. 1972 *Mon. Not. R. Astron. Soc.* **160**, 99.
Gabriel, A. H. & Jordan, C. 1969 *Mon. Not. R. Astron. Soc.* **145**, 241.
Giacconi, R., Gursky, H., Paolini, F. & Rossi, B. B. 1962 *Phys. Rev. Lett.* **9**, 439.
Kahn, S. M. *et al.* 2001 *Astron. Astrophys.* **365**, L312.
Kinkhabwala, A. *et al.* 2002. *Astrophys. J.* **575**, 732.
Kinkhabwala, A., Behar, E., Sako, M., Gu, M. F., Kahn, S. M. & Paerels, F. 2003. *Astrophys. J.* in press, astro-ph/0304332.
Liedahl, D. A. 1999 In *X-ray Spectroscopy in Astrophysics* (ed. J. van Paradijs & J. A. M. Bleeker), p. 189. Heidelberg: Springer.
Mazzotta, P., Mazzitelli, G., Colafrancesco, S. & Vittorio, N. 1998 *Astrophys. J. Suppl.* **133**, 403.
Mitchell, R. J., Culhane, J. L., Davison, P. J. & Ives, J. C. 1976 *Mon. Not. R. Astron. Soc.* **175**, 29.
Ness, J.-U. *et al.* 2001 *Astron. Astrophys.* **367**, 282.
Porquet, D. & Dubau, J. 2000 *Astrophys. J. Suppl.* **143**, 495.
Pradhan, A. K. 1982 *Astrophys. J.* **263**, 477.
Pravdo, S. H. *et al.* 1976 *Astrophys. J.* **206**, L41.
Rasmussen, A. P., Behar, E., Kahn, S. M., den Herder, J. W. & van der Heyden, K. 2001 *Astron. Astrophys.* **365**, L231.
Sanford, P., Mason, K. O. & Ives, J. 1975 *Mon. Not. R. Astron. Soc.* **173**, 9.
Van der Heyden, K. J. *et al.* 2002 *Astron. Astrophys.* **392**, 955.

3

X-rays from stars

BY MANUEL GÜDEL

Paul Scherrer Institut

Introduction

There is little *a priori* reason for outer atmospheres of stars to engage in high-energy processes. Yet, over the past decades it has become clear that only a few stellar classes abstain from doing so. First discovered in 1948 (Burnight 1949), *solar* X-rays originate in the outer, extended atmosphere of the Sun, the corona. More recent imagery shows it to be a highly structured, variable and dynamic envelope of the Sun, with measured temperatures of up to a few million K (MK; see Fig. 3.1). Its confinement and structure are basically due to the presence of magnetic fields that are anchored in surface active regions. Understanding the mechanisms that heat the coronal plasma has been one of the great challenges of solar physics, driving numerous theoretical, numerical and observational studies.

One of the basic questions of stellar X-ray astronomy is whether coronae like the Sun's exist on other stars, and if so, in which types of star we detect them, and how they compare with the solar example, both phenomenologically and physically. By studying stellar samples of different ages, we can trace the high-energy evolution of the Sun and the solar system back in time. Further, we would like to understand why hot, massive stars that are not supposed to possess magnetic fields are nevertheless luminous X-ray sources. A look at the Hertzsprung–Russell diagram (HRD) of detected X-ray stars in Fig. 3.2, compiled from selected catalogues of survey programs (Alcalá *et al.* 1997; Berghöfer *et al.* 1996; Hünsch *et al.* 1998a,b, 1999; Lawson *et al.* 1996) shows all basic features expected from an optical HRD (we plot each star at the locus of the optically determined absolute magnitude M_V and the colour index $B-V$ regardless of possible unresolved binarity). Although the samples used for the figure are in no way 'complete' (in volume or brightness), the main sequence is clearly evident, and so is the giant branch. The top right part

Frontiers of X-Ray Astronomy, ed. A.C. Fabian, K.A. Pounds and R.D. Blandford. Published by Cambridge University Press.

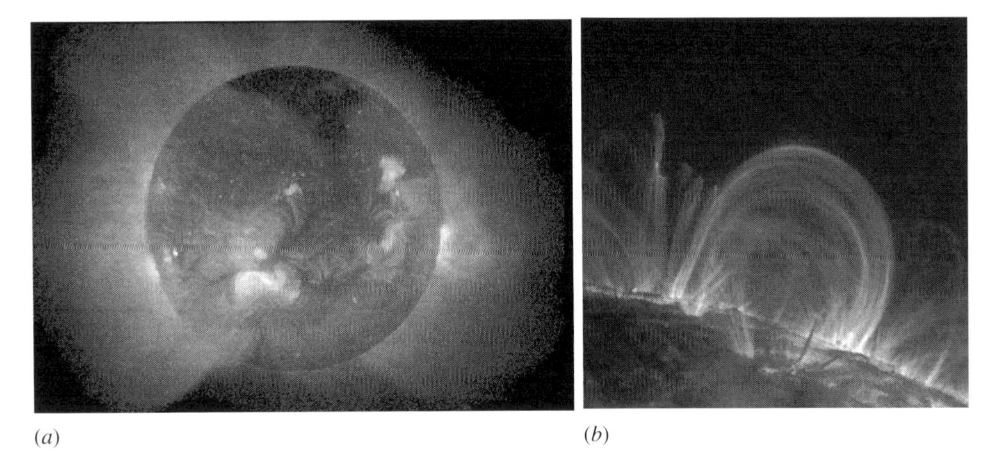

(*a*) (*b*)

Figure 3.1. (*a*) The Sun observed in X-rays by the Yohkoh satellite. Inhomogeneities associated with magnetic active regions are evident (picture from the Yohkoh mission of ISAS, Japan). (*b*) Active region showing a sample of complex coronal loops in the extreme ultraviolet range, as observed by the TRACE satellite (picture from the NASA Trace project).

of the diagram, comprising cool giants, is almost devoid of detections, however. The so-called corona-wind dividing line (dashed in Fig. 3.2; after Linsky & Haisch (1979)) separates coronal giants and supergiants to its left from stars with massive winds to its right. It is unknown whether the wind giants possess magnetically structured coronae at the base of their winds – the X-rays may simply be absorbed by the overlying wind material. The few residual detections may at least partly be attributed to low-mass companions. The large remaining area from spectral class M up to at least mid-F comprises stars that are – in the widest sense – solar-like. Their X-rays are produced in a hot corona similar to the solar corona, i.e. an ensemble of tangled magnetic structures in the form of loops connecting photospheric and chromospheric active regions (Fig. 3.1).

The underlying mechanism that generates the magnetic fields in cool stars, the 'magnetic dynamo', operates through an interaction between rotation (more precisely, differential rotation) and convection. The X-ray luminosity L_X is indeed strongly correlated with the rotation period P in otherwise similar stars, approximately fulfilling $L_X \propto P^{-2}$ (Pallavicini *et al.* 1981). In the most rapid rotators with periods of less than 1–2 days, however, L_X empirically 'saturates' around a value of $\approx 10^{-3} L_{bol}$. Since L_{bol} increases toward earlier main-sequence spectral types and toward giants, the upper limit to the coronal X-ray luminosity also increases. This effect is clearly visible in Fig. 3.2. Saturation may result from the coverage of the entire surface by active regions. Alternatively, it may be the internal dynamo itself that saturates once the rotation period is sufficiently short.

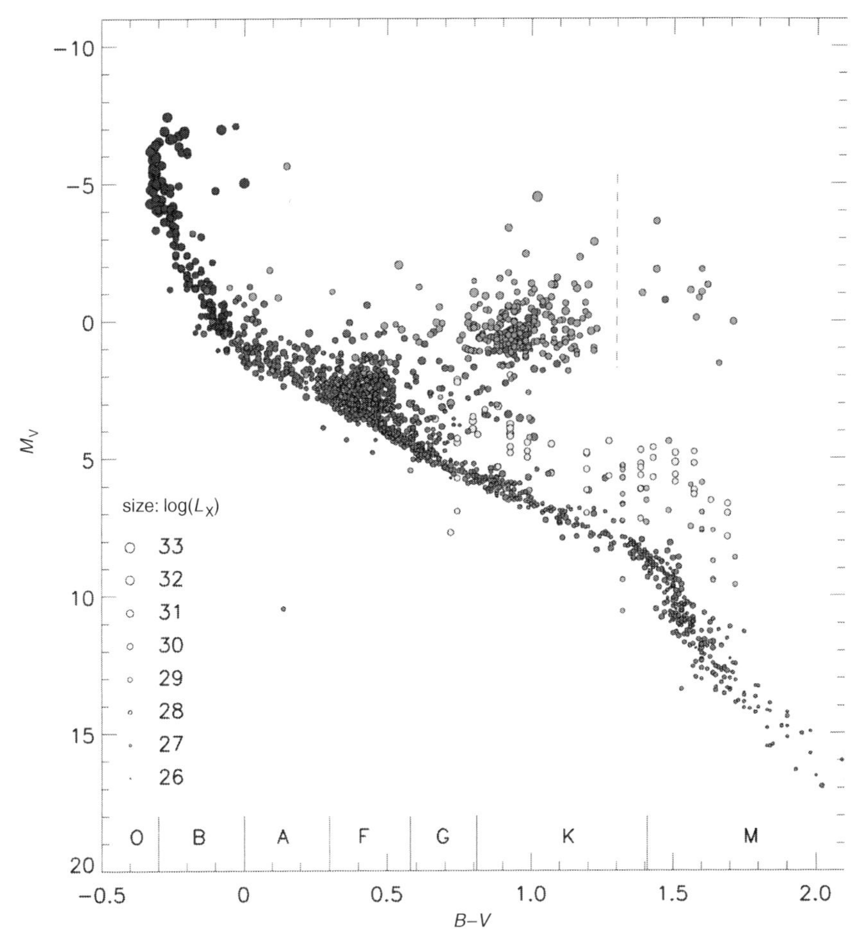

Figure 3.2. Hertzsprung-Russell Diagram HRD based on about 2000 X-ray detected stars extracted from the catalogs by Berghöfer *et al.* (1996) and Hünsch *et al.* (1998a,b, 1999). Where missing, distances from the Hipparcos catalogue (ESA 1997) were used to calculate the relevant parameters. The low-mass pre-main sequence stars are from studies of the Chamaeleon I dark cloud (Alcalá *et al.* 1997; Lawson *et al.* 1996) and are representative of other star formation regions. Different greyscales are used for different catalogues without signifying any physical distinction. The size of the circles characterizes $\log L_X$ as indicated in the panel at lower left.

In any case, the picture employing an increasing coronal volume for increasing stellar size fails to explain another puzzling feature of cool star X-ray emission. The characteristic coronal temperature rises with increasing X-ray luminosity or, using a measure that in this picture is less dependent on stellar size, with increasing L_X/L_{bol} (see e.g. Schrijver *et al.* (1984)). If explicit measurements of coronal temperatures are not available, spectral hardness ratios are commonly used as a

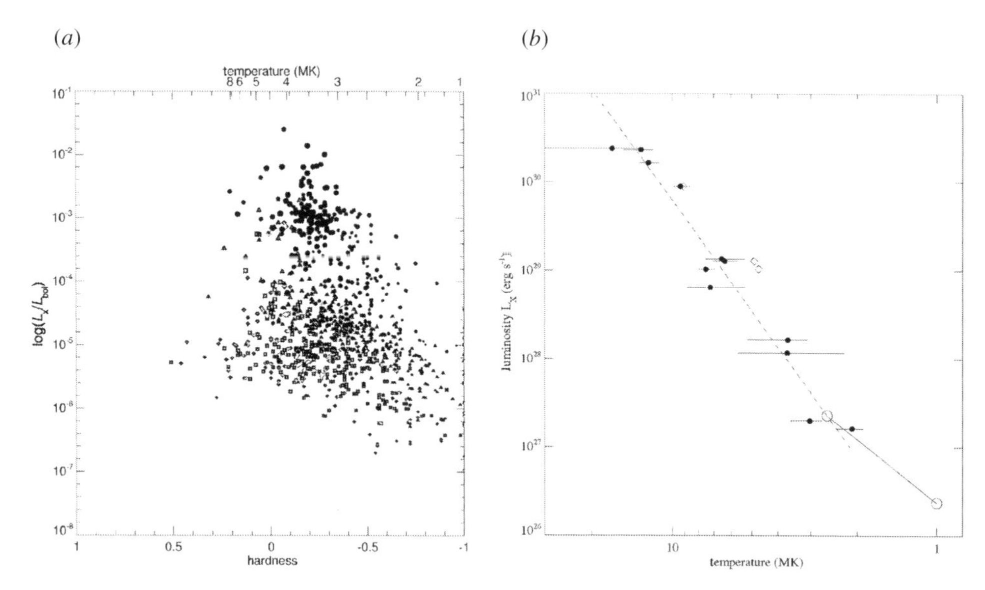

Figure 3.3. (*a*) Spectral hardness as defined for ROSAT PSPC survey spectra (see Hünsch *et al.* (1998b) for a definition) for stars in Fig. 3.2 closer than 50 pc, divided into the classes of main-sequence mid-K to late-M dwarfs (circles), main-sequence G to mid-K dwarfs (triangles), main-sequence F stars (quadrangles), F- and G-type subgiants (diamonds), and late-type giants (crosses). The symbol sizes characterize $\log L_X$. Increasing hardness is to the left, corresponding to increasing characteristic coronal temperature. The top axis gives corresponding temperatures for an isothermal plasma between 1 and 8 MK. (*b*) Dependence of the coronal temperature of solar analogues (ROSAT sample from Güdel *et al.* (1997)) on the total non-flaring X-ray luminosity. The circles give the range for the solar corona after Peres *et al.* (2000). The squares refer to emission-measure weighted average coronal temperatures of three solar analogs as derived from archival XMM-Newton data of the G stars EK Dra, π^1 UMa, and χ^1 Ori.

temperature indicator (see Hünsch *et al.* (1998b) for a definition used for ROSAT PSPC observations). Figure 3.3(a) shows the loci of the cool stars from the sample in Fig. 3.2 on a hardness versus L_X/L_{bol} plot, distinguishing between five stellar classes: mid-K to late M dwarfs, G to mid-K dwarfs, F dwarfs, F/G-type subgiants, and late-type giants. The hardness axis is inverted, with increasing hardness (roughly corresponding to increasing coronal temperature) toward the left, somewhat analogous to the optically defined HRD. Since the more massive and more evolved classes are rarer and on average farther away, their X-ray spectra may be more absorbed, resulting in a larger hardness. To limit such selection bias, Fig. 3.3(a) shows stars within 50 pc only. Although flaring sources have not been flagged, the five classes show different trends on this diagram, with the coolest dwarfs displaying the steepest and the giants displaying the flattest distribution. One obvious bias may be introduced by undetected low-mass companions of larger stars, in which case L_{bol} is overestimated for the relevant star, thus making the trend flatter. Also, L_{bol} depends on the effective photospheric temperature; its magnitude per unit surface area is thus larger for earlier-type stars. The least biased sample is clearly provided by the coolest main-sequence stars. There is thus little doubt that 'more active stars' display hotter coronae. This is supported by a dedicated study of solar analogs for which the coronal temperatures were determined from spectral fits to ROSAT data (Fig. 3.3(b); the hotter of the usually two dominant temperatures has been used (Güdel *et al.* 1997)).

Because the outer convection zone is very shallow or non-existent in early-F and A-type stars, the generation of magnetic fields and hence of X-ray luminous coronae is supposed to break down, a picture that is well supported by observations (many of the detected stars within spectral class A in Fig. 3.2 are actually known binaries, with a lower-mass star probably being responsible for the X-rays). Moving toward early-type stars of spectral classes O and B, a completely different picture emerges, with stars now mostly following a trend $L_X \approx 10^{-7} L_{\mathrm{bol}}$ (Pallavicini *et al.* 1981). In single hot stars, the X-rays are thought to be produced in shocks developing in line-driven winds that are intrinsically unstable, with shocks being distributed throughout the stellar wind. Stellar mass loss increases toward hotter stars, and therefore the most luminous X-ray sources are found among early O stars. In hot-star binaries, winds may collide between the two companions, producing a shock zone that becomes the dominant source of X-rays. The most extreme levels of X-ray emission are attained in Wolf–Rayet (WR) stars that suffer heavy mass loss. X-ray detections include both single and binary WR stars.

The research discipline of high-energy processes in stars has been boosted by the satellite missions Chandra and XMM-Newton. Apart from addressing persistent problems (e.g. on coronal heating, on wind-mass loss, on coronal magnetic structure), the new facilities uncover a growing number of physical issues not anticipated before. X-rays (and non-thermal radio emissions) have been sought in quite unlikely places such as brown dwarfs and planets, the earliest stages of stellar evolution, and in stellar jets/outflows from young stars, and they have been detected in all of them. The discipline of stellar high-energy astrophysics has also been spreading out into new, even multidisciplinary fields: for example, we are recognizing the importance of high-energy processes in young stars for their molecular environment; the ionization of the infalling molecular material by the star itself; production of spallation products in the coalescing protoplanetary disc material by stellar flares, etc. The following sections summarize various X-ray findings from the areas of cool stars (problems related to energy release and flares), hot stars (wind shocks and colliding winds), pre-main sequence stars (protostars, T Tau objects, and jets), and brown dwarfs, with emphasis on the common physical mechanisms that may be responsible for the observed phenomena. We will focus on advances

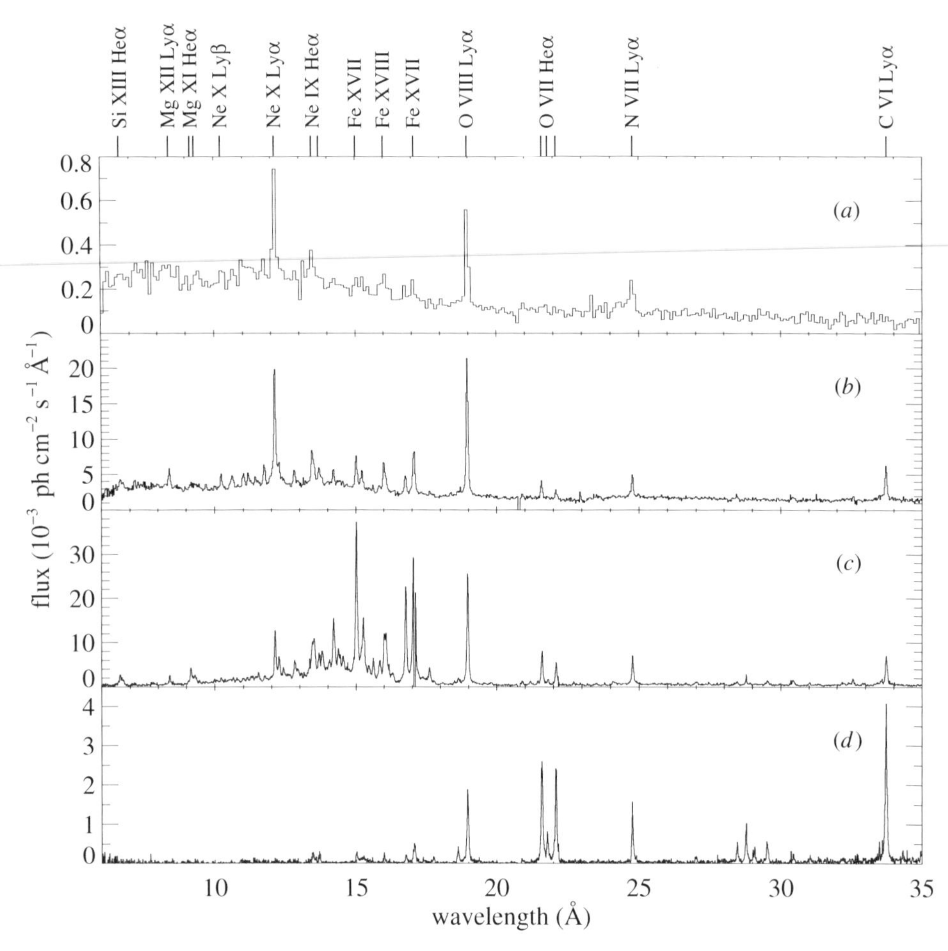

Figure 3.4. Examples of XMM-Newton reflection grating spectrometer spectra covering the temperature ranges observed in quiescent stellar coronae. (*a*) YY Men, an FK Com-type giant, with a dominant temperature of c.30 MK; (*b*) HR 1099, an RS CVn-type binary with $T_{\rm peak} \approx$ 10–30 MK; (*c*) Capella, $T_{\rm peak} \approx$ 7 MK; and (*d*) Procyon (F5 IV–V), $T_{\rm peak} \approx$ 1–2 MK. Note the different continua and line systems.

due to XMM-Newton and Chandra, in particular on results from high-resolution spectroscopy. For an overview of the pre-Chandra/XMM-Newton view of stellar X-ray astronomy, see, for example, Rosner *et al.* (1985), Pallavicini (1989), and Haisch & Schmitt (1996).

The question of coronal heating in cool stars

The ultimate cause of coronal heating remains elusive, even in the case of the Sun. A look at some high-resolution X-ray spectra (Fig. 3.4) of stellar coronae reveals a wide range of plasma conditions. Optically thin spectral lines of various elements, in particular of Fe (L-shell lines around 1 keV), are used together with the continuum to disentangle the temperature structure of coronae (see, for example, the detailed discussion in Behar *et al.* (2001)). The characteristic measured temperatures range between 1–3 MK in rather cool, inactive coronae like the Sun's, and some 30–40 MK in extremely X-ray luminous, magnetically active stars. What causes this difference? Why do stars maintain outer atmospheres 2–3 orders of magnitude hotter than their photospheres?

Coronal heating physics not only defines one of the more fascinating problem areas in astrophysics, with applications far beyond stellar atmospheres, the question of what heats cosmic plasmas like those around stars, including the solar and stellar winds, has also been a driver for excellent studies of theoretical plasma physics. A review of proposed mechanisms, including dissipation of Alfvén waves, acoustic waves, and current dissipation, is beyond the scope of the present chapter. The interested reader is referred to reviews such as those by Narain & Ulmschneider (1990), Ulmschneider *et al.* (1991), or Zirker (1993). Solar observations have added momentum to the hypothesis that a large number of small flares ultimately release the required coronal energy explosively (e.g., Krucker & Benz (1998)). Flare statistics has also been extended to magnetically active stars using long time series from various satellites. Flares appear to be distributed in (radiative) energy E as a power law of the form

$$\frac{dN}{dE} = kE^{-\alpha}; \tag{3.1}$$

the constant k, the exponent α, and possible energy cut-offs of the power-law distribution determine the fraction of coronal heating by flares. In particular, if $\alpha \geq 2$, the extrapolation to small flare energies requires a lower energy cut-off $E_0 > 0$ since otherwise the energy integration diverges. In this case, there is evidently no shortage of energy if E_0 is sufficiently low. Results from magnetically active stars consistently imply best-fit values of $\alpha > 2$ (Audard *et al.* 2000; Kashyap *et al.* 2002; Güdel *et al.* 2003), with cut-offs E_0 in the domain of small solar flares.

Supporting evidence for coronal heating by a large number of flares is found in: coronal temperatures that increase with increasing magnetic activity (as measured, for example by the photospheric spot coverage (Schrijver *et al.* 1984; Güdel *et al.* 1997, also Fig. 3.3); a flare rate that increases proportionally with the X-ray luminosity (Audard *et al.* 2000); and a coronal emission measure distribution compatible with the continuous heating and cooling of a large number of small flares (Güdel *et al.* 1997). It has also been suggested that anomalous coronal abundances found in very active stars are due to flares (Brinkman *et al.* 2001), although the relevant mechanisms are not yet understood. If coronal (or chromospheric) plasma is ultimately heated in flares, then the question on the heating physics reduces to understanding flare physics – a seemingly complex task involving a closer understanding of the magnetic field energy buildup in the corona, the development of instabilities, the mode of explosive energy release (e.g. the debated type of magnetic reconnection), the dominant channel of energy transformation (electric currents, accelerated particle populations, propagation of shocks), and the relevant processes of relaxation (radiation, conduction, expansion). X-ray astronomy shows its strengths in contributions to the last two topics.

Flares as probes of the coronal energy release

The operation of plasma heating is best seen in individual large flares. A standard flare scenario developed in solar physics assumes that high-energy particles (electrons, protons, ions) are accelerated up to hundreds of MeV in regions of reconnecting coronal magnetic fields (Dennis 1988; Hudson & Ryan 1995). The particles, bound to magnetic field lines, travel toward lower atmospheric layers, where they collisionally heat the cooler chromospheric gas. The resulting overpressure drives heated material into the corona ('chromospheric evaporation'), building up the X-ray source. A large hot component with maximum temperatures up to 100 MK is added to the total stellar emission measure distribution, while the cooler plasma remains nearly unchanged (Audard *et al.* 2001b; Huenemoerder *et al.* 2001). When electrons impact at chromospheric levels, a prompt optical enhancement appears predominantly in the optical U band, the 'U-band flare', which is probably related to enhanced Balmer continuum or to black-body emission from the heated surface (see e.g. Hawley *et al.* (1995)). Since the optical luminosity $L_{\rm O}$ roughly traces the injected power while the X-ray luminosity $L_{\rm X}$ scales approximately with the total accumulated energy in the coronal plasma, one expects

$$L_{\rm O} \propto \frac{\rm d}{{\rm d}t} L_{\rm X}, \tag{3.2}$$

analogous to the 'Neupert effect' known from solar observations (Neupert 1968; Dennis & Zarro 1993). Solar observers normally use radio or hard X-ray (greater than 10 keV) observations instead of optical monitoring because the same high-energy particles also emit synchrotron (radio) emission and, when they impact on the chromosphere, non-thermal hard X-ray bremsstrahlung. This important diagnostic for the chromospheric evaporation scenario has stubbornly eluded stellar observers, with few exceptions (e.g., Kahler *et al.* 1982; de Jager *et al.* 1986, 1989). A quantitative analysis of the stellar Neupert effect was given by Hawley *et al.* (1995) and Güdel *et al.* (1996).

Figure 3.5(*a*) shows an example observed by XMM-Newton during a large flare on the M dwarf Proxima Centauri. The correspondence between the U-band light

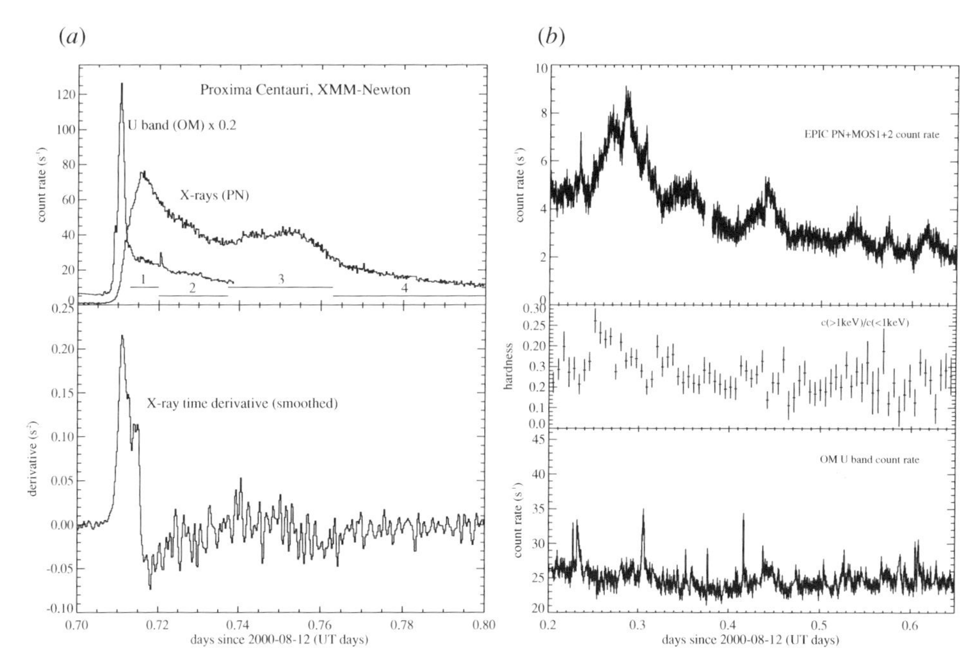

Figure 3.5. (*a*) An example of the Neupert effect seen by XMM-Newton on Proxima Centauri. The top panel shows the EPIC PN X-ray light curve (gradual decay) and the OM U-band light curve (fast decay). The bottom panel shows the time derivative of the X-ray light curve which closely correlates with the U-band curve in the early flare phase. (*b*) Example of low-level behaviour of Proxima Centauri. From top to bottom: XMM-Newton EPIC light curve showing small flares and modulations; hardness (count rate above 1 keV/count rate below 1 keV); U band light curve measured by the OM. Many optical and X-ray flares correlate in time.

curve and the X-ray time derivative is compelling, suggesting an important role of high energy particles in the initial heating and evaporation of chromospheric material.

To return to the overall coronal X-ray emission of stars outside obvious, strong flaring, Fig. 3.5(b) shows an episode on Proxima Centauri that would principally be recognized – by all previous satellites – as slowly modulated ‘emission in quiescence’. There are several surprising aspects of this observation: (i) the presence of continuous low-level X-ray variability, accompanied by a large number of small optical flares; (ii) several optical and X-ray enhancements fulfill (3.2), i.e. they give indirect evidence for frequent operation of chromospheric evaporation during low-level emission; (iii) the spectral hardness is continuously variable, indicating the presence of heating events; and (iv) the densities derived from O VII emission lines (see below) vary as well, indicating the presence of dynamic processes involving coronal magnetic fields.

Coronal structure and composition

Structure and size of stellar coronae are strongly constrained by the distribution of photospheric magnetic fields and the coronal temperature structure. Interferometric observations at radio wavelengths give evidence for extended magnetic structures with sizes up to several stellar radii (Benz *et al.* 1998; Mutel *et al.* 1998) and with a highly ordered, large-scale magnetic topology (Beasley & Güdel 2000). While radio-emitting high-energy electrons tend to prefer low-density, preferably extended magnetic regions where collisional losses are minimized, X-rays originate predominantly from high-density volumes owing to the n_e^2 dependence of the radiative losses. X-ray observations recording stellar eclipses and rotational modulation suggest much more compact structures in magnetically active stellar coronae than seen at radio wavelengths, with rather small coronal heights but appreciable densities of several times 10^{10} cm^{-3} (White *et al.* 1990; Schmitt & Kürster 1993). Eclipse modelling of the M dwarf binary YY Gem with XMM-Newton shows an overall distribution of coronal plasma with a scale height of $(1\text{–}4)\times 10^{10}$ cm $\approx (0.25\text{–}1)R_*$, compatible with the pressure scale height derived from spectroscopic temperature measurements. The inferred coronal densities of up to a few times 10^{10} cm^{-3} are also compatible with direct spectroscopic measurements (Güdel *et al.* 2001b; Stelzer *et al.* 2002).

Because the X-ray luminosity L_X is proportional to the volume emission measure $\mathrm{EM} = \int n_e^2 dV$, flux measurements alone provide ambiguous information on source volume V and electron density n_e. Spectroscopic observations help resolve the degeneracy. XMM-Newton and Chandra provide access to He-like line triplets of C V, N VI, O VII, Ne IX, Mg XI, Si XIII, and S XV. The triplets consist of three closely separated lines due to the resonance transition (r) $1s^2\ ^1S_0\text{–}1s2p\ ^1P_1$, the intercombination transition (i) $1s^2\ ^1S_0\text{–}1s2p\ ^3P_1$, and the forbidden transition (f) $1s^2\ ^1S_0\text{–}1s2s\ ^3S_1$. The flux ratio $R = \mathrm{f/i}$ is density-sensitive (larger ratios for smaller densities (Gabriel & Jordan 1969; see also Chapter 2)), and also slightly dependent on the temperature (see e.g. Porquet *et al.* (2001)). Figure 3.6 shows the He-like triplet of N VI in the quiescent corona of the inactive F subgiant Procyon together with the theoretical prediction for the f/i flux ratio, including effects due to the photospheric ultraviolet radiation field (Ness *et al.* 2001). In this example, electron densities around 10^{10} cm^{-3} are indicated. Figure 3.7 reveals dramatic changes in the electron density during the flare on Proxima Centauri shown in Fig. 3.5(*a*), providing direct evidence for changes in the mass distribution in the emitting coronal volume (the plasma probed by the O VII and the Ne IX triplets is relatively cool, covering the 1–5 MK range).

Density measurements tend to show distinct differences between active and inactive stars. While low densities are found in weakly active or inactive coronal

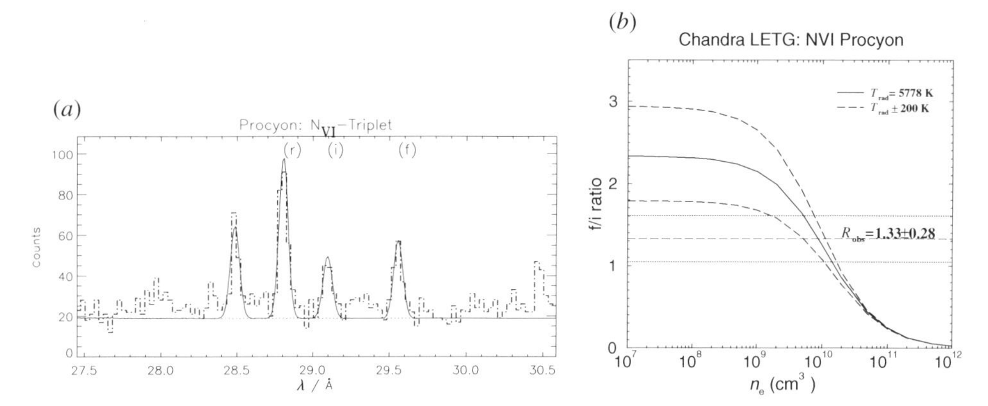

Figure 3.6. (*a*) Example of a density-sensitive line triplet of He-like N VI, measured on the inactive mid-F star Procyon. (*b*) The solid line illustrates the flux ratio between the f and i lines as a function of electron density for different photospheric ultraviolet radiation fields; the observed f/i ratio indicates densities at or below the low-density limit accessible by this triplet ($\approx 10^{10}$ cm^{-3}; from Ness *et al.* (2001)).

stars ($10^9 - 10^{10}$ cm^{-3} (Audard *et al.* 2001a; Brinkman *et al.* 2000; Canizares *et al.* 2000; Mewe *et al.* 2001; Ness *et al.* 2001; Phillips *et al.* 2001; Raassen *et al.* 2002)), electron densities up to several times 10^{10} cm^{-3} are found in magnetically very active stars during 'quiescence' (Güdel *et al.* 2001a,b; Huenemoerder *et al.* 2001; Ness *et al.* 2002), although the correlation is not unique (Audard *et al.* (2001b) and Ayres *et al.* (2001) report low densities in the active RS CVn binary HR 1099). In the context of the flare-heating hypothesis for coronal heating, higher densities are expected in more active stars as the high cadence of flares drives more high-density material into the corona by virtue of chromospheric evaporation, thus building up a corona that is, on average, denser than an inactive corona. Owing to the n_e^2 dependence of the X-ray luminosity, the X-ray losses from dense coronae should be very large. This interpretation is in line with the trend shown in Fig. 3.3 and can be summarized in the working hypothesis of statistical flare heating: more X-ray luminous coronae of otherwise similar stars are hotter because their more frequent large flares increase the coronal density and thus the X-ray luminosity while at the same time keeping the coronal temperatures high.

Elemental abundances in the corona may be an interesting tracer for the coronal heating mechanism since the latter is ultimately responsible for the mass transport into the corona. X-ray spectroscopy can be used to measure coronal abundance anomalies in magnetically active stars. In the solar corona and in the solar wind, elements with a first ionization potential (FIP) below 10 eV, such as Mg, Si, Ca, Fe, Ni, are overabundant by a factor of a few compared with the photospheric abundances, whereas high-FIP (greater than 10 eV) elements, such as C, N, O, Ar, Ne, are closer

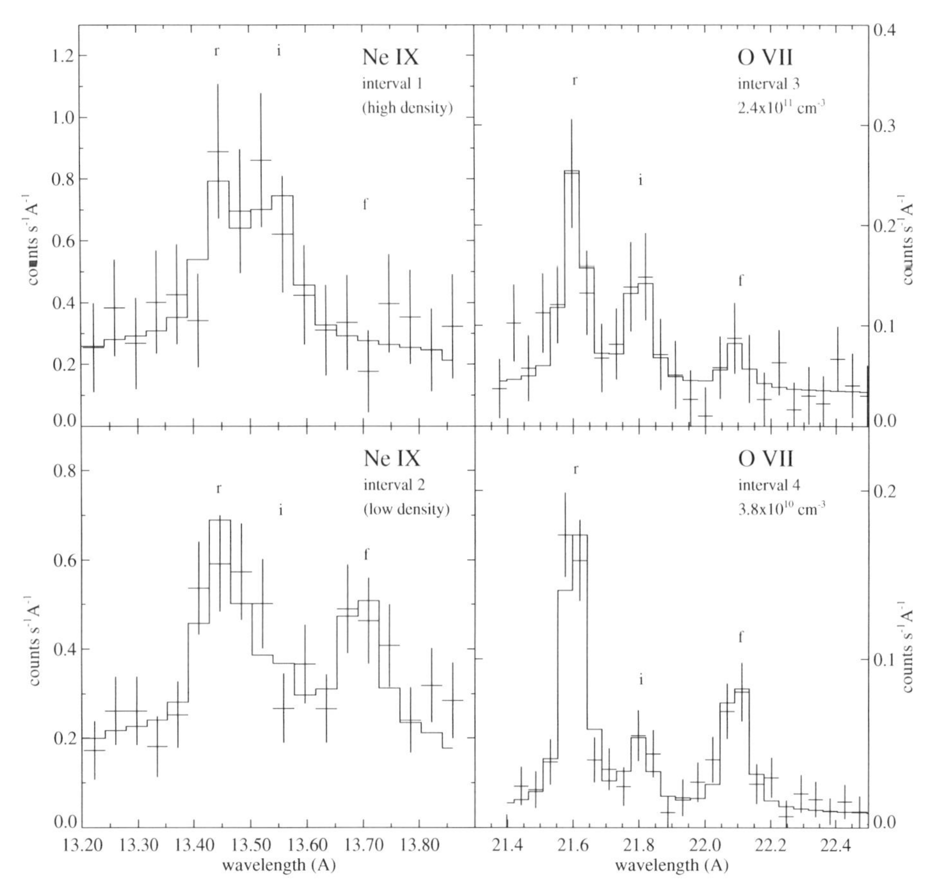

Figure 3.7. Density-sensitive triplet of Ne IX (left) and O VII (right) obtained during four episodes of the large flare shown in Fig. 3.5(*a*) (see marked intervals in that figure). The small f/i ratios (upper panels) indicate high densities ($\approx 4 \times 10^{11}$ cm^{-3} for O VII) during the primary and secondary flare peaks, while the large f/i ratios (lower panels) indicate a rapid return to low densities ($\approx 2 \times 10^{10}$ cm^{-3} for O VII) during the flare decays. The pre-flare densities are similar to the latter.

to their respective photospheric values. This abundance pattern is known as the FIP effect. Physical models involve element fractionation in the chromosphere along pressure and temperature gradients. The chromosphere has the right temperature (4000–10000 K) for high-FIP elements to be predominantly neutral and for low-FIP elements to be considerably ionized. Many models therefore require the presence of electric or magnetic fields that affect ions and neutrals differently (for reviews, see Feldman (1992) and Hénoux (1995)). Abundance anomalies in stars have been reported since the ASCA and EUVE eras but the detected patterns were at variance with solar findings. Very active stars reveal low abundances for all measured

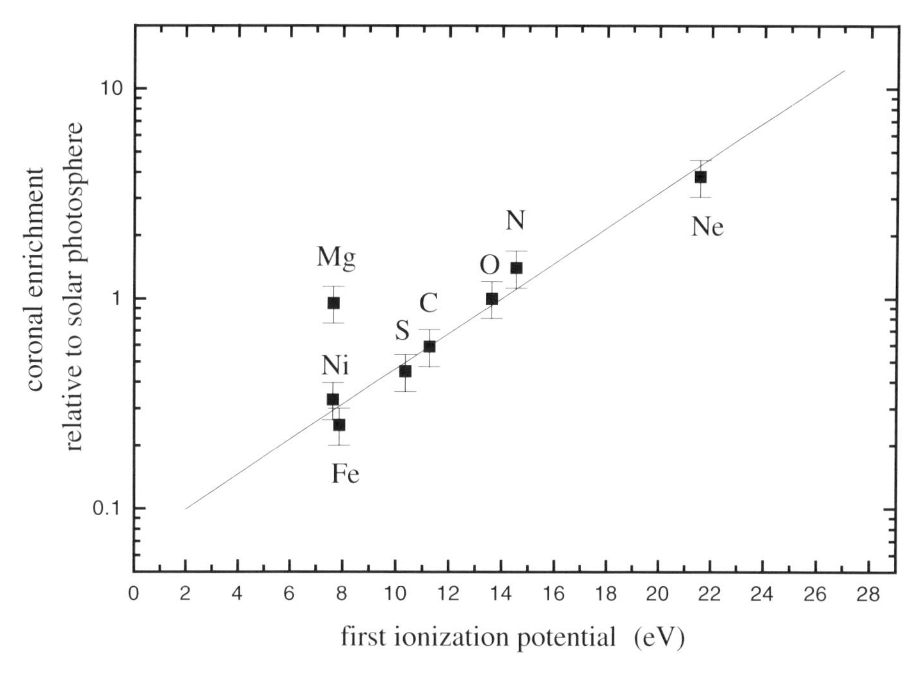

Figure 3.8. Elemental abundance pattern in the active RS CVn-type binary HR 1099, measured by XMM-Newton. The coronal abundances are given relative to solar photospheric abundances, with O normalized to unity (reproduced from Brinkman *et al.* (2001)).

elements (White 1996; Drake 1996; Singh *et al.* 1996; Schmitt *et al.* 1996, Mewe *et al.* 1997), while less active stars show trends toward an FIP effect, or they reveal a coronal composition identical to the photospheric composition (Drake *et al.* 1995, 1997; Laming *et al.* 1996). High-resolution spectroscopy with XMM-Newton and Chandra accesses a larger number of individual elements. In very active stars, an inverse FIP effect is now regularly found, with low-FIP elements being underabundant relative to high-FIP elements (Fig. 3.8; Brinkman *et al.* 2001; Audard *et al.* 2001b; Güdel *et al.* 2001a), resulting in a large Ne/Fe abundance ratio (Drake *et al.* 2001); it seems that this effect weakens toward less active stars and may switch over to the solar-type FIP effect in inactive stars. The physical cause of this behaviour and its relation to the heating mechanism are presently unclear.

On the other hand, elemental abundances may partially reflect the abundances of the stellar photosphere, and since heavy elements fundamentally determine the radiative losses of a coronal gas, the metallicity of a star may partially determine its X-ray coronal properties. A study of the metal-poor open cluster NGC 2516 and a comparison with the co-eval but metal-rich Pleiades cluster indicates that the activity level of M dwarfs is insensitive to stellar metallicity, whereas the median X-ray luminosity of G–K dwarfs was found to be lower in NGC 2516 (Sciortino *et al.* 2001; Harnden *et al.* 2001 – but see the new metallicity determination of NGC 2516 by Sung *et al.* 2002).

Toward younger stars: From discs and jets to astro-ecology

Energetically violent processes occur in a star's life at its birth and during its evolution to a normal, hydrogen-burning star. This is somewhat perplexing as stars form in the coldest places of the universe, namely inside dense, massive molecular clouds with temperatures of 10–30 K. Star formation proceeds through four phases, with different signs of high-energy mechanisms (for reviews, see Shu *et al.* (1987) and Feigelson & Montmerle (1999)): (i) As a molecular cloud core contracts and accumulates matter in its centre, a cool protostar forms while most of the future mass of the star is still infalling. These so-called class 0 protostars are mainly accessible by millimeter and radio observing techniques since optical light and X-rays are blocked by the dense molecular envelope. Class 0 sources often drive massive outflows. The total time spent in this phase is of order 10^4 y. (ii) Once the star has accreted most of its mass and is still deeply embedded in its molecular envelope, it strongly interacts with its accretion disc; magnetic activity, located on the star, on the accretion disc, or in magnetic fields between the star and its disc, becomes evident. Well-developed jets and outflows indicate an important role of mass ejection while the star accretes. These class I objects are usually detected as infrared sources, as thermal radio jet sources, and now also as energetic X-ray stars. This phase of stellar evolution lasts approximately 10^5 y. (iii) As the molecular envelope is dispersed with time, the star becomes fully optically revealed while still retaining its accretion disc. Such class II or classical T Tau stars (CTTS) are sources of strong X-ray emission, probably due to the presence of solar-like coronae but possibly also owing to interactions between magnetic fields on the star and on the disc. (iv) When the accretion disc also disperses, the contracting star spins up to very high rotation rates (rotation periods of order 1 day or less), inducing a strong magnetic dynamo. In this weak-lined T Tau phase (class III, WTTS), magnetic activity is thought to be entirely due to a magnetic corona. Both the CTTS and the WTTS phases last for about 10^6–10^7 y, after which time the star approaches the Zero-Age Main Sequence and starts spinning down due to angular momentum loss via a magnetized wind.

An extensive review of high-energy aspects of star formation and early evolution covering the pre-Chandra/XMM-Newton era is given in Feigelson & Montmerle (1999). While X-ray (and radio) emission from CTTS and WTTS was abundantly detected in young stellar clusters and star-forming regions by the early X-ray satellites, the strongly absorbing environments of class 0 and I protostars made the identification of X-rays – and hence of magnetic activity in general – more of a

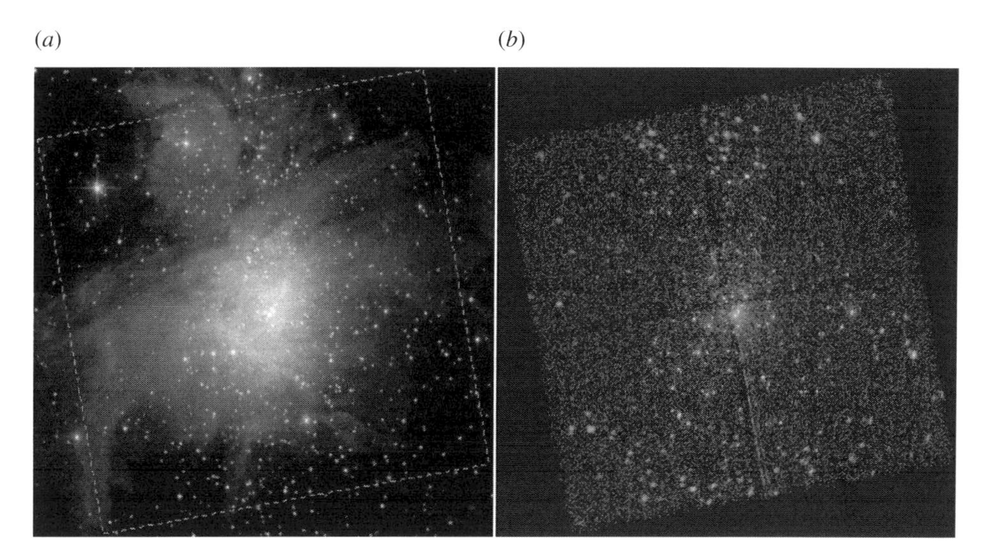

Figure 3.9. The ONC seen (*a*) at near infrared wavelengths (2MASS) and (*b*) in X-rays. The Chandra image reveals at least 1000 pre-main sequence X-ray sources (Garmire *et al.* 2000; Feigelson *et al.* 2002b; figures reproduced courtesy of N. Grosso).

challenge. First detections of class I protostars with ROSAT and ASCA include sources in the ρ Oph molecular cloud (Casanova *et al.* 1995; Kamata *et al.* 1997) and in Corona Australis (Koyama *et al.* 1996). Flares with outstanding total radiative energies ($L_X \approx 10^{33}$–10^{36} erg s^{-1}) indicate that a large volume around the star, of order 0.1 AU, must be involved in the explosive release of magnetic energy (Grosso *et al.* 1997). The magnetic environment of such young stars may thus be very different from a solar-like corona. Magnetic geometries in young stars potentially include the following: solar-type multipolar fields anchored in the stellar photosphere, some of them possibly being of 'fossil' nature, i.e. left-over magnetic fields from the collapsing molecular cloud; magnetic fields linking the stellar surface with the inner border of the accretion disc (typically near the corotation radius); magnetic fields above the disc, resulting from mass ejection after reconnection events; fields that are eventually carried away by jets and outflows; or fields anchored completely on the accretion disc, induced by convection and differential rotation in the disc itself (see Feigelson & Montmerle (1999) for further discussion).

Two of the most prolific nearby star-formation regions (SFR) have obtained extensive observations with Chandra and XMM-Newton, namely the Orion Nebula Cluster (ONC) and the ρ Ophiuchi Dark Cloud. The ONC image from Chandra contains at least 1000 stars between 0.1 My and a few My of age (Garmire *et al.* 2000; Feigelson *et al.* 2002b; Flaccomio *et al.* 2002a,b; Fig. 3.9). Many of them are in the classical (disc-surrounded) or weak-lined (discless) T Tau stage of evolution.

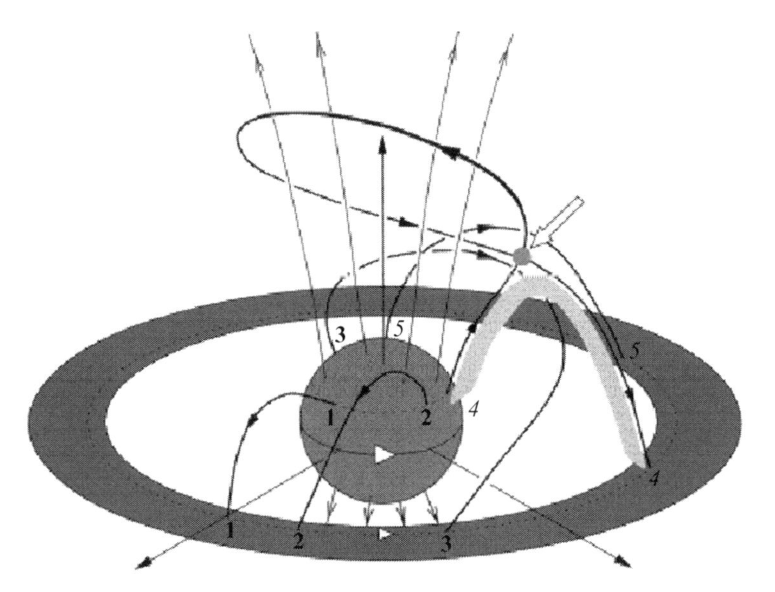

Figure 3.10. Sketch showing wound-up magnetic fields between a protostar and its accretion disc. If the star and the disc do not rotate synchronously, magnetic fields are wound up (labels 1→5). Interacting magnetic field lines may release energy by reconnection for example in class I protostars (from Montmerle *et al.* (2000)).

The observed X-ray spectra point at the presence of very hot plasma, with a median of *c.* 30 MK (Feigelson *et al.* 2002b). The most outstanding aspect of the sample is its high rate of variability, including X-ray flares from massive stars in the Trapezium. The X-ray luminosity is found to be maintained at a high level as the stars contract down the Hayashi track, i.e. during the first 2 My for a solar-like star. At later stages, during the radiative track (2–10 My), part of the stellar sample declines in L_X (Garmire *et al.* 2000). However, there appears to be no discernible effect on the X-ray luminosity due to the presence or absence of an accretion disc across a large range of ages (Feigelson *et al.* 2002a; see also Preibisch & Zinnecker 2001).

Discs may be more important in younger stars for which the inner disc and the stellar surface are not rotating in phase. The star-disc magnetic fields may wind up, producing reconnection and consequently releasing large amounts of energy (Montmerle *et al.* 2000; Fig. 3.10). Such non-synchronous rotation is probably present in deeply embedded class I protostars that are now regularly detected as X-ray sources despite their large hydrogen absorption column densities (Imanishi *et al.* 2001a; Feigelson *et al.* 2002b). The characteristic plasma temperatures increase toward more embedded stars, pointing to a higher efficiency of magnetic energy release. Large flares are also regularly detected among these objects; they suggest mechanisms that involve reconnection in large-scale magnetic fields

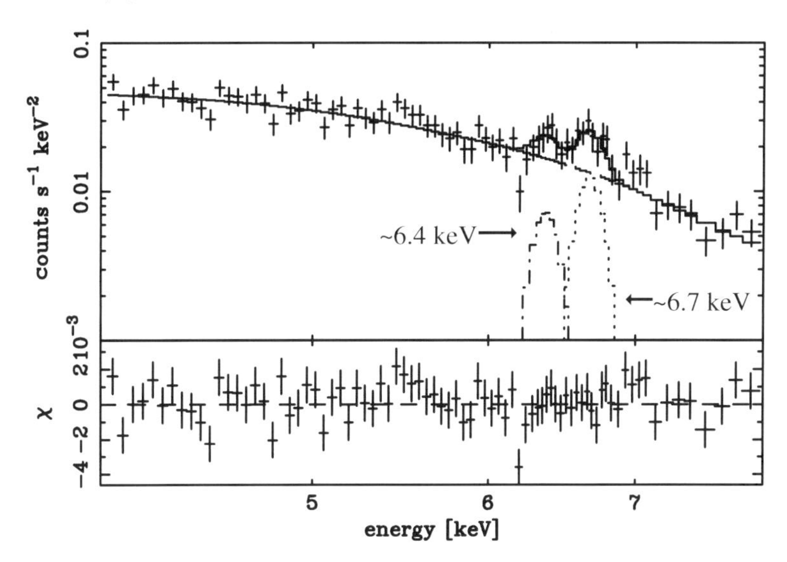

Figure 3.11. Possible fluorescence on a protostellar accretion disc is seen in the 6.4 keV 'cold' Fe line next to the Fe XXV complex at 6.7 keV during a strong flare on the class I object YLW 16A (from Imanishi *et al.* (2001a)).

between the star and the inner accretion disc (Montmerle *et al.* 2000) although their overall characteristics (peak temperature, peak emission measure) fit well into an extrapolation of solar flares to larger energies (Imanishi *et al.* 2001a).

X-rays also give direct evidence for the presence of accretion discs. Kastner *et al.* (2002) report unusually small f/i ratios for He-like Ne IX lines in the CTTS TW Hya, indicating plasma densities up to *c.* 10^{13} cm^{-3}. They interpret the X-rays as coming from the base of dense accretion columns on the stellar surface. Alternatively, the R ratio of a low-density plasma could also, or additionally, be reduced by a strong UV radiation field – see the section on X-rays from hot stars. Direct evidence for inner accretion discs may now be available through fluorescence lines formed by irradiation of an extended surface, for example an accretion disc, by hard emission from flares. Imanishi *et al.* (2000a) report both the Fe XXV 6.7 keV line and a fluorescence 6.4 keV line during a strong flare on the class I protostar YLW 16A (Fig. 3.11).

Some objects detected in the ONC display externally illuminated protoplanetary discs (proplyds) that harbour young, forming stars. Chandra detects very hard and strongly variable X-rays at the location of several of these objects (Schulz *et al.* 2001). The variability time scales (10 min–2 h) suggest that the X-rays are formed in the immediate stellar environment, perhaps in magnetic coronae, rather than at the shock interface between the envelope and the wind from the nearby O star θ^1 Ori C.

The considerable X-ray variability observed in class I protostars and T Tau stars suggests the presence of magnetic fields. However, going back in time to class 0 sources, we hardly expect to find copious sources of X-rays around these nascent stars embedded in dense, cool molecular material. Nature seems to teach us otherwise. Tsuboi *et al.* (2001) detected several X-ray sources coincident with class 0 condensations identified at millimetre wavelengths in Orion OMC-3. Although systematic confirmation is badly needed, the tentative properties of the sources are surprising: their luminosities of 10^{30} erg s^{-1} are very similar to those of more evolved T Tau stars!

Protostars and some T Tau stars drive prominent bi-polar outflows and optical and radio jets that often terminate in Herbig–Haro objects. The latter represent shock regions between the outflowing jet material and the circumstellar material, plus shocks internal to the jet itself. Although shocks are expected to heat the gas to temperatures of *c.* 10^6 K, protostellar jets have only very recently been detected in X-rays. Pravdo *et al.* (2001) find a weak but definitive X-ray source at the leading edge of the HH2 knot. Similarly, Favata *et al.* (2002) report a clear X-ray source at the position of the L1551 IRS5 jet. Temperatures range up to several MK. While of considerable interest by themselves, the jet X-ray sources may play an important role by irradiating the star's accretion disc with high-energy photons along a low-absorption path from above the disc, thus efficiently ionizing the disc molecular material out to large distances from the star, and thereby coupling the disc to magnetic fields (Favata *et al.* 2002).

Young stars may also affect their physical and chemical environments by high-energy particles (in particular, protons) ejected during flares or accelerated by coronal mass ejections (see Glassgold *et al.* (2000) for a review). Meteorites show evidence for violent high-energy interactions in the young solar system, e.g. anomalously high abundances of daughter products of short-lived nuclids like ^{41}Ca, ^{26}Al, or ^{53}Mn, or chondrules requiring flash melting of solid material to *c.* 2000 K (see Feigelson *et al.* (2002a)). The required high-energy protons are amply present in young stellar systems. From scaling arguments based on the surprisingly high rate of very large flares in pre-main sequence stars, Feigelson *et al.* (2002a) suggest a 10^5-fold enhancement in energetic protons ejected into the stellar environment compared to the contemporary Sun! 'Astro-ecology' is seemingly becoming a veritable new field of stellar research.

X-rays from brown dwarfs

Brown dwarfs (BD) are stars with masses of less than 0.07–0.08 $M_{\odot}$. They are unable to sustain hydrogen burning in their core, although those with $M > 0.01\ M_{\odot}$ can burn deuterium during a short episode lasting *c.* 10^7 y. For reviews see, for example, Kulkarni (1997) and Reid & Hawley (2000). In the context of magnetic activity, BDs are interesting objects. First, as they descend the Hayashi track, they essentially resemble very late M dwarfs (M5 and later) that are fully convective, or contracting T Tau stars that also lack a stable nuclear energy source (Neuhäuser 1997). The issue of magnetic dynamo generation in a fully convective star is controversial. Since the radiative core is missing, we expect at least a qualitative change – perhaps a complete breakdown – of the magnetic dynamo action across the border from partially to fully convective stars, but detailed statistical studies indicate no such change in the overall X-ray luminosities of M dwarfs (Fleming *et al.* 1993). Several very late-type M dwarfs have been detected during appreciable X-ray flares (Giampapa *et al.* 1996; Fleming *et al.* 2000). One might therefore expect BDs to engage in extensive magnetic activity as well, possibly depending on their rotation rates, *unless* the putative breakdown of magnetic activity occurs at the hydrogen burning mass limit. As in more massive stars, magnetic activity may be relevant for the spin-down as a consequence of angular momentum loss in a magnetized wind. Neuhäuser *et al.* (1999), however, argue that low-mass objects probably have only weak or no winds; their magnetic braking time scales could thus be much longer than for main-sequence stars. Neuhäuser *et al.* attribute the non-detection of Pleiades-age BDs and candidate BD (CBDs) rather to the weakening of the second dynamo ingredient, namely internal convection. As a BD continuously cools after the deuterium-burning stage, the temperature gradient between the stellar core and the surface decreases, and so does the convection velocity.

First detections of BDs as X-ray sources date back only a few years (see Neuhäuser & Comerón (1998) and references therein). As argued above, the most likely places to find 'active' BDs are star-formation regions. Nearby star-formation regions indeed show rich samples of BDs or CBDs, and those were the sites where initial X-ray detections were reported (Neuhäuser *et al.* 1999). New searches for BDs benefit from the increased sensitivity of XMM-Newton and Chandra. Imanishi *et al.* (2001b) studied the ρ Oph cloud with Chandra to find two out of eight BDs and five out of ten CBDs. Similarly, Feigelson *et al.* (2002b) report 30 BDs/CBDs in the ONC, and Preibisch & Zinnecker (2001) found four BDs and three CBDs in IC 348. The detections show rather high characteristic temperatures, $kT \approx 0.9$–2.5 keV and sometimes even higher, with X-ray luminosities similar to active, low-mass pre-main-sequence stars, $L_X \approx 10^{28}$–10^{30} erg s^{-1}. The ratio $L_X/L_{bol} \approx 10^{-5}$–$10^{-3}$ is again similar to T Tau stars or active low-mass M dwarfs. Preibisch & Zinnecker (2001) argue that the observed X-rays are compatible with coronal emission. The most outstanding property, however, is their rather high flaring rate.

The latter property of BDs/CBDs is reminiscent of very-low-mass main-sequence stars. It has been noted before that stars at the lower-mass end are more prone to flaring, whereas signatures of steady magnetic activity become difficult to detect (see the review in Rutledge *et al.* (2000)). Although some of the detected BDs are still accreting mass, no correlation between the X-ray luminosity and the presence of discs is found (Imanishi *et al.* 2001b), parallelling the absence of such a correlation in disc-surrounded CTTS and discless WTTS stars (Preibisch & Zinnecker 2001; Feigelson *et al.* 2002a; see also the preceeding section), and suggesting the presence of solar-like magnetic coronae.

Owing to the decreasing internal convection in ageing BDs, we expect little magnetic activity in *older*, non-accreting and collapsed BDs. Initial searches in X-rays (Neuhäuser *et al.* 1999 and references therein) and at radio wavelengths (Krishnamurthi *et al.* 1999) revealed a dearth of detections, even at the modest age of the Pleiades. Chandra (Krishnamurthi *et al.* 2001) and XMM-Newton (Briggs and Pye 2003) observations of the Pleiades have detected late M dwarfs but so far also failed to detect X-ray emission from substellar members of this cluster, pushing the limit of the quiescent X-ray luminosity of (C)BDs at ages of 100 My down to $L_X \approx 3 \times 10^{27}$ erg s^{-1}. But once again, nature provided surprises. Rutledge *et al.* (2000) observed the nearby (5 pc), old (500 My), non-accreting BD LP 944-20 with Chandra during 44 ks and identified one rather strong X-ray flare ($L_{X,peak} \approx 1.2 \times 10^{26}$ erg s^{-1}, duration *c.* 1 h), although there was no discernible quiescent emission. The same star was found as a copious flaring gyrosynchrotron source at radio wavelengths (Berger *et al.* 2001). While it does seem that steady X-ray emission is dying at the bottom of the main sequence, magnetic activity continues to reveal itself in the form of episodic energy release.

If strong solar-type coronal magnetic activity plays an important role in evolved BDs, then, in analogy to more massive main-sequence stars, one would expect considerable spin-down due to angular momentum loss via a magnetized wind (Berger 2002). This has obviously been prevented in LP 944-20 since its rotation period is smaller than 4.4 h (Rutledge *et al.* 2000). Either, such stars do indeed not shed appreciable steady winds, or their intermittent magnetic activity is not sufficient to transport away angular momentum at a sufficiently fast rate.

X-rays from hot stars

Hot stars (spectral classes B and earlier) shed winds involving mass loss up to several times $10^{-5} M_\odot$ y^{-1}. It is a matter of ongoing debate whether or not stellar winds produce X-rays through a universal mechanism, or whether various processes compete. A basic model applied to winds of single OB stars is based on plasma heating in shocks that form as a result of line-driven instabilities (Lucy & White 1980; Lucy 1982; Feldmeier *et al.* 1997). The temperature of an adiabatic shock is related to the shock velocity v by $kT = (3/16)\bar{m}v^2$, with $\bar{m} \approx 1.3m_p$ being the mean particle mass and m_p being the mass of the proton. The modest shock

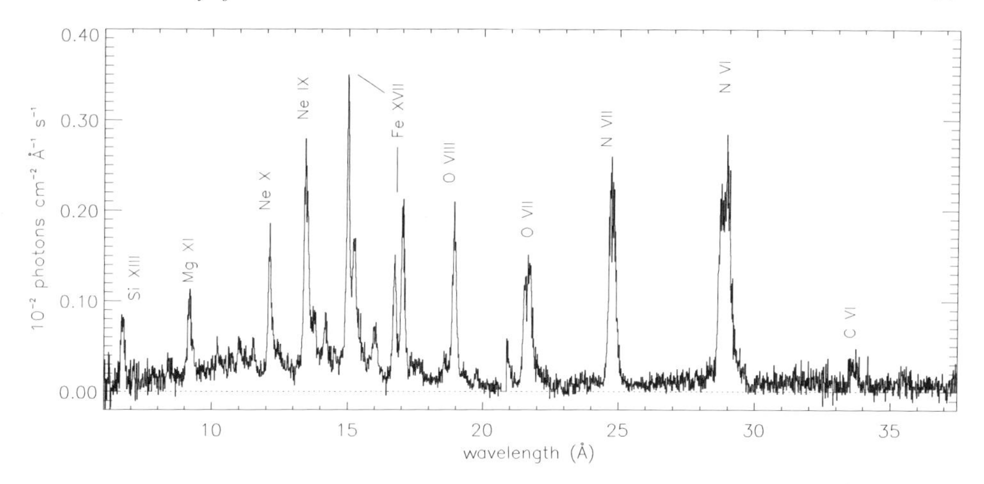

Figure 3.12. XMM-Newton reflection grating spectrum of ζ Pup. All lines are significantly broadened, and some are blueshifted. The f lines in the He-like triplets are strongly suppressed due to the ultraviolet radiation field. Note the very strong lines of nitrogen, in agreement with a *c*. 10-fold abundance enhancement relative to the solar abundance (from Kahn *et al.* (2001)).

velocities of a few 100 km s^{-1} thus imply rather soft X-rays, corresponding to temperatures of order a few MK. One usually assumes that the same mechanism also applies to Wolf–Rayet (WR) stars that are subject to stronger mass loss rates, but WR stars reveal intriguing differences to OB stars. For example, the standard hot-star relation $L_X \approx 10^{-7} L_{\rm bol}$ does not apply to WR stars (Wessolowski 1996). Extremely hot temperatures (see e.g. Cassinelli *et al.* 1994, and also below) also challenge the wind-instability model as they require very large wind velocities. In such cases, colliding winds between the components of close binaries may be the cause of hard X-rays. Alternatives, such as magnetically confined coronae at the base of the winds, are problematic since the consequent strong absorption of the X-ray emission by the overlying wind is not observed; it requires clumping of the winds. Non-thermal mechanisms such as inverse Compton radiation have been considered for the X-ray emission as well, but observations remain ambiguous.

We first discuss observational results related to the standard wind-shock picture. The X-ray sources are assumed to be distributed across the stellar wind out to large radii. Owocki & Cohen (2001) present detailed calculations of X-ray line profiles modified by the wind velocity distribution (line broadening) and selective wind attenuation; in particular, absorption of the emission from the far side of the star results in a suppression of the red wings of the lines such that the lines become asymmetric and apparently shifted to the blue. This standard wind model is well supported by observations of ζ Pup. The line broadening (Figs. 3.12 and 3.13) seen by XMM-Newton and Chandra corresponds to velocities (*c*. 900–1400 km s^{-1})

(*a*)

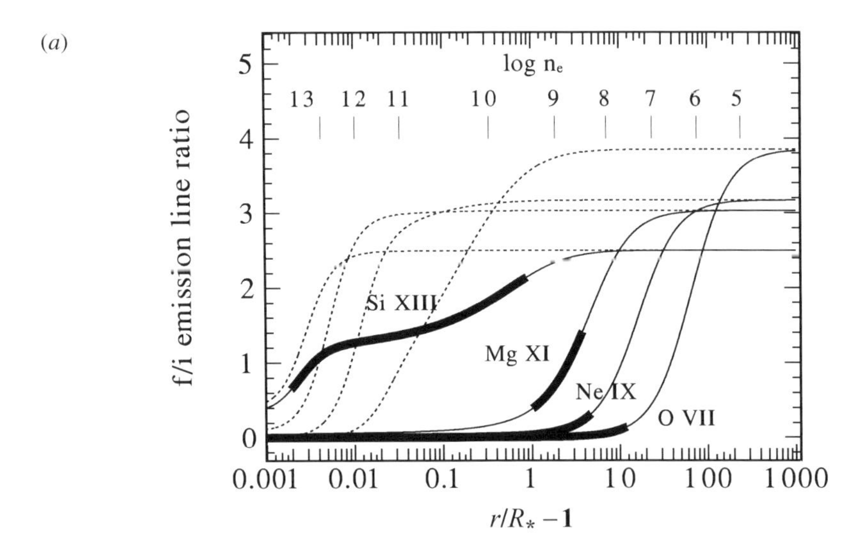

(*b*)

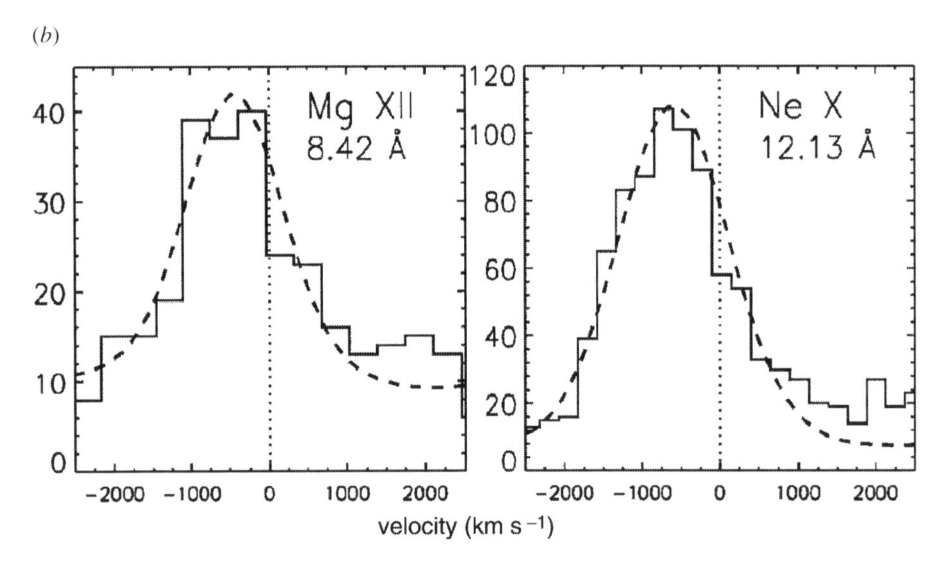

Figure 3.13. (*a*) Calculated f/i flux ratios for different He-like ions in a model atmosphere of ζ Ori. The bold lines indicate the allowed ranges from observations (after Waldron & Cassinelli (2001); updated calculations from W. Waldron (2002), private communication). (*b*) Asymmetric and blueshifted line profiles of Mg XII and Ne X from ζ Pup. The dotted lines indicate the rest wavelengths. The dashed curves are Gaussian fits (Chandra HETGS observation, after Cassinelli *et al.* (2001), with permission).

somewhat less than the terminal wind velocity (Kahn *et al.* 2001; Cassinelli *et al.* 2001), and it increases for decreasing atomic number Z, indicating that the low-Z, cooler lines are formed further out in the wind. This view is consistent with optical depth calculations. The $\tau = 1$ surfaces in a wind for He-like ions are closer to the star for harder emission, i.e. for higher-Z ions. Calculations using realistic wind models place the $\tau = 1$ surfaces at 22, 23, 9, 3.5, and 2.5 stellar radii for N VI, O VII, Ne IX, and Mg XI (Kahn *et al.* (2001), and similar values in Cassinelli *et al.* (2001)). Since the emissivity scales with n_e^2 and n_e drops off with radius in simple models, the observed emission is supposed to originate from close to these radii.

Upper limits to the formation distances can be deduced from the He-like line triplets (see the section on coronal structure and composition and also Chapter 2). Even if the source density is below the low-density limit of the R-ratio (maximizing the ratio), the ultraviolet radiation field of the hot star can significantly alter the ratio by photoexciting the $2\,^3S_1$ to the $2\,^3P_1$ state, thus decreasing R. Since the radiation field dilutes with increasing stellar distance, the observed R-ratio contains information on the formation distance in a standard wind model. The results are fully compatible with the $\tau = 1$ lower limits (Kahn *et al.* 2001; Cassinelli *et al.* 2001).

Contrary to expectation, some other O stars reveal no line asymmetries, nor any blueshifts despite appreciable line broadening. This novel result made possible by the high energy resolution available from gratings suggests that the X-ray sources are immersed in an optically thin wind, which is difficult to explain with the currently known wind parameters (Waldron & Cassinelli 2001; Schulz *et al.* 2000). Possibly, wind clumping or non-symmetric outflows may be the answer (Waldron & Cassinelli 2001). Additionally, standard wind calculations indicate rather small formation radii for S XV (*c.* $1.2R_*$) or Si XIII, suggesting that very hot plasma ($T \approx 10^7$ K) exists close to the stellar surface (Fig. 3.13(a) for ζ Ori) where no strong shocks are expected to develop. Unusually hot X-ray emitting plasma has in particular also been detected in the Orion Trapezium O stars, with temperatures ranging up to about 30 MK (Schulz *et al.* 2001). These measurements confirm earlier ASCA observations (Yamauchi *et al.* 1996) and defy an explanation in terms of wind shocks in single stars.

An even more provocative example among hot stars is τ Sco (Cassinelli *et al.* 1994). Not only are its emission lines unshifted, they are also unbroadened and again indicate the presence of very hot (*c.* 20 MK) plasma (Mewe *et al.* 2002). The hard emission cannot be explained by the standard radiative wind-shock model. Howk *et al.* (2000) proposed a mechanism in which shocked wind clumps are immersed in the outflowing wind while falling back toward the stellar surface at large speeds relative to the wind. The leading bow shock induces heating to several tens of MK. Since these structures are formed close to the base of the wind and attain modest velocities relative to the star, no significant line broadening and blueshift is seen.

Single WR stars have been much more challenging for X-ray observers. A high-quality observation of the putative single WR star WR 110 has been presented by Skinner *et al.* (2002a). Apart from a plasma component with temperatures around a few MK as expected from the wind-shock model, the spectrum indicates the presence of a rather hot plasma with temperatures of at least 16 MK (as inferred from the presence of S XV lines). The filling factor of the X-ray emitting material is rather small, $f \approx 10^{-7}$. Although a non-thermal mechanism such as inverse Compton scattering of ultraviolet photons off relativistic shock-accelerated electrons (Chen & White 1991) remains a viable possibility, the predicted spectral index of the power-law component is not found; also, evidence of magnetic fields is yet to be discovered. A likely alternative is the presence of an unknown companion that also sheds a wind, giving rise to a wind–wind collision zone in which much higher temperatures are attained (Skinner *et al.* 2002a; see next section).

A surprisingly similar spectrum was observed from EZ CMa, including a hot component with temperatures around 40 MK (Skinner *et al.* 2002b). In this case, a companion is more suggestive since the star shows periodic optical and ultraviolet variability. The hard X-ray component may be due to the WR wind shocking directly onto the companion surface.

Radically different models involving compact, closed magnetic structures closer to the stellar surface have been suggested. In the 'magnetically confined wind-shock model' initially proposed by Babel & Montmerle (1997), the radiatively driven wind flows along dipolar magnetic field lines from both hemispheres to a shock zone in the equatorial plane, where dense, high-temperature material forms a prominent X-ray source. Although attractive to explain hot, high-density plasma, the velocities required to explain the *symmetric* line broadening (1000 km s^{-1}, exceeding the free-fall velocity!) are challenging, since both upflows and downflows must be present. On the other hand, colliding gas streams of such velocities should induce much hotter plasma still (*c.* 10^8 K; Owocki & Cohen 2001), and this could well explain the excessively hot plasma detected in some massive stars (Schulz *et al.* 2000; Skinner *et al.* 2002a,b).

X-rays from colliding winds

In binaries of massive, wind-shedding stars, X-ray emission originates predominantly from a hydrodynamic wind collision shock between the two stars (Prilutskii & Usov 1976), although shocks in the winds of the individual stars may also contribute. A simulation of a wind-collision binary is shown in Fig. 3.14. The outside observer may see X-ray variability due to at least two effects: Firstly, the wind collision shock may vary in strength (and thus heating) due to orbit eccentricity, the X-ray maximum typically being expected during periastron passage where the wind

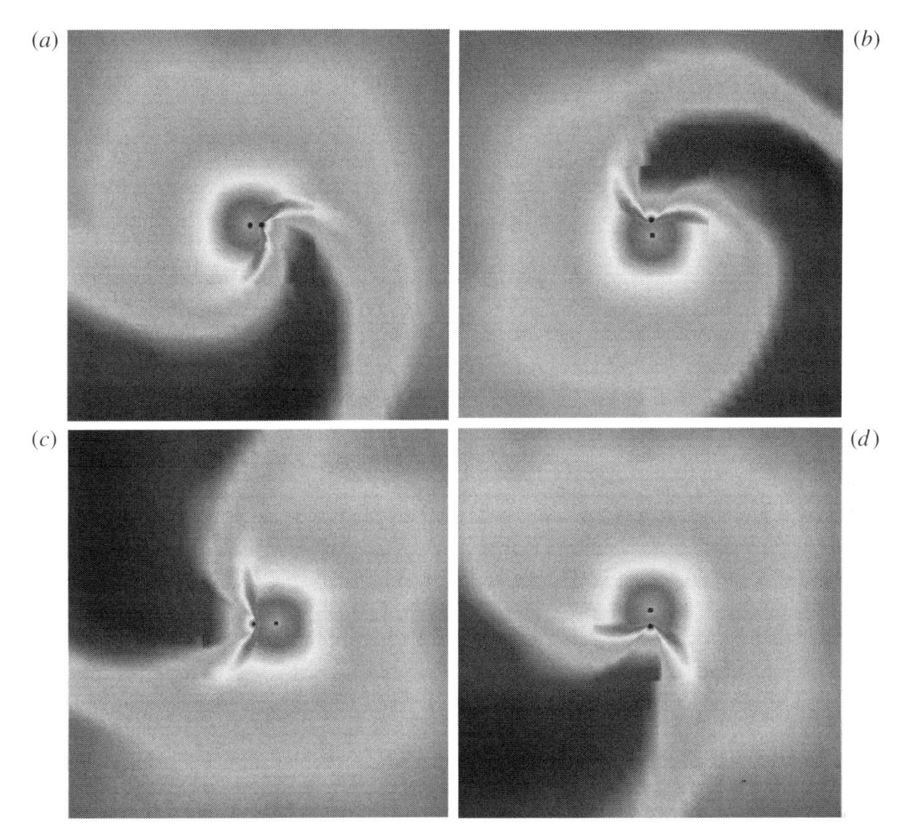

Figure 3.14. Hydrodynamic simulation of the γ^2 Vel binary system as seen from above the orbital plane. Greyscale levels give the gas density in the plane. The two stars are shown as black dots, with the WR star being surrounded by its strong wind, while the outflowing shock zone (the two 'antennae') wraps around the O star. The four phases shown refer to: (*a*) periastron, (*b*) first quarter, (*c*) apastron, and (*d*) third quarter. (From Walder *et al.* (1999), with permission.)

density is highest. Secondly, absorption effects due to variable line-of-sight wind column densities in the course of one orbital revolution may modulate the intrinsic X-ray emission. The former behaviour is expected for small line-of-sight inclinations of the orbital axis, while the latter behaviour is common for large inclination angles. We discuss two cases.

The prototypical WC8+O7.5 wind-collision binary γ^2 Vel ($P_{\rm orb} = 78.5$ d) has proven to be an ideal object to study wind–wind-collision shocks in X-rays. Given the large inclination angle of its orbit ($\approx 65^{\rm o}$), its X-ray flux is strongest when the observer looks down along the low-density cavity carved out of the massive WR wind behind the O star (Willis *et al.* 1995). The Chandra HETG spectrum (1–15Å) obtained during this phase is rich in emission lines suggesting plasma between 4 and 25 MK, mostly attributed to the wind-collision shock zone (Skinner *et al.* 2001).

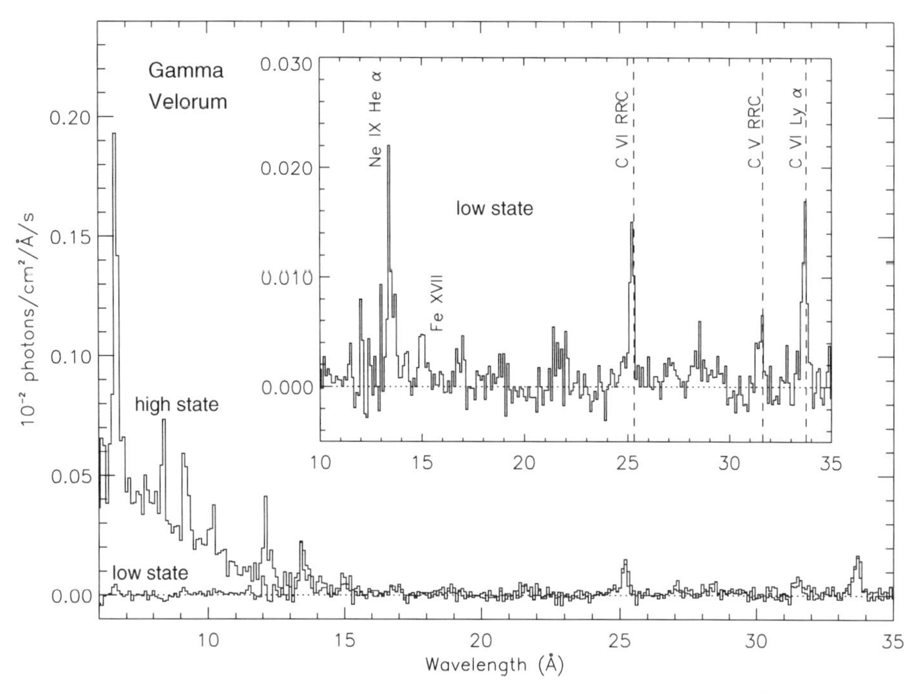

Figure 3.15. XMM-Newton reflection grating spectra of γ^2 Vel obtained at X-ray maximum (weak absorption, 'high state'), and at X-ray minimum (strong absorption, 'low state'). The long-wavelength part is shown again in the inset, emphasizing the radiative recombination continua (Schmutz *et al.* 2002).

Despite high intrinsic wind velocities, no line shifts are detected, while symmetric broadening to *c.* 1000 km s^{-1} is present. The He-like triplets of Si XIII, Mg XI and Ne IX show no effect from photoexcitation by the O star photosphere, placing their formation distances beyond $3R_O$, $9R_O$, and $30R_O$, where R_O is the O star radius (Skinner *et al.* 2001). The long-wavelength spectrum seen by the XMM-Newton reflection grating spectroment reveals completely different characteristics (Fig. 3.15). While shortward of 12Å, much of the emission is strongly suppressed for viewing angles away from the O star wind cavity, the spectrum longward of 12Å reveals no change, suggesting that it is formed far out in the wind. A closer examination reveals radiative recombination continua of C V and C VI with widths compatible with wind temperatures away from the WR star (a few 10^4 K; Schmutz *et al.* 2002). Presumably, the hard photons emitted by the hot wind-collision shock travel outward along the wind cavity and photoionize distant regions, possibly the cavity walls, to produce a recombination spectrum.

The O5.5 I+O7 V eccentric binary HD 93403 ($P_{\rm orb}$ = 15.1 d) has been observed by XMM-Newton at various phases along its orbit (Rauw *et al.* 2002). Given its low inclination angle (*c.* 30°) and its large eccentricity (0.234), X-ray variability in the harder band is expected to be caused predominantly by the changing colliding-wind shock characteristics since the X-ray luminosity is expected to scale as $L_X \propto v^{-3.2}r^{-1}$ (Stevens *et al.* 1992), i.e. the luminosity should be at maximum near periastron. A 20% increase at that phase is indeed observed although the spectral characteristics remain similar. An uncorrelated modulation of the softest flux is ascribed to some additional phase-dependent line-of-sight absorption.

In the standard picture of colliding winds, the individual stellar winds should produce modest temperatures of a few MK only, while it is the wind–wind collision shocks that produce the hotter component. Clear proof requires high-spatial resolution observations that resolve the stars from the wind shock. Such may have been achieved for the first time with Chandra observing the WR binary WR 147 in which the components are separated by about 0.6″ (Pittard *et al.* 2002). Positional analysis and comparison with numerical modelling indicate that the dominant source of X-rays is indeed the wind–wind collision shock, while there may be some contamination by soft emission from the WR wind itself.

One of the most outstanding high-energy sources in the galaxy is coincident with the massive star η Car. This star or stellar system has a unique history of optical activity, with an outburst in the middle of the nineteenth century that made η Car the second-brightest star in the night sky. It is believed to be one or several high-mass objects obscured by extensive dust additional to a bipolar nebula and an equatorial disc. η Car seems to be in a Luminous Blue Variable phase in which it loses large amounts of mass in a short time on its way to becoming a WR star. High-resolution Chandra imagery resolves a hard central component from a wide, rather soft ring-like feature (Seward *et al.* 2001). The latter may be related to the hot gas behind the blast wave from the eruption about 160 y ago (Seward *et al.* 2001), while the hard central source is ascribed to a wind–wind collision shock in a massive binary. Grating spectroscopy with Chandra reveals extremely hot temperatures (up to 100 MK, (Corcoran *et al.* 2001)) which again favour an interpretation in terms of wind–wind collision. A fluorescent line of cold Fe was resolved at 6.4 keV, roughly consistent with fluorescence of the absorbing cold material along the line of sight (Corcoran *et al.* 2001). Extensive hydrodynamic modelling of the wind–wind collision agrees with a companion possibly being an O-type supergiant or a WR star with a very high terminal wind velocity (3000 km s^{-1} (Pittard & Corcoran 2002)) that leads to dominant X-ray emission from the *companion* wind.

Conclusions

Spectroscopic studies available from Chandra and XMM-Newton play a pivotal part in the understanding of the physical processes in stellar (magnetic and non-magnetic) atmospheres. It is now routinely possible to derive densities and to study the influence of ultraviolet radiation fields, both of which can be used to infer the geometry of the radiating sources. Line profiles provide important information on bulk mass motions and attenuation by neutral matter, e.g. in stellar winds. The increased sensitivity has revealed new types of X-ray sources in systems that were thought to be unlikely places for X-rays: flaring brown dwarfs, including rather old, non-accreting objects, and terminal shocks in jets of young stars are important examples. New clues concerning the role of stellar high-energy processes in the modification of the stellar environment (ionization, spallation, etc.) contribute significantly to our understanding of the 'astro-ecology' in forming planetary systems.

Technological limitations are evident. The spectral resolution has not reached the level where bulk mass motions in cool stars become easily measurable. Higher resolution would also be important to perform X-ray 'Doppler imaging' in order to reconstruct the 3-D distribution of the X-ray sources around a rotating star. Higher sensitivity will be required to perform high-resolution spectroscopy of weak sources such as brown dwarfs or embedded pre-main-sequence sources. A new generation of satellites such as *Constellation-X* or *XEUS* should pursue these goals.

I thank M. Audard, K. R. Briggs, J. P. Cassinelli, F. Favata, E. Feigelson, D. Folini, N. Grosso, D. Huenemoerder, K. Imanishi, J. Kastner, M. Leutenegger, T. Montmerle, J.-U. Ness, G. Rauw, R. Rutledge, S. Skinner, Y. Tsuboi, R. Walder, and W. Waldron for helpful discussions and figure material presented in this chapter. Work at PSI is financially supported by the Swiss National Science Foundation (grant no. 2000-058827.99).

References

Alcalá, J. M., Krautter, J., Covino, E., Neuhäuser, R., Schmitt, J. H. M. M. & Wichmann, R. 1997 *Astron. Astrophys.* **319**, 184–200.
Audard, M., Güdel, M., Drake, J. J. & Kashyap, V. L. 2000 *Astrophys. J.* **541**, 396–409.
Audard, M., Behar, E., Güdel M., Raassen, A. J. J., Porquet, D., Mewe, R., Foley, C. R. & Bromage, G. E. 2001a *Astron. Astrophys.* **365**, L329–335.
Audard, M., Güdel, M. & Mewe, R. 2001b *Astron. Astrophys.* **365**, L318–323.
Ayres, T. R., Brown, A., Osten, R. A., Huenemoerder, D. P., Drake, J. J., Brickhouse, N. S. & Linsky, J. L. 2001 *Astrophys. J.* **549**, 554–577.
Babel, J. & Montmerle, T. 1997 *Astron. Astrophys.* **323**, 121–138.
Beasley, A. J. & Güdel, M. 2000 *Astrophys. J.* **529**, 961–967.
Behar, E., Cottam, J. & Kahn, S. M. 2001 *Astrophys. J.* **548**, 966–975.
Benz, A. O., Conway, J. & Güdel, M. 1998 *Astron. Astrophys.* **331**, 596–600.
Berger, E. *et al.* 2002 *Astrophys. J.*, **572**, 503–513.
Berger, E., *et al.* 2001 *Nature* **410**, 338–340.
Berghöfer, T. W., Schmitt, J. H. M. M. & Cassinelli, J. P. 1996 *Astron. Astrophys. Suppl. Ser.* **118**, 481–494.
Briggs, K. R. & Pye, J. P. 2003 *Mon. Not. R. Astron. Soc.*, in press.
Brinkman, A. C., *et al.* 2000 *Astrophys. J.* **530**, L111–114.
Brinkman, A. C., *et al.* 2001 *Astron. Astrophys.* **365**, L324–328.
Burnight, T. R. 1949, *Phys. Rev.* A **76**, 165.
Canizares, C. R., *et al.* 2000 *Astrophys. J.* **539**, L41–44.

Casanova, S., Montmerle, T., Feigelson, E. D. & André, P. 1995 *Astrophys. J.* **439**, 752–770.
Cassinelli, J. P., Cohen, D. H., MacFarlane, J. J., Sanders, W. T. & Welsh, B. Y. 1994 *Astrophys. J.* **421**, 705–717.
Cassinelli, J. P., Miller, N. A., Waldron, W. L., MacFarlane, J. J. & Cohen, D. H. 2001 *Astrophys. J.* **554**, L55–58.
Chen, W. & White, R. L. 1991 *Astrophys. J.* **366**, 512–528.
Corcoran, M. F., *et al.* 2001 *Astrophys. J.* **562**, 1031–1037.
de Jager, C. *et al.* 1986 *Astron. Astrophys.* **156**, 95–100.
de Jager, C. *et al.* 1989 *Astron. Astrophys.* **211**, 157–172.
Dennis, B. R. 1988 *Solar Phys.* **118**, 49–94.
Dennis, B. R. & Zarro, D. M. 1993 *Solar Phys.* **146**, 177–190.
Drake, J. J., Laming, J. M. & Widing, K. G. 1995 *Astrophys. J.* **443**, 393–415.
Drake, J. J., Laming, J. M. & Widing, K. G. 1997, *Astrophys. J.* **478**, 403–416.
Drake, J. J., *et al.* 2001 *Astrophys. J.* **548**, L81–85.
Drake, S. A. 1996 In *6th Annual Astrophysics Conference*, Maryland (eds,. S. S. Holt & G. Sonneborn) pp. 215–226. San Francisco, CA:ASP.
ESA 1997 *The Hipparcos and Tycho Catalogues*, European Space Agency SP-1200.
Favata, F., Fridlund, C. V. M., Micela, G., Sciortino, S. & Kaas, A. A. 2002 *Astron. Astrophys.* **386**, 204–210.
Feigelson, E. D. & Montmerle, T. 1999 *A. Rev. Astron. Astrophys.* **37**, 363–408.
Feigelson, E. D., *et al.* 2002b *Astrophys. J.* **574**, 258.
Feigelson, E. D., Garmire, G. P. & Pravdo, S. H. 2002a *Astrophys. J.* **572**, 335–349.
Feldman, U. 1992 *Phys. Scripta* **46**, 202–220.
Feldmeier, A., Kudritzki, R.-P., Palsa, R., Pauldrach, A. W. A. & Puls, J. 1997 *Astron. Astrophys.* **320**, 899–912.
Flaccomio, E., *et al.* 2002a *Astrophys. J.* **582**, 382.
Flaccomio, E., *et al.* 2002b *Astrophys. J.* **582**, 398.
Fleming, T. A., Giampapa, M. S., Schmitt, J. H. M. M. & Bookbinder, J. A. 1993 *Astrophys. J.* **410**, 387–392.
Fleming, T. A., Giampapa, M. S. & Schmitt, J. H. M. M. 2000 *Astrophys. J.* **533**, 372–377.
Gabriel, A. H. & Jordan, C. 1969 *Mon. Not. R. Astron. Soc.* **145**, 241–248.
Garmire, G., *et al.* 2000 *Astron. J.*, **120**, 1426–1435.
Giampapa, M. S., *et al.* 1996 *Astrophys. J.* **463**, 707–725.
Glassgold, A. E., Feigelson, E. D. & Montmerle, T. 2000 In *Protostars and Planets IV* (ed. V. Mannings, A. P. Boss & S. S. Russell), p. 429, Tucson, AZ: University of Arizona Press.
Grosso, N., Montmerle, T., Feigelson, E. D., André, P., Casanova, S. & Gregorio-Hetem, J. 1997 *Nature* **387**, 56–58.
Güdel, M., Benz, A. O., Schmitt, J. H. M. M. & Skinner, S. L. 1996 *Astrophys. J.* **471**, 1002–1014.
Güdel, M., Guinan, E. F. & Skinner, S. L. 1997 *Astrophys. J.* **483**, 947–960.
Güdel, M., *et al.* 2001a *Astron. Astrophys.* **365**, L336–343.
Güdel, M., *et al.* 2001b *Astron. Astrophys.* **365**, L344–352.
Güdel, M., Audard, M., Kashyap, V. L., Drake, J. J. & Guinan, E. F. 2003 *Astrophys. J.*, **582**, 423.
Hawley, S. L., *et al.* 1995 *Astrophys. J.* **453**, 464–479.
Haisch, B. & Schmitt, J. H. M. M. 1996 *PASP* **108**, 113–129.
Harnden Jr, F. R., *et al.* 2001 *Astrophys. J.* **547**, L141–145.
Hénoux, J.-C. 1995 *Adv. Space Res.* **15**, 23–32.

Howk, J. C., Cassinelli, J. P., Bjorkman, J. E. & Lamers H. J. G. L. M. 2000 *Astrophys. J.* **534**, 348–358.
Hudson, H. & Ryan, J. 1995 *A. Rev. Astron. Astrophys.* **33**, 239–282.
Huenemoerder, D. P., Canizares, C. R. & Schulz, N. S. 2001 *Astrophys. J.* **559**, 1135–1146.
Hünsch, M., Schmitt, J. H. M. M. & Voges, W. 1998a *Astron. Astrophys. Suppl. Ser.* **127**, 251–255.
Hünsch, M., Schmitt, J. H. M. M. & Voges, W. 1998b *Astron. Astrophys. Suppl. Ser.* **132**, 155–171.
Hünsch, M., Schmitt, J. H. M. M., Sterzik, M. F. & Voges, W. 1999 *Astron. Astrophys. Suppl. Ser.* **135**, 319–338.
Imanishi, K., Koyama, K., & Tsuboi, Y. 2001a *Astrophys. J.* **557**, 747–760.
Imanishi, K., Tsujimoto, M. & Koyama, K. 2001b *Astrophys. J.* **563**, 361–366.
Kahler, S., *et al.* 1982 *Astrophys. J.* **252**, 239–249.
Kahn, S. M., *et al.* 2001 *Astron. Astrophys.* **365**, L312–317.
Kamata, Y., Koyama, K., Tsuboi, Y. & Yamauchi, S. 1997 *PASJ* **49**, 461–470.
Kashyap, V. L., Drake, J. J., Güdel, M. & Audard, M. 2002 *Astrophys. J.* **580**, 1118.
Kastner, J. H., Huenemoerder, D. P., Schulz, N. S., Canizares, C. R. & Weintraub D. A. 2002 *Astrophys. J.*, **567**, 434–440.
Koyama, K., Hamaguchi, K., Ueno, S., Kobayashi, N. & Feigelson, E. D. 1996 *PASJ* **48**, L87–92.
Krishnamurthi, A., Leto, G. & Linsky, J. L. 1999 *Astron. J.* **118**, 1369–1372.
Krishnamurthi, A., Reynolds, C. S., Linsky, J. L., Martín, E. & Gagné, M. 2001 *Astron. J.* **121**, 337–346.
Krucker, S. & Benz, A. O. 1998 *Astrophys. J.* **501**, L213–216.
Kulkarni, S. R. 1997 *Science* **276**, 1350–1354.
Laming, J. M., Drake, J. J. & Widing K. G. 1996, *Astrophys. J.* **462**, 948–959.
Lawson, W. A., Feigelson, E. D. & Huenemoerder, D. P. 1996 *Mon. Not. R. Astron. Soc.* **280**, 1071–1088.
Linsky, J. L. & Haisch, B. M. 1979 *Astrophys. J.* **229**, L27–32.
Lucy, L. B. & White, R. L. 1980 *Astrophys. J.* **241**, 300–305.
Lucy, L. B. 1982 *Astrophys. J.* **255**, 286–292.
Mewe, R., Kaastra, J. S., van den Oord, G. H. J., Vink, J. & Tawara, Y. 1997 *Astron. Astrophys.* **320**, 147–158.
Mewe, R., Raassen, A. J. J., Drake, J. J., Kaastra, J. S., van der Meer, R. L. J. & Porquet, D. 2001 *Astron. Astrophys.* **368**, 888–900.
Mewe, R., Raassen, A. J. J., Cassinelli, J. P., van der Hucht, K. A., Miller, N. A. & Güdel, M. 2002 *Astron. Astrophys.*, **398**, 203.
Montmerle, T., Grosso, N., Tsuboi, Y. & Koyama, K. 2000 *Astrophys. J.* **532**, 1097–1110.
Mutel, R. L., Molnar, L. A., Waltman, E. B. & Ghigo, F. D. 1998 *Astrophys. J.* **507**, 371–383.
Narain, U. & Ulmschneider, P. 1990 *Space Sci. Rev.* **54**, 377–445.
Ness, J.-U., *et al.* 2001 *Astron. Astrophys.* **367**, 282–296.
Ness, J.-U., Schmitt, J. H. M. M., Burwitz, V., Mewe, R. & Predehl P. 2002 *Astron. Astrophys.* **387**, 1032–1046.
Neuhäuser, R. 1997 *Science* **276**, 1363–1370.
Neuhäuser, R. & Comerón, F. 1998 *Science* **282**, 83–85.
Neuhäuser, R. *et al.* 1999 *Astron. Astrophys.* **343**, 883–893.
Neupert, W. M. 1968 *Astrophys. J.* **153**, L59–64.
Owocki, S. P. & Cohen, D. H. 2001 *Astrophys. J.* **559**, 1108–1116.
Pallavicini, R. 1989. *A&A Rev.* **1**, 177–207.

Pallavicini, R., Golub, L., Rosner, R., Vaiana, G. S., Ayres, T. & Linsky, J. L. 1981 *Astrophys. J.* **248**, 279–290.
Peres, G., Orlando, S., Reale, F., Rosner, R. & Hudson, H. 2000 *Astrophys. J.* **528**, 537–551.
Phillips, K. J. H., Mathioudakis, M., Huenemoerder, D. P., Williams, D. R., Phillips, M. E. & Keenan, F. P. 2001 *Mon. Not. R. Astron. Soc.* **325**, 1500–1510.
Pittard, J. M. & Corcoran, M. F. 2002 *Astron. Astrophys.* **383**, 636–647.
Pittard, J. M., *et al.* 2002 *Astron. Astrophys.* **388**, 335–345.
Porquet, D., Mewe, R., Dubau, J., Raassen, A. J. J. & Kaastra, J. S. 2001 *Astron. Astrophys.* **376**, 1113–1122.
Pravdo, S. H., Feigelson, E. D., Garmire, G., Maeda, Y., Tsuboi, Y. & Bally J. 2001 *Nature* **413**, 708–711.
Preibisch, T. & Zinnecker H. 2001 *Astrophys. J.* **122**, 866–875.
Prilutskii, O. F. & Usov, V. V. 1976 *Sov. Astron.* **20**, 2.
Raassen, A. J. J., *et al.* 2002 *Astron. Astrophys.* **389**, 228.
Rauw, G., *et al.* 2002 *Astron. Astrophys.* **388**, 552–562.
Reid, N. & Hawley, S. L. 2000 *New Light on Dark Stars: Red Dwarfs, Low Mass Stars, Brown Dwarfs*. New York: Springer.
Rosner, R., Golub, L. & Vaiana G. S. 1985 *A. Rev. Astron. Astrophys.* **23**, 413–452.
Rutledge, R. E., Basri, G., Martín, E. L. & Bildsten, L. 2000 *Astrophys. J.* **538**, L141–144.
Schmitt, J. H. M. M. & Kürster, M. 1993 *Science* **262**, 215–218.
Schmitt, J. H. M. M., Stern, R. A., Drake, J. J. & Kürster, M. 1996 *Astrophys. J.* **464**, 898–909.
Schmutz, W. *et al.* 2002, in preparation.
Schrijver, C. J., Mewe, R. & Walter, F. M. 1984 *Astron. Astrophys.* **138**, 258–266.
Schulz, N. S., Canizares, C. R., Huenemoerder, D. & Lee, J. C. 2000 *Astrophys. J.* **545**, L135–139.
Schulz, N. S., Canizares, C., Huenemoerder, D., Kastner, J. H., Taylor, S. C. & Bergstrom, E. J. 2001 *Astrophys. J.* **549**, 441–451.
Sciortino, S. *et al.* 2001 *Astron. Astrophys.* **365**, L259–266.
Seward, F. D., Butt, Y. M., Karovska, M., Prestwich, A., Schlegel, E. M. & Corcoran, M. 2001 *Astrophys. J.* **553**, 832–836.
Shu, F. H., Adams, F. C. & Lizano, S. 1987 *A. Rev. Astron. Astrophys.* **25**, 23–81.
Singh, K. P., Drake S. A., White N. E. & Simon T. 1996 *Astron. J.* **112**, 221–229.
Skinner, S. L., Güdel, M., Schmutz, W. & Stevens, I. R. 2001 *Astrophys. J.* **558**, L113–116.
Skinner, S. L., Zekhov, S. A., Güdel, M. & Schmutz, W. 2002a *Astrophys. J.*, **572**, 477–486.
Skinner, S. L., Zekhov, S. A., Güdel, M. & Schmutz, W. 2002b *Astrophys. J.* **579**, 764.
Stelzer, B., *et al.* 2002 *Astron. Astrophys.* **392** 585.
Stevens, I. R., Blondin, J. M. & Pollock, A. M. T. 1992 *Astrophys. J.* **386**, 265–287.
Sung, H., Bessell, M. S., Lee, B. & Lee, S. 2002 *Astron. J.* **123**, 290–303.
Tsuboi, Y., *et al.* 2001 *Astrophys. J.* **554**, 734–741.
Ulmschneider, P., Priest, E. R. & Rosner, R. (eds.) 1991 *Mechanisms of Chromospheric and Coronal Heating*. Proceedings of the International Conference, Heidelberg, Heidelberg: Springer, 649 pp.
Walder, R., Folini, D. & Motamen, S. 1999 In *Wolf-Rayet Phenomena in Massive Stars and Starburst Galaxies*, Proc. IAU Symposium 193 (eds. K. A. van der Hucht, G. Koenigsberger & P. R. J. Eenens) pp. 298–305, San Francisco, CA: ASP.
Waldron, W. L. & Cassinelli J. P. 2001 *Astrophys. J.* **548**, L45–48.

Wessolowski, U. 1996 In *Röntgenstrahlung from the Universe* (eds. H. U. Zimmermann, J. Trümper & H. Yorke) MPE Report 263, pp. 75–76.
White, N. E. 1996 In *9th Cambridge Workshop on Cool Stars, Stellar Systems, and the Sun* (eds. R. Pallavicini & A. K. Dupree), pp. 193–202, San Francisco: ASP.
White, N. E., Shafer, R. A., Parmar, A. N., Horne, K. & Culhane, J. L. 1990 *Astrophys. J.* **350**, 776–795.
Willis, A. J., Schild, H. & Stevens, I. R. 1995 *Astron. Astrophys.* **298**, 549–566.
Yamauchi, S., Koyama, K., Sakano, M. & Okada, K. 1996 *PASJ* **48**, 719–737.
Zirker, J. B. 1993 *Solar Phys.* **148**, 43–60.

4

X-ray observations of accreting white-dwarf systems

BY M. CROPPER, G. RAMSAY

University College London

C. HELLIER

Keele University

K. MUKAI

Laboratory for High Energy Astrophysics

C. MAUCHE

Lawrence Livermore National Laboratory

D. PANDEL

University of California

Introduction

Accreting white dwarfs are found in a number of types of binary systems. Accretion can take place via a wind from the companion (as in Mira, where the white dwarf Mira B accretes from the red giant Mira A, and in other symbiotic stars) but most commonly it occurs via an accretion stream because the companion overfills its Roche lobe. Such semi-detached systems are called cataclysmic variables (CVs). The secondary may be a red giant (as in some recurrent novae), a lower main-sequence star (most CVs) or a degenerate white dwarf (the AM CVn systems). A comprehensive review of CVs can be found in Warner (1995).

CVs produce a major component of the total galactic X-ray emission. Their wide range of temporal behaviours on accessible time-scales, and their emission at wavelengths from the infrared to X-rays, has made them a well-studied class, particularly as regards the understanding of accretion discs and magnetically mediated accretion. Here we review the results from XMM-Newton and Chandra X-ray observations of these objects, and new directions regarding ultrashort-period systems with white-dwarf secondaries.

Frontiers of X-Ray Astronomy, ed. A.C. Fabian, K.A. Pounds and R.D. Blandford. Published by Cambridge University Press.

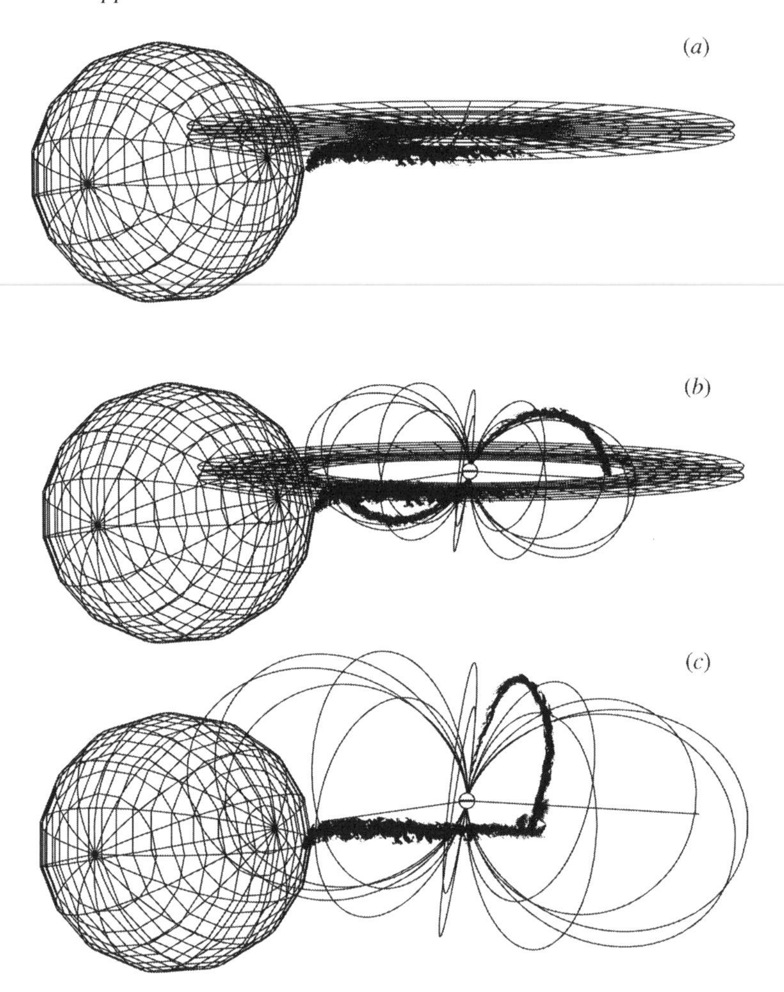

Figure 4.1. Schematic illustrating the different types of accretion flow in CVs: (*a*) non-magnetic systems; (*b*) intermediate polars; and (*c*) polars.

Accretion in CVs

The mass transfer from the secondary results in an initially ballistic stream from the L1 point, falling into the Roche lobe of the primary. The magnetic field on the primary plays a strong role in determining the behaviour of these systems (see Fig. 4.1). For magnetic fields up to *c.* 1 MG, the stream forms an accretion disc, with the material proceeding through the disc onto the surface of the white dwarf via a boundary layer. For magnetic fields between 1 and 10 MG, the inner part of the accretion disc is hollowed out, and the accreting material follows the field lines to accrete quasi-radially near the two magnetic poles. Magnetic fields stronger than *c.* 10 MG prevent the formation of a disc entirely, and the accretion stream threads directly onto the magnetic field. The material again follows the field lines, accreting near one, or in some cases, both magnetic poles (depending on the orientation of the magnetic axis). These strong field systems are called AM Her systems, or polars, while the intermediate case is known as DQ Her systems or intermediate polars (IPs). Generally, the magnetic field in polars is sufficiently strong to synchronize (or nearly synchronize) the white-dwarf spin rotation with the binary orbital period.

In the non-magnetic CVs the accretion disc emits over a range of temperatures, increasing towards the centre. Much of the gravitational potential energy is radiated in X-rays and the extreme ultraviolet (EUV) at the boundary layer interfacing the disc to the white dwarf. There is an extensive body of work on the emission from the boundary layer: see Collins *et al.* (2000) and references therein. These works generally adopt assumptions such as optically thick conditions and the absence of any magnetic field, although some models do assume optically thin conditions (e.g. Narayan & Popham (1993)). The boundary layer is optically thick for moderate and high transfer rates (Pringle & Savonije 1979). This condition is generally fulfilled when the system is in a high optical state or outburst. Until recently, observations were of insufficient quality to demonstrate that X-rays originate in the boundary layer in *quiescent* systems.

In the magnetic CVs, the accreting material falling down the field lines encounters a shock near the surface of the white dwarf. This converts the kinetic energy of the infalling material (approximately the escape velocity, or a few 10^3 km s^{-1}) into thermal velocities: $kT \sim 50$ keV for a $1M_\odot$ white dwarf. The hot plasma emits bremsstrahlung radiation at these X-ray temperatures and cyclotron emission at optical or infrared wavelengths (depending on the magnetic field strength), settling as it cools. The surface of the white dwarf is heated by this emission, giving rise to a third spectral component at soft X-ray wavelengths (*c.* 40 eV). Brief and comprehensive overviews are given in Cropper *et al.* (2000) and Wu (2000), respectively.

The multi-temperature nature of the region renders single-temperature bremsstrahlung fits to X-ray observations of limited use if physical parameters are to be extracted from them. We have developed (Cropper *et al.* 1999) a description for the hydrodynamical variables (temperature, pressure, density and flow velocity) within the post-shock flow in a semi-analytic treatment to include cyclotron emission and the variation of the gravitational potential within the post-shock region, important if the shock is tall. Once a density and temperature structure is known, optically thin spectra (for example, those generated using the Mewe, Kaastra & Leidahl (MEKAL) model (Mewe *et al.* 1995)) can be summed in strata with the appropriate emitting volume. These stratified accretion-column models run sufficiently fast for fits to be made to the X-ray spectra. The inputs to the model are the mass of the white dwarf, the mass-transfer rate per unit area, the ratio of cyclotron to bremsstrahlung cooling at the shock (related to the magnetic field) and the metal abundance relative to solar (affecting the line strengths).

Non-magnetic systems

OY Car

OY Car is an eclipsing, non-magnetic CV (dwarf nova). It was observed twice by XMM-Newton, in June and August 2000, during a normal (non-outbursting) state. Figure 4.2 (from Ramsay *et al.* (2001a)) shows the X-ray, optical (B-band) and UV (240–340 nm) light curves of the eclipse in this system. The X-ray eclipse is sharp, indicating that the X-ray-emitting boundary layer is confined close to the surface of the white dwarf. The ingress and egress may be slightly shorter at harder energies. Their duration indicates that the X-ray emission is from smaller regions than the entire white dwarf, and may be from higher latitudes (Wheatley & West 2002). Ramsay *et al.* (2001a) also found evidence for low levels of X-ray emission throughout the eclipse, which they attribute to an extended corona (which is seen when the system is in outburst). Ramsay *et al.* (2001b) found that the boundary-layer emission was optically thin, and multi-temperature spectral fits were required in order to fit the observed Fe and S emission-line strengths. Optically thin models for the boundary layer were also required in Advanced Satellite for Cosmology and Astrophysics (ASCA) observations of high-accretion-rate systems (Mauche & Mukai (2002) and references therein.)

The UV and optical eclipse ingresses and egresses are more gradual, consistent with the diameter of a white dwarf. The eclipse of the disc is also seen in these bands, and it is total in the case of the UV, providing a limit to the extent of the UV-emitting disc. In the optical range there is significant contribution from the hot spot, where the stream impacts on the edge of the disc.

The XMM-Newton observations may have detected the rotation period of the white dwarf in OY Car. Ramsay *et al.* (2001a) found evidence for a period of 2240 s, which was especially strong in soft X-rays. Such a modulation could be due to a quasi-stable bright spot on the photosphere of the white dwarf. Alternatively, it could be due to energy-dependent absorption, perhaps in disc material circulating above the orbital plane.

V603 Aql

V603 Aql is a bright old nova (Nova Aql 1918, when it reached a visual magnitude of about 1.1). The orbital period is 199 min, but the photometric period is slightly longer, probably from a precessing disc. The system was observed with Chandra

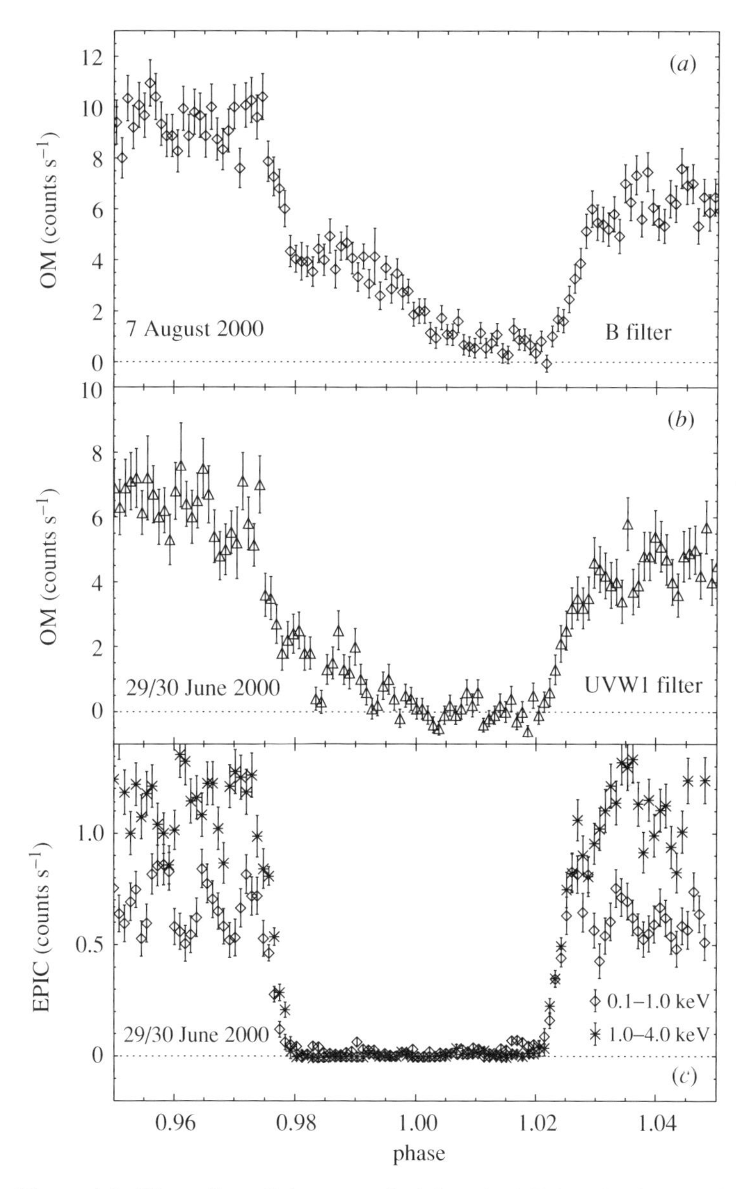

Figure 4.2. The eclipse light curves in (*a*) optical (B band), (*b*) UV (240–340 nm) and (*c*) X-ray spectra of OY Car taken from Ramsay *et al.* (2001a).

in April 2001 for 64 ks, with the High Energy Transmission Grating Spectrometer (HETG).

The spectrum is shown in Fig. 4.3. This shows strong photoionized lines of Fe, Si, Mg, Ne and O. The X-rays are probably emitted from optically thin regions of the boundary layer (possibly with an optically thick region emitting in the EUV). There

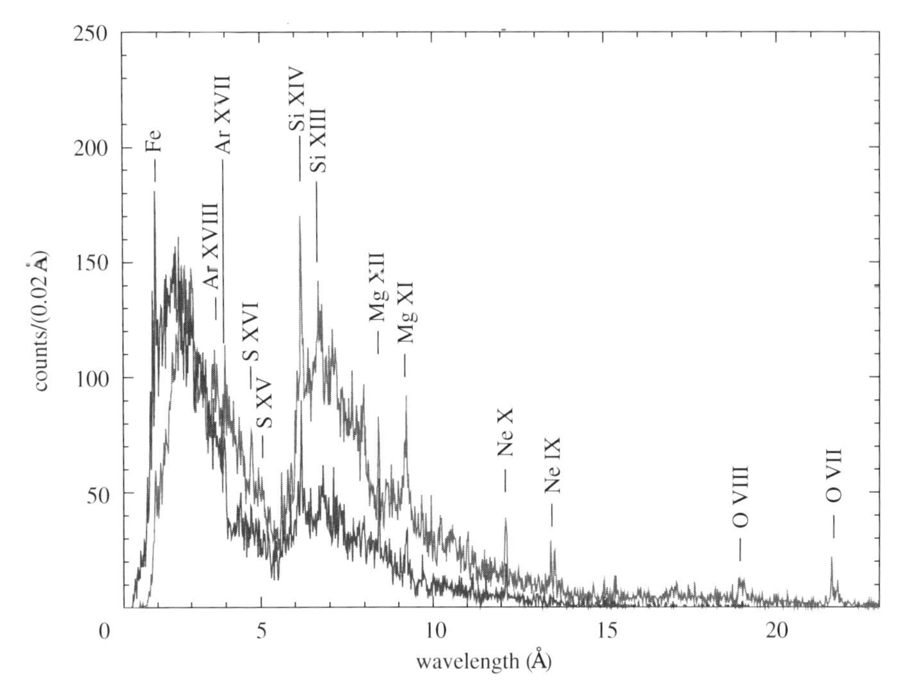

Figure 4.3. The high-energy (lower) and medium-energy (upper) HETG spectra of V603 Aql. The continuum shape is due mostly to the instrumental response function. Unlabelled lines are mostly due to Fe L (Mukai *et al.* 2002).

is a density gradient even within this region, with the lower-density region emitting line-rich spectra, and higher-density regions mostly emitting higher-temperature, continuum-rich spectra. There are also forbidden lines (for example, in the Mg-XI triplet). Upper limits to the density of these indicate that their emitting volume is significantly larger than that of the continuum.

Intermediate polars

EX Hya

EX Hya is a bright, well-studied intermediate polar with a short orbital period (98 min) and a white-dwarf spin period (67 min) which is in a (2:3) resonance with the orbital period. A 100 ks XMM-Newton observation was made in July 2000. There is a prominent X-ray modulation on the spin period, decreasing with increasing energy (Fig. 4.4): this was known from earlier observations (e.g. Rosen *et al.* (1988)). A phase-resolved spectral analysis shows for the first time (Cropper *et al.* 2002) that this modulation is primarily due to changes in the absorption. The data folded on the *orbital* period show a grazing eclipse of varying depth (dependent on

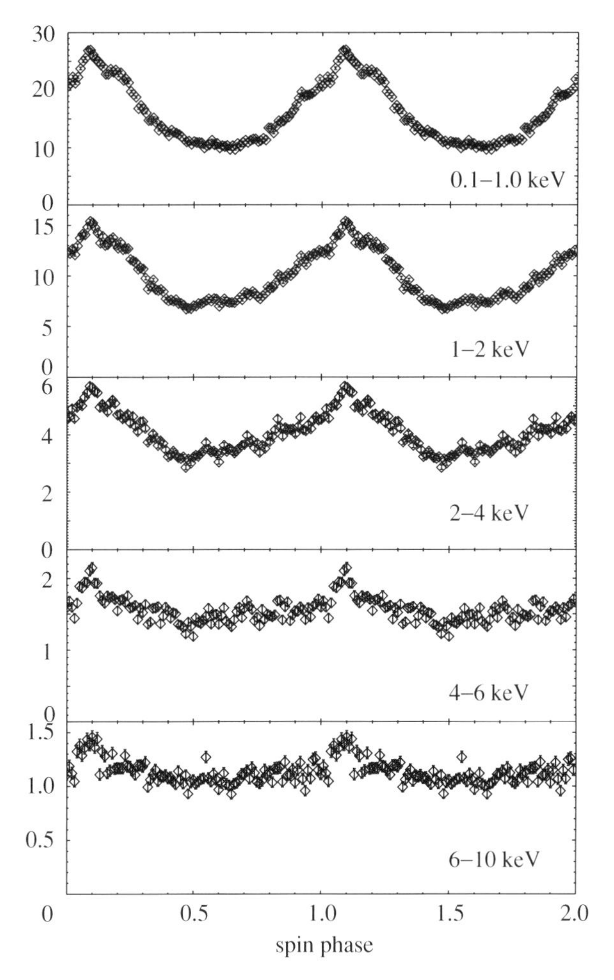

Figure 4.4. The light curve of EX Hya folded on the 67-min spin period in different energy bands. The modulation on the 98-min orbital period has been removed from the data. Reproduced with permission from Cropper *et al.* (2002).

spin phase). There was also significant variability at orbital phases just before the eclipse, again due to changes in absorption.

We extracted European Photon Imaging Camera (EPIC)-pn spectra from phases near the spin maximum (see Fig. 4.5), and fitted our stratified accretion column model. The fit implied a white-dwarf mass of $0.52 \pm 0.03 M_{\odot}$. It can be seen from the residuals in Fig. 4.5 that the stratified models provide a good fit to the data.

A much more demanding test of the models is the fit to the reflection-grating spectrometer (RGS) spectrum. The spectrum is shown in Fig. 4.6(*a*). The RGS spectrum is rich in emission lines of O, Ne, Mg, Si and Fe. Fitting a stratified model to these data results in a mass of 0.51 $M_{\odot}$: the fit is shown superimposed and the

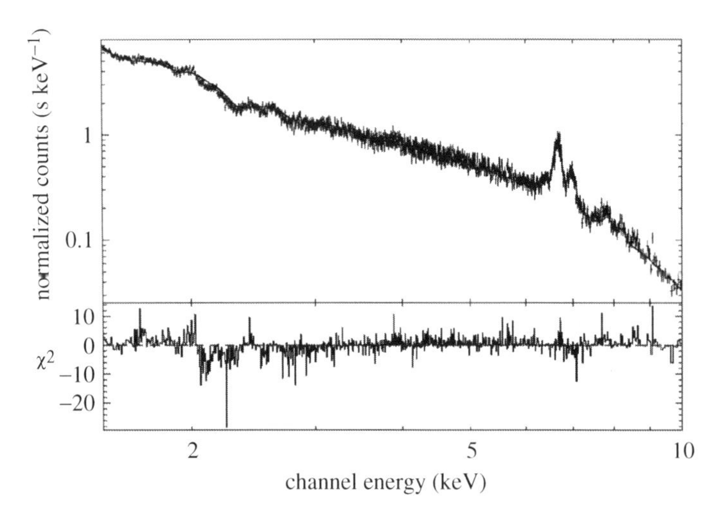

Figure 4.5. The EPIC-MOS spectrum of EX Hya from phases around the spin maximum, together with a stratified model fit and the residuals from the fit. Reproduced with permission from Cropper *et al.* (2002).

residuals from the fit are shown in (*d*). It is clear that the model accounts for both the line emission and continuum slope. Much poorer fits are obtained if a single-temperature MEKAL model, or even a three-temperature MEKAL model ((*b*) and (*c*)), is used. This indicates that the stratified models provide a good description of the physical parameters in the post-shock region.

The fits indicate that almost all the line emission is optically thin emission from the hot post-shock plasma. The residuals in Fig. 4.6 are nevertheless significant, and may be due to photoionized emission from the irradiated pre-shock flow.

AO Psc and FO Aqr

AO Psc and FO Aqr are well-observed IPs with orbital periods of *c*. 3.6 and *c*. 4.8 h, respectively. They were observed with XMM-Newton in May and June 2001, with *c*. 50 ks exposures. Although these systems are generally considered to be similar, their spectra (Figs. 4.7 and 4.8) are very different: FO Aqr is strongly absorbed, while AO Psc is less so. This was seen from fits to earlier ASCA data (see, for example, Mukai *et al.* (1994)), but now is much more evident with the low-energy response of XMM-Newton. It indicates that the treatment of the absorption in these systems is critical to an understanding of their spectra: current simple models using partial covering or even warm absorbers are insufficient, and an absorbing structure more appropriate to the dimensions of, and physical conditions in, the intervening material will need to be developed.

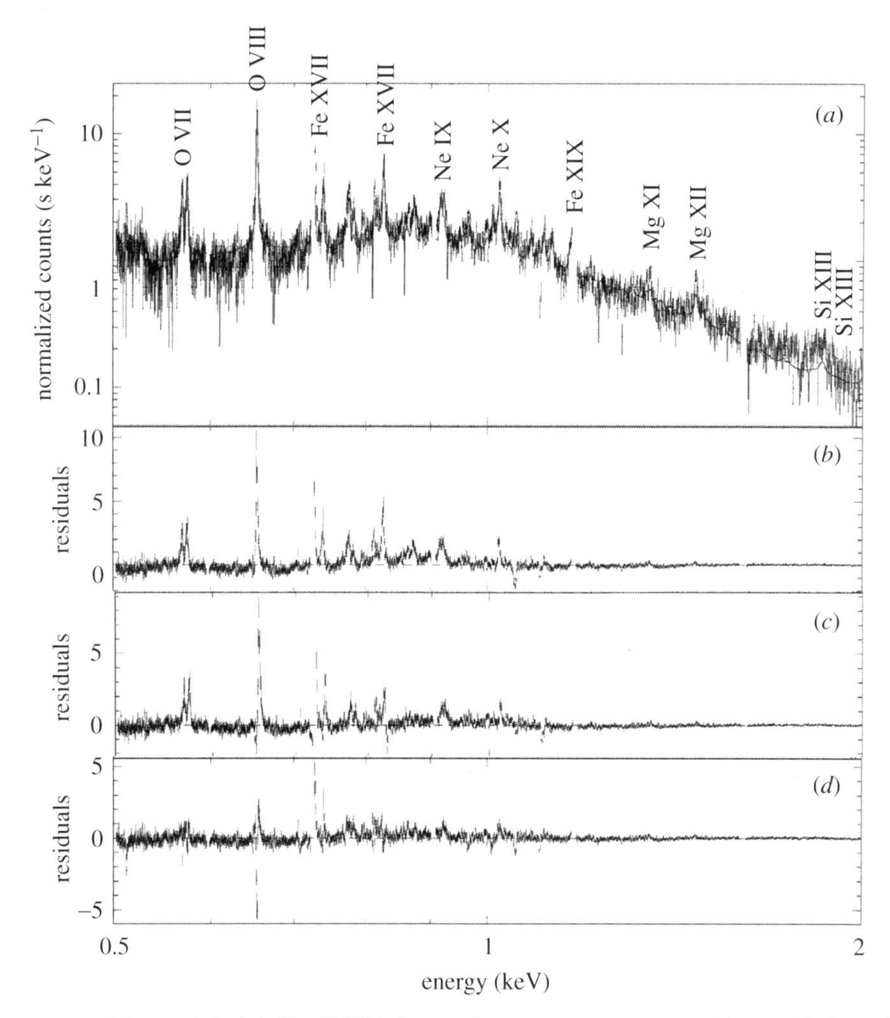

Figure 4.6. (*a*) The RGS1 first-order-average spectrum with stratified model fit, most evident at higher energies. Residuals from the fit are shown in (*d*), $\chi^2 = 2.8$, while residuals from single-temperature and three-temperature MEKAL-model fits are shown in (*b*) MEKAL 1T, $kT = 1.6$ keV, $\chi_n^2 = 6.1$ and (*c*) MEKAL 3T, $\chi_n^2 = 4.0$. Reproduced with permission from Cropper *et al.* (2002).

The spectra show a strong Fe Kα line complex, with lines at 6.4 keV from cool fluorescence (probably from the white-dwarf surface), and at 6.7 and 6.97 keV from Fe XXV and XXVI. There is also evidence for a strong O VII emission line at 0.56 keV in the spectra, particularly in FO Aqr.

The long-duration observations in both X-ray and UV show a strong modulation on both the orbital period and the spin period of the white dwarf. In AO Psc (Fig. 4.9) the amplitude of the orbital modulation is confined mainly to soft X-rays. Over the

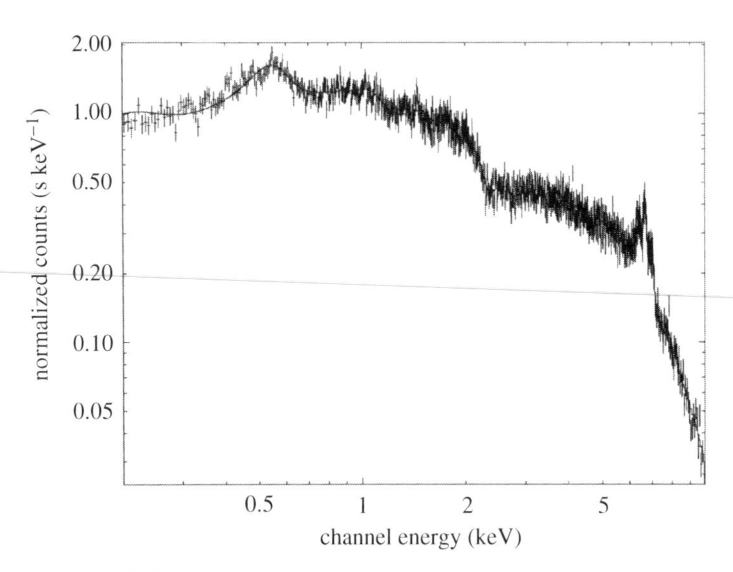

Figure 4.7. The integrated XMM-Newton EPIC-pn spectrum of AO Psc. Reproduced with permission from Hellier *et al.* (2002).

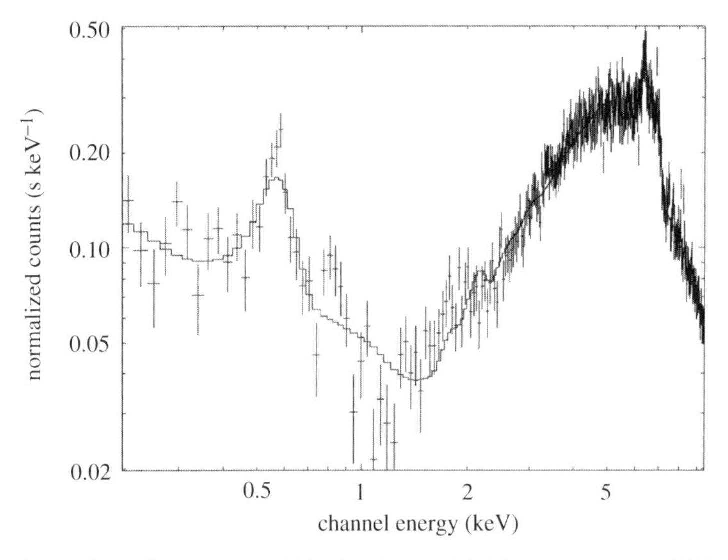

Figure 4.8. The integrated XMM-Newton EPIC-pn spectrum of FO Aqr. Reproduced with permission from Hellier *et al.* (2002).

orbital period, the UV spin-modulation amplitude grows when the soft-X-ray spin-modulation amplitude decreases. The X-rays and the UV are in-phase, indicating that the UV emission is most likely to be the emission from the heated white-dwarf surface at the time that the X-ray-emitting region is seen closest to the line of sight.

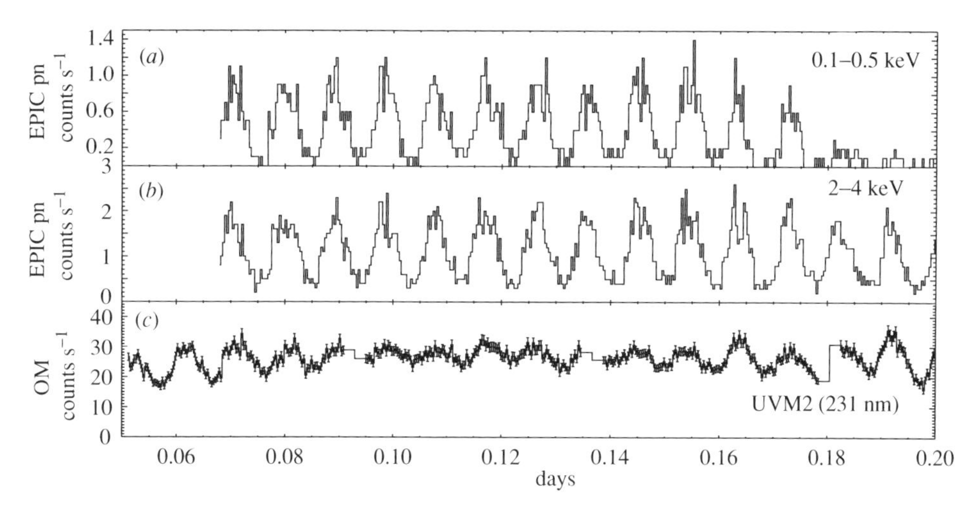

Figure 4.9. Time series of one orbit duration of AO Psc XMM-Newton in (*a*) soft and (*b*) medium X-ray bands together with (*c*) the 230-nm UV band. Reproduced with permission from Hellier *et al.* (2002).

V1223 Sgr

V1223 Sgr was observed for 50 ks with Chandra HETG in April 2000. The integrated spectrum is shown in Fig. 4.10. This shows strong, narrow lines with line ratios and inferred temperatures characteristic of photoionized plasma. These lines probably originate in the relatively cool magnetically threaded material before it encounters the accretion shock.

Polars

WW Hor

WW Hor is a relatively faint eclipsing polar observed with XMM-Newton for 23 ks in December 2000. The spectral and temporal studies for this system have been reported in Ramsay *et al.* (2001c) and Pandel *et al.* (2002), respectively.

Figure 4.11 shows the light curves folded on the 115-min orbital period from Pandel *et al.* (2002). The system is essentially undetectable in X-rays for half of the orbital period. This is the result of the accretion region being out of view behind the limb of the white dwarf at these orbital phases. At a phase of around 0.75 this region rotates over the limb, passes across the face of the white dwarf and disappears over the limb again at a phase around 0.25. Around phase 0.0, the secondary eclipses the white dwarf and the accretion region, causing the dip at the centre of the bright phase. In the optical-B band, a sinusoidal variation is seen, which can be explained as the flux from a large heated region around the accretion spot on the surface of the

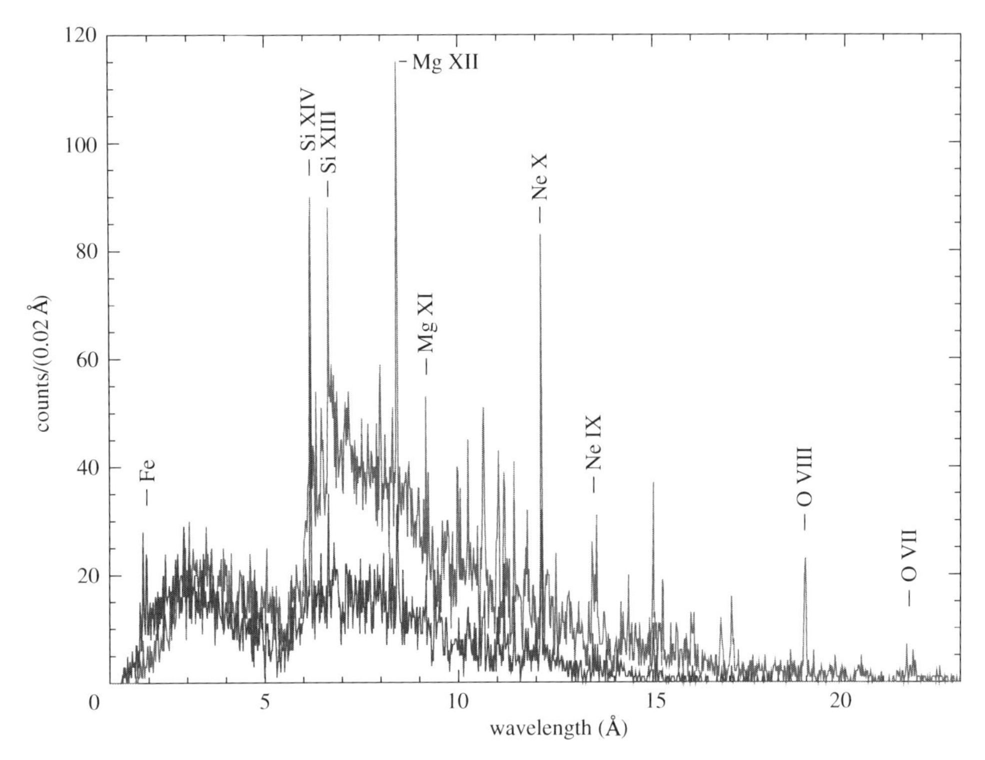

Figure 4.10. The Chandra high-energy (lower) and medium-energy (upper) HETG spectra of V1223 Sgr. Reproduced with permission from Mukai *et al.* (2002).

white dwarf. The contribution from the cyclotron flux can be seen in the R band: this is beamed perpendicular to the magnetic field lines, so is seen at its brightest when the accretion region is close to the limb.

The most interesting aspect of the data in Fig. 4.11 is that the bright-phase duration (measured at the points halfway to maximum brightness) is slightly longer for hard X-rays than for soft X-rays; hence the accretion column is hotter at the top. This indicates that we are seeing the temperature stratification of the accretion region directly, as predicted by the stratified models (see Cropper *et al.* (2000)).

Ramsay *et al.* (2001c) found that the EPIC spectrum could be fitted with the stratified models, without any need for a distinct soft X-ray component from the heated white dwarf. This is surprising in the light of the body of previous observations of polars, in which the soft component is generally found to be dominant.

BY Cam

BY Cam is one of the four known polars in which the white-dwarf period is slightly asynchronous with the orbital period (the synodic period is 14.5 days). It was

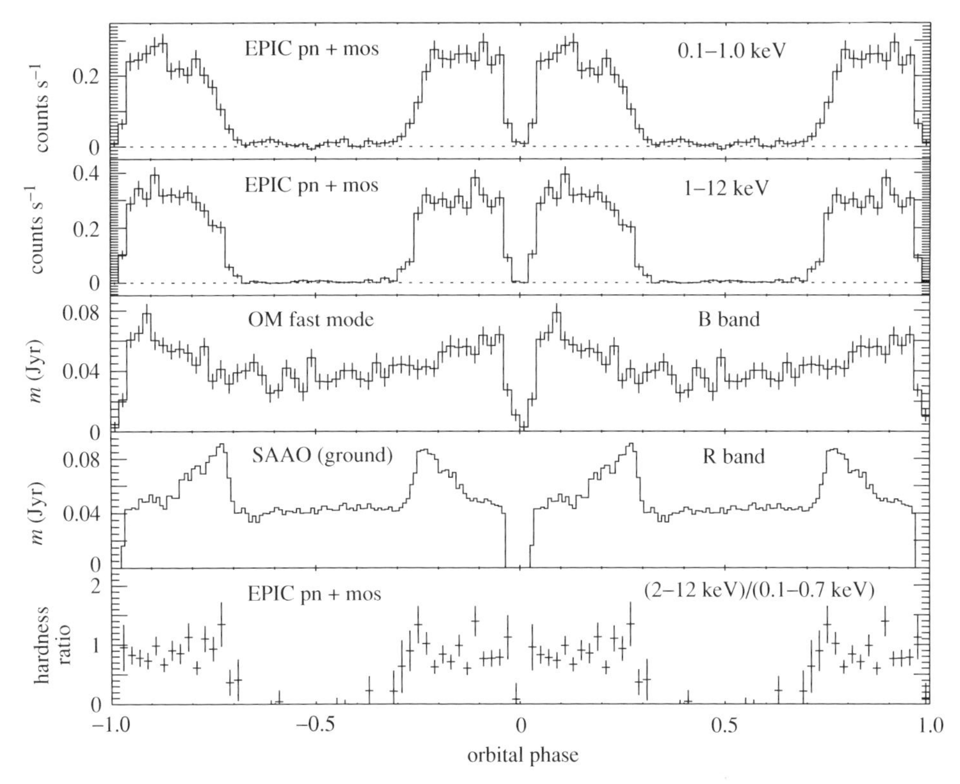

Figure 4.11. The light curve of WW Hor folded on the orbital period in different X-ray energy bands (top two panels), and the optical-B and R bands from the XMM Optical Monitor and SAAO, respectively (third and fourth panels). The bottom panel shows the X-ray hardness ratio. Adapted from Pandel *et al.* (2002).

observed in August 2001 as part of a large programme of quick-look (6 ks) observations of polars and a fuller report is available in Ramsay & Cropper (2003).

The X-ray light curves show a steep change in brightness one-third of the way through the observation, particularly in the harder band. Spectra accumulated from data after this increase are significantly different from those before it, exhibiting a strong, soft, black-body component. Ramsay & Cropper (2003) interpreted this to be the result of a second accretion region rotating into view. In contrast, the emission from the first accretion region did not require a soft component. After subtraction of a 20 000 K black body representing the white-dwarf photospheric emission, using the UV-flux measurement from the XMM Optical Monitor, they suggested that the reprocessed component could have reduced in temperature, so that it falls in the EUV, as a result of the lower heating rate at this pole. Allowing the mass to be a free parameter, the stratified models predict similar white-dwarf masses, but different accretion rates and normalizations, at each pole.

Ramsay & Cropper (2003) investigated the flaring at hard and soft energies during the time that the second accretion region was in view. They found that the hard X-ray flares *lagged* the soft X-ray flares by *c.* 20 s and were anticorrelated: this surprising finding was attributed to denser 'blobs' of material causing local heating in the white-dwarf atmosphere, which then released sufficient optically thick material to obscure the harder X-rays.

Early lessons learned from XMM-Newton and Chandra

The new XMM-Newton data have provided demanding tests of our understanding of the accretion process in magnetic CVs. In non-magnetic systems the X-ray eclipse data provide strong constraints on the radial and latitudinal extent of the boundary layer between the disc and white-dwarf surface. In magnetic systems, the success of the stratified accretion models in describing not only the continuum emission but also the line emission from these systems (for example EX Hya) indicates that we have a relatively good understanding of the physical processes taking place in the post-shock accretion region, and of its temperature and density structure. This means that we can have some confidence in the fitted values of our input parameters, such as the mass of the white dwarf and the local accretion rate; these are important for our understanding of these systems. XMM-Newton has also allowed us to observe the temperature stratification in the post-shock accretion region for the first time in WW Hor. This is a direct confirmation of the temperature structure in these accretion shocks.

Photoionized X-ray lines, which probably have their origin in the wind from the accretion disc, or from the irradiated pre-shock gas in magnetic systems, are seen. These need to be included in the X-ray modelling. It has also become clear (for example from the FO Aqr data) that, for some intermediate polars in particular, absorption and photoionizing effects can be problematic and need to be taken into account more appropriately in the model fits.

During the last decade there has been significant discussion regarding the relative strengths of the soft reprocessed component in polars compared with the optically thin emission from the multi-temperature post-shock region (the soft/hard ratio). This has been referred to as the 'soft X-ray problem'. In BY Cam we have seen that the soft/hard ratio is significantly different in the two accretion regions, and have inferred from the UV data that this is the result of the heated region at the less-active pole being shifted into the unobservable EUV. It has been a surprising result that, in several of the systems we have observed with XMM-Newton, the soft component is absent or negligible at energies higher than the interstellar absorption cut-off. This indicates that the strength of this component in the X-ray band is very sensitive to the local accretion rate.

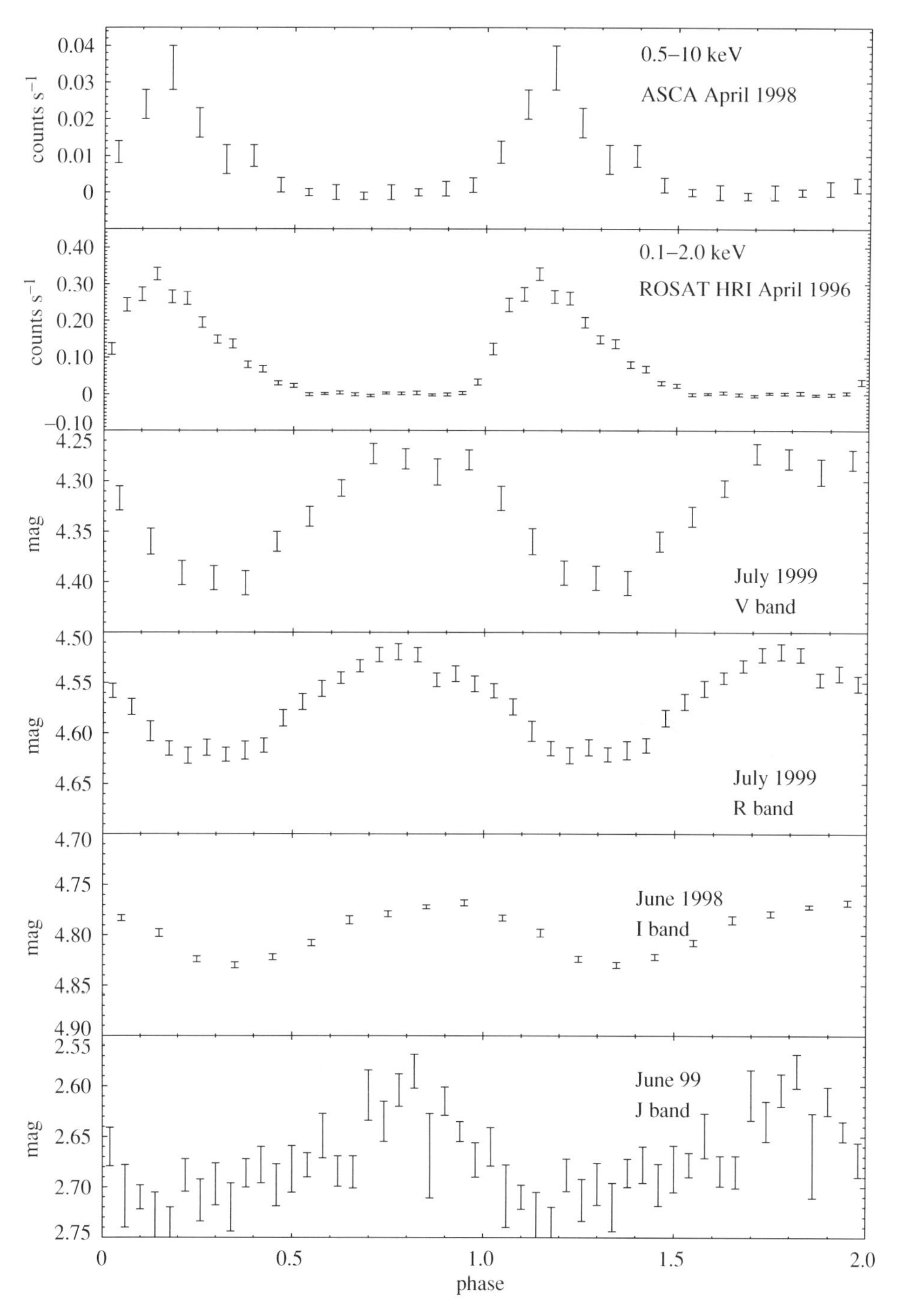

Figure 4.12. The light curve of RXJ1914+24 folded (twice) on the 569-s period. The X-ray light curves are in the top two panels, followed by three optical and one infrared band. Reproduced with permission from Ramsay *et al.* (2002b).

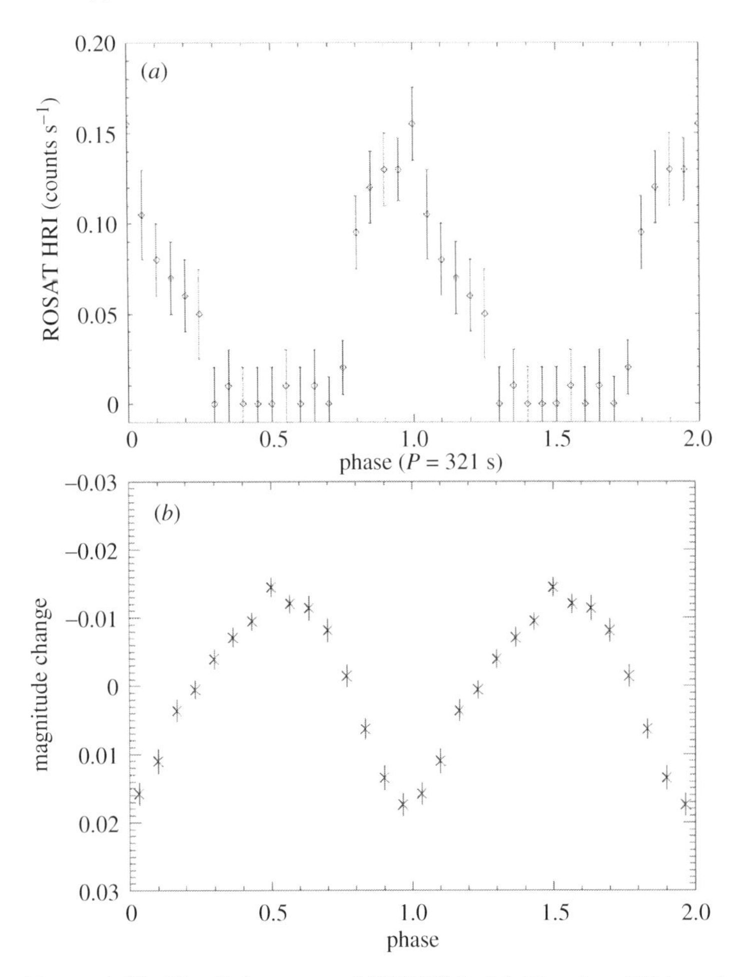

Figure 4.13. The light curve of RXJ0806+15 (October 1994 to April 1995) in X-rays (*a*) from Burwitz & Reinsch (2001) and in the optical (*b*) from Ramsay *et al.* (2002a) folded on the 321-s period. Note that the relative phasing is as yet unknown.

Double-degenerate systems

An interesting new development arising from the ROSAT survey data is the discovery of two ultrashort-period binaries, RXJ1914+24 and RXJ0806+15, with *orbital* periods of 569 and 321 s, respectively (Cropper *et al.* 1998; Israel *et al.* 1999). In this case, the secondary stars must be degenerate He cores, and indeed this is consistent with their faint, hot, photometric colours. That they are binaries is suggested by their antiphased X-ray and optical light curves, probably due to reprocessing. Slightly longer-period systems of this type are called AM CVn systems (see Warner (1995)), but it is clear now that these new systems are different in important rcspccts.

The difference is most evident in the light curves (Figs. 4.12 and 4.13), where the X-ray light curve is consistent with zero counts for half of the cycle. At least for RXJ1914+24, the X-ray and optical variation is antiphased. This is most naturally explained if the optical light originates from the X-ray-heated secondary. There are no strong emission lines in RXJ1914+24, and the optical radiation is unpolarized. Weak He lines have been seen in RXJ0806+15 (Israel *et al.* 2002).

Three explanations have been put forward to explain these properties. Cropper *et al.* (1998) suggest they might be double-degenerate polars, Ramsay *et al.* (2002b) and Marsh & Steeghs (2002) consider that they may be double-degenerate counterparts of Algol systems, and Wu *et al.* (2002) suggest that they may be 'electric stars', scaled-up versions of the Jupiter–Io unipolar inductor. In this case the energy release is not by accretion at all: the X-ray emission results from resistive heating in the white-dwarf photosphere by large currents driven between the two stars. At this stage it is unclear whether any of these models is correct.

In any case, these are fascinating systems, and with the shortest orbital periods known will be the strongest continuous sources of gravitational radiation in the sky, and thus important targets for gravitational-wave detectors.

References

Burwitz, V. & Reinsch, K. 2001 In *AIP Conference Proceedings Series* (ed. N. White, G. Malagoti, G. Palumbo), vol. 599, p. 522, Maryland: AIP.

Collins, T. J. B., Helfer, H. L. & Van Horn, H. M. 2000 *Astrophys. J.* **543**, 934.

Cropper, M., Harrop-Allin, M. K., Mason, K. O., Mittaz, J. P. D., Potter, S. B. & Ramsay, G. 1998 *Mon. Not. R. Astr. Soc.* **293**, 222.

Cropper, M., Wu, K., Ramsay, G. & Kocabiyik, A. 1999 *Mon. Not. R. Astr. Soc.* **306**, 684.

Cropper, M., Wu, K. & Ramsay, G. 2000 *New Astron. Rev.* **44**, 57.

Cropper, M. *et al.* 2002 XMM-Newton observations of EX Hya. (In preparation.)

Hellier *et al.* 2002 XMM-Newton observations of the Intermediate Polars AO PSc and FO Agr. (In preparation.)

Israel, G. L. *et al.* 1999 *Astron. Astrophys.* **349**, L1.

Israel, G. L. *et al.* 2002 *Astron. Astrophys.* **386**, L13.

Marsh, T. R. & Steeghs, D. 2002 *Mon. Not. R. Astr. Soc.* **331**, L7.

Mauche, C. W. & Mukai, K. 2002 *Astrophys. J.* **566**, L33.

Mewe, R., Kaastra, J. S. & Liedahl, D. A. 1995 *Legacy* **6**(6), 16.

Mukai, K., Ishida, M. & Osborne, J. P. 1994 *Publ. Astr. Soc. Jpn* **46**, L87.

Mukai, K., Kinkhabwala, A., Peterson, J. R., Kalin, S. & Paerels, F. 2003 *Astrophys. J.* **586**, L77.

Narayan, R. & Popham, R. 1993 *Nature* **362**, 820.

Pandel, D. *et al.* 2002 *Mon. Not. R. Astr. Soc.* **332**, 116.

Pringle, J. E. & Savonije, G. J. 1979 *Mon. Not. R. Astr. Soc.* **187**, 777.

Ramsay, G. & Cropper, M. 2003 *Mon. Not. R. Astr. Soc.* **338**, 219.

Ramsay, G. *et al.* 2001a *Astron. Astrophys.* **365**, L288.

Ramsay, G. *et al.* 2001b *Astron. Astrophys.* **365**, L294.

Ramsay, G. 2001c *Mon. Not. R. Astr. Soc.* **326**, L27.

Ramsay, G., Hakala, P. & Cropper, M. 2002a *Mon. Not. R. Astr. Soc.* **332**, L7.
Ramsay, G., Wu, K., Cropper, M., Schmidt, G., Iwamuro, K., Sekiguchi, K. & Maihara, T. 2002b *Mon. Not. R. Astr. Soc.* **333**, 575.
Rosen, S., Mason, K. O. & Córdova, F. A. 1988 *Mon. Not. R. Astr. Soc.* **231**, 549.
Warner, B. 1995 *Cataclysmic Variable Stars*. Cambridge: Cambridge University Press.
Wheatley, P. & West, R. 2002 In *The Physics of Cataclysmic Variables and Related Objects*. ASP Conference Proceedings, vol. 261, p. 433 (eds. B. T. Gänsicke, K. Bauermann & K. Reinsch). San Francisco, CA: ASP.
Wu, K. 2000 *Space Sci. Rev.* **93**, 611.
Wu, K., Cropper, M., Ramsay, G. & Sekiguchi, K. 2002 *Mon. Not. R. Astr. Soc.* **331**, 221.

5

Accretion flows in X-ray binaries

BY CHRIS DONE

University of Durham

Introduction

One of the key puzzles in X-ray astronomy is understanding accretion flows in a strong gravitational field. This applies to both active galactic nuclei (AGN) and quasars, where the accretion is onto a supermassive black hole, and to the stellar-mass galactic black holes (GBH) and even the neutron-star systems. Neutron-star radii are of the order of three Schwarzchild radii, i.e. the same as that for the last stable orbit of material around a black hole. Thus, they have very similar gravitational potentials so should have very similar accretion flows, though of course with the major difference that neutron stars then have a solid surface, so can have a boundary layer and a stellar magnetic field.

The main premise throughout this chapter is that progress in understanding accretion in *any* of these objects should give us some pointers to understanding accretion in *all* of them. Galactic sources are intrinsically less luminous, but a great deal closer than the AGN, so generally are much brighter. The variability time-scales also scale with the mass of the central object, so it is much easier to study *changes* in the accretion flow onto galactic sources. Thus we can use the galactic sources as a laboratory for understanding accretion processes and then see how much of this can be transferred to AGN.

Instabilities in accretion flows

Accretion onto a black hole via an optically thick disc has been known for decades to produce rather robust spectral predictions (hereafter SS from Shakura & Sunyaev (1973)). The emission should consist of a sum of quasi-black-body spectra, peaking at a maximum temperature of *c.* 0.6 keV for accretion rates around Eddington onto a *c.* 10 $M_\odot$ GBH, or at *c.* 0.3 keV for accretion rates at *c.* 1% of Eddington. The

Frontiers of X-Ray Astronomy, ed. A.C. Fabian, K.A. Pounds and R.D. Blandford. Published by Cambridge University Press.

whole accretion-disc structure is determined mainly by the luminosity (i.e. mass-accretion rate) as a fraction of the Eddington luminosity, $\dot{m}$, and is only weakly dependent on the mass of the black hole.

However, these steady-state accretion-disc solutions are generically unstable. At low mass-accretion rates then, the point at which hydrogen makes the transition between being mostly neutral to being mostly ionized gives a dramatic instability. When a part of the disc starts to reach temperatures at which the Wien photons can ionize hydrogen, these photons are absorbed. This energy no longer escapes from the disc, so it heats up, which produces higher temperatures and more photons which can ionize hydrogen, leading to more absorption and still higher temperatures. The runaway heating only stops when hydrogen is mostly ionized.

If the temperature crosses the hydrogen-ionization instability point at *any* radius in the disc, then the *whole disc* is unstable. This classic disc-instability model (DIM) is responsible for a huge variety of variability behaviours. In the neutron stars and black holes, the DIM is made more complex by X-ray irradiation: the outburst is triggered by the DIM, then irradiation contributes to the ionization of hydrogen, so controls the evolution of the disc, while it can also enhance the mass transfer from the companion star (see, for example, the review by Lasota (2001)).

There may also be a further instability at high mass-accretion rates, where the pressure in the disc is dominated by radiation. If the heating-rate scales with the total pressure, then this scales as T^4 where radiation pressure dominates. A small increase in temperature leads to a large increase in heating rate, and so to a bigger increase in temperature. The runaway heating only stabilizes when the time-scale for the radiation to diffuse out of the disc is longer than the accretion time-scale for it to be swallowed by the black hole (optically thick advection (Abramowicz *et al.* 1988)). The spectrum of these slim discs is slightly different from those of standard SS discs, as the energy generated in the innermost orbits is preferentially advected rather than radiated (Watarai & Mineshige 2001). At these high mass-accretion rates the disc can also overheat, causing the inner disc to become strongly Comptonized (Shakura & Sunyaev 1973; Beloborodov 1998).

However, there is much uncertainty about the disc structure at such high mass-accretion rates. The association of the viscosity with the magnetic dynamo means that the viscous heating may scale only with the gas pressure, rather than the total pressure, which removes the radiation pressure instability (Stella & Rosner 1984). Even the advective cooling may be circumvented if the disc becomes clumpy, so that the disc material becomes less efficient in trapping the radiation (Turner *et al.* 2002; Gammie 1998; Krolik 1998).

Figure 5.1(a) shows the equilibrium solutions (i.e. places where heating balances cooling) of an SS disc at a given radius and viscosity. The disc surface

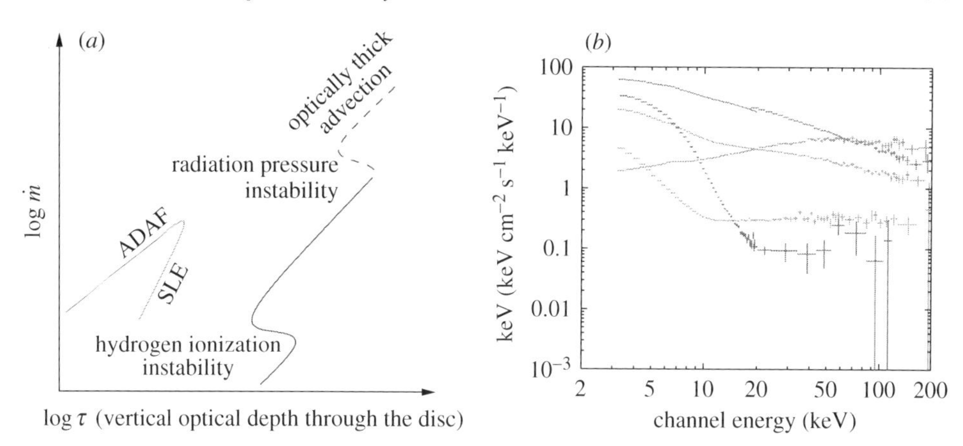

Figure 5.1. (*a*) The multiple solutions to the accretion flow equations. The right-hand line is the SS disc, modified by advective cooling at the highest mass-accretion rates and by atomic opacities at the lowest mass-accretion rates. The left-hand line is the optically thin flows, the advective flows (ADAF) and Shapiro–Lightman–Eardley (SLE) solutions (after Chen *et al.* (1995)). (*b*) RXTE PCA and HEXTE data from RXTE J1550-564 showing the low state (peaking at 100 keV) and a variety of high mass-accretion-rate spectra, all of which peak at *c.* 1 keV. Also shown are the high state (disc sharply peaking at 1 keV), ultrasoft spectrum (disc showing rounded peak at 1 keV) and two high states (disc strongly Comptonized to give a smooth, steep spectrum).

density,

$$\Sigma = \int \rho \, \mathrm{d}z \propto \tau,$$

where τ is the optical depth of the disc, generally increases as the mass-accretion rate increases, but the hydrogen ionization and radiation pressure instabilities are so strong that the disc surface density actually decreases at these points. The dashed line shows where the disc behaviour becomes very uncertain.

Observations of accretion flows in galactic black holes

Most black-hole binaries are in systems where the outer edge of the disc is cool enough to dip below the hydrogen-ionization point. They are generically transient and show large variability. These give us a sequence of spectra at differing mass-accretion rates onto the central object, allowing us to test accretion models.

What is seen is generally very different from the simple SS disc-emission ideas outlined above. At high mass-accretion rates (approaching Eddington) the spectra are dominated by a soft component at $kT \sim 1$ keV, which is strongly (very-high-state, VHS) or weakly (high-state, HS) Comptonized by low-temperature thermal (or quasi-thermal) electrons with $kT \sim$ 5–20 keV (Zycki *et al.* 1998; Gierlinski *et al.* 1999; Kubota *et al.* 2001). There is also a rather steep power-law tail ($\Gamma \sim$ 2–3), which extends out beyond 511 keV in the few objects with good high-energy data (Grove *et al.* 1998). At lower mass-accretion rates, below *c.* 2–3% of Eddington, there is a rather abrupt transition when the soft component drops in temperature and luminosity. Instead this (low-state, LS) spectrum is dominated by thermal Comptonization, with $\Gamma < 1.9$, rolling over at energies of *c.* 150 keV (see, for example, the reviews by Tanaka & Lewin (1995), van der Klis (1995) and Nowak (1995)). This spectral form seems to continue even down to very low luminosities (*c.* 0.01% of Eddington: the quiescent or off state (Kong *et al.* 2000)). Figure 5.1(*b*) shows a selection of VHS, HS and LS spectra from the GBH transient RXTE J1550-564.

While XMM-Newton and Chandra have opened up new windows in high-resolution X-ray imaging and spectroscopy, the Rossi X-ray Timing Explorer (RXTE) gives an unprecedented volume of data on these sources. To get a broad idea of the range of spectral states, quantity as well as quality is important! But this also means that plotting individual spectra is too time consuming. Colour–colour and colour–intensity diagrams have long been used for neutron-star X-ray binaries to get an overview of source behaviour (see, for example, the review by van der Klis (2000)). The problem is that they often depend on the instrument response (counts within a certain energy range) and on the absorbing column towards the source. To get a measure of the source behaviour we want to plot *intrinsic* colour, i.e. unabsorbed flux ratios over a given energy band. To do this we need a physical model. Plainly, there can be emission from an accretion disc, together with a higher-energy component from Comptonization. Reflection of this emission from the surface of the accretion disc can also contribute to the spectrum. Thus we use a model consisting of a multicoloured accretion disc, Comptonized emission (which is *not* a power law at energies close to either the seed photon temperature or the mean electron energy), with a Gaussian line and a smeared edge to roughly model the reflected spectral features, with galactic absorption.

We use this model to fit the RXTE PCA data from several different black holes from archival RXTE data to follow their broadband spectral evolution. We choose four energy bands, 3–4 keV, 4–6.4 keV, 6.4–9.7 keV and 9.7–16 keV, and integrate the unabsorbed model over these ranges to form *intrinsic* colours and use the generally fairly well-known distance to convert the extrapolated bolometric flux to total luminosity. Again, since the mass of the central object is fairly well known, we can translate the bolometric luminosity into a fraction of the Eddington luminosity.

Figure 5.2(*a*) shows a hard colour–luminosity plot for the black holes Cyg X-1 (diamonds), GX339-4 (squares) and J1550-564 (circles). The general trend is for spectra with $L/L_{\rm Edd} < 0.03$ to be hard, while those at higher luminosities are soft. The hard spectra from J1550-564 at 10% of Eddington are from the extremely

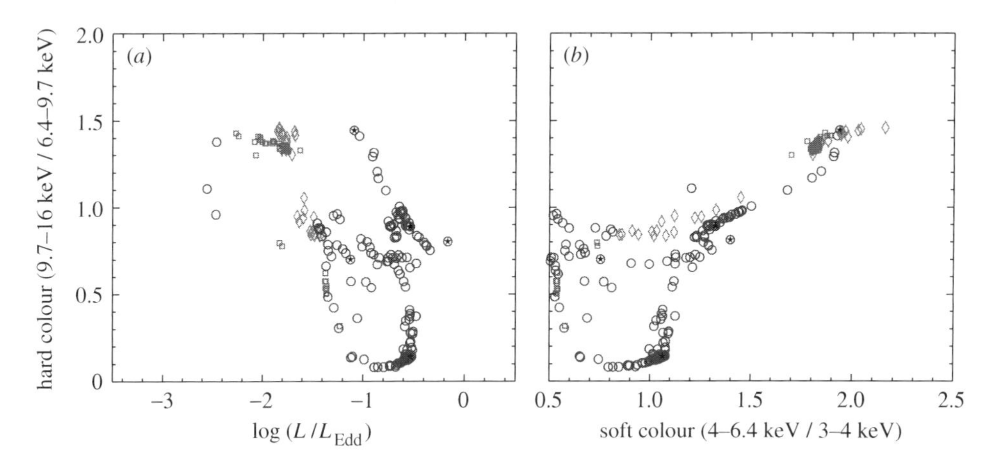

Figure 5.2. (*a*) Luminosity–colour and (*b*) colour–colour plots for the RXTE PCA data on the galactic black holes GX339 (squares), Cyg X-1 (diamonds) and J1550-564 (circles). Points corresponding to the spectra in Fig. 5.1(*b*) are marked with stars.

rapid *rise* to outburst (see Wilson & Done (2001)), where the accretion flow was presumably far from steady state. Apart from this, it is clear that there is a transition from hard to soft spectra at a few per cent of Eddington.

The huge range in spectral behaviour is brought out more clearly on a colour–colour plot (Fig. 5.2(*b*)). The hardest spectra (low/hard state) form a well-defined diagonal track, where hard and soft colour change together while the soft spectra show an amazing variety of shapes. The points corresponding to the spectra in Fig. 5.1(*b*) are (2,1.5) for the low/hard, (0.7,0.7) for the HS, (1.4,0.8) for the VHS and (1,0.1) for the ultrasoft spectra.

Observations of accretion flows around neutron stars

Neutron stars without a strong magnetic field ($B < 10^{12}$ G) come in two flavours, named atolls and Z sources. Z sources are named after a Z-shaped track they produce on an X-ray colour–colour diagram, while atolls are named after their C- (or atoll-)shaped track. These differences between the two low-mass X-ray binary (LMXB) categories probably reflect differences in both mass-accretion rate, $\dot{M}$, and magnetic field, B, with the Z sources having high luminosity (typically more than 50% of the Eddington limit) and magnetic field ($B \geq 10^9$ G), while the atolls have lower luminosity (generally less than 10% of Eddington) and low magnetic field ($B \ll 10^9$ G) (Hasinger & van der Klis 1989).

Most of these systems are *stable* to the disc instability (neutron stars round the same mass companion need to be closer than a black hole for the star to fill its

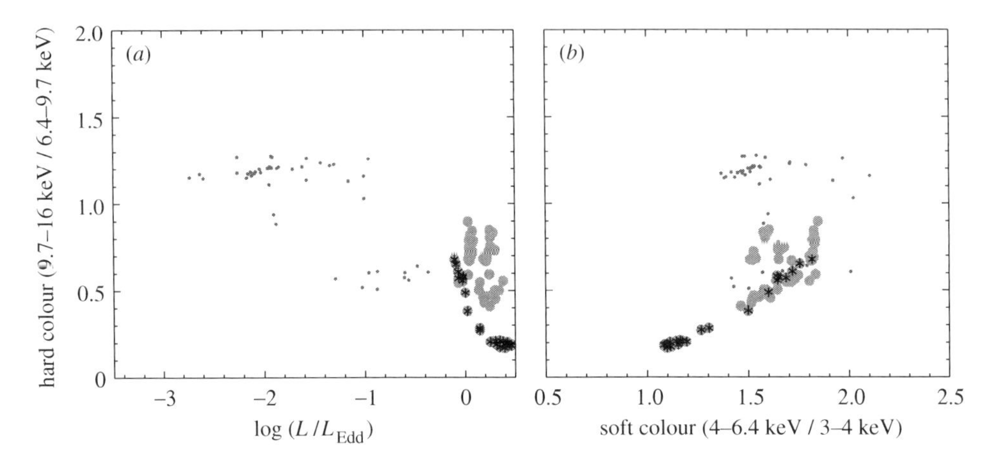

Figure 5.3. (*a*) Luminosity–colour and (*b*) colour–colour plots for the RXTE PCA data on the transient atolls (low-magnetic-field disc-accreting neutron stars, small dots). Again there is a hard-to-soft spectral switch at *c*. 10% of Eddington. The standard Z sources Cyg X-2, Sco X-1 and GX 17 + 2 are shown as large circles, while stars show Cir X-1.

Roche lobe). The disc is smaller and hotter, so is less likely to trigger the hydrogen ionization instability (King & Ritter 1998)). However, there are a few (probably evolved) atoll systems which are transient; Fig. 6.3(*a*) shows the colour–luminosity plot for these systems (small filled circles), while Fig. 6.3(*b*) shows their colour–colour diagram. Plainly there is again a switch from a well-defined hard state (which looks very similar to the brightest low-state spectra from the black holes (Yoshida *et al.* 1993; Barret & Vedrenne 1994; Barret *et al.* 2000)) to a softer spectrum at *c*. 10% of Eddington, although here the soft state also forms a single well-defined track. Because these are *intrinsic* colours, this diagram can be overlaid on that for the black holes. It is clear that the neutron-star spectra evolve in very different ways with mass-accretion rate, and that the softest black-hole spectra (HS and ultrasoft) are not seen in these atoll sources.

However, the 'ultrasoft' spectrum is not a unique black-hole signature, as the odd neutron-star binary Cir X-1 (tentatively classed as a Z source) has been known for a long time to show such a spectrum. Parts (*a*) and (*b*) of Fig. 5.3 also include data from Cir X-1 (stars) and standard Z sources (large circles). It is immediately clear that the standard Z sources do not vary by much from Eddington and that their spectral variability along their Z is also rather small (the Z track in Fig. 5.3(*b*) is mostly masked by the symbols' size!). While Cir X-1 is not convincingly like either class of system, the important point here is that Cir X-1 certainly ends up at the same 'ultrasoft' spectrum as the black holes on a colour–colour diagram. Perhaps when the mass-accretion rate is extremely high, the flow is so optically and geometrically thick that it completely swamps the central object, so that the nature of the central object becomes unimportant. However, *no* neutron-star systems show the colours associated with the HS spectra. HS spectra, or more specifically the steep power-law tails extending to high energies seen in the HS spectra, have been suggested as a unique black-hole signature (Barret & Vedrenne 1994; Laurent & Titarchuk 1999), although observations show such components in the Z sources (see, for example, di Salvo *et al.* (2001a,b) and references therein). The change in colour between the black holes and neutron stars can be explained even if the accretion flow itself is identical as the neutron stars have additional emission from a boundary layer which will give a higher temperature (harder) component (Popham & Sunyaev 2001).

X-ray emission in the low/hard state

Hard X-rays are a generic feature of accretion onto a black hole in all spectral states, yet the standard SS disc models cannot produce such emission. Even worse, the observations show that we have to explain different sorts of hard X-ray emission, with a fairly well-defined spectral switch between the hard and soft spectra at luminosities of *c.* 3% of Eddington. This mechanism has to work also in the low-magnetic-field neutron stars at low mass-accretion rates.

To get hard X-rays, a large fraction of the gravitational energy released by accretion must be dissipated in an optically thin environment where it does not thermalize, and so is able to reach much higher temperatures. For the SS disc, an obvious candidate is magnetic flares above the disc, generated by the Balbus–Hawley magnetohydrodynamic (MHD) dynamo responsible for the disc viscosity (Balbus & Hawley 1991). Buoyancy could cause the magnetic-field loops to rise up to the surface of the disc, so they can reconnect in regions of fairly low particle density. This must be happening at some level, and is shown by numerical simulations, but current models (although these are highly incomplete, as in general the simulated discs are not radiative) do not carry enough power out from the disc to reproduce the observed low/hard state (Miller & Stone 2000).

If such mechanisms produce the LS spectra, then we also need a mechanism for the transition. The obvious candidate is the radiation pressure instability. However, the disc is no longer a standard SS disc, as much of the power is dissipated in a magnetic corona rather than in the optically thick disc. The disc is cooler and denser (Svensson & Zdziarski 1994), which means that it is gas-pressure-dominated up to higher mass-accretion rates than a standard SS disc. The disc just starts to hit the radiation pressure instability at a few per cent of Eddington when the fraction of power dissipated in the corona is *c.* 60% but, in the limit where all the power is dissipated in the corona, the disc is *stable* at all mass-accretion rates below the Eddington limit (Svensson & Zdziarski 1994).

An alternative to the SS disc models is that the inner disc is replaced by an optically thin, X-ray-hot accretion flow. This flow still has to dissipate angular momentum, so magnetic reconnection is still the source of heating, but the assumption here is that the accretion energy is given mainly to the protons, and that the electrons are only heated via Coulomb collisions. The proton temperature approaches the virial temperature, so pressure support becomes important and the flow is no longer geometrically thin. The electrons cool by radiating, while the protons cool only by Coulomb collisions, so the flow is intrinsically a two-temperature plasma. Where the electrons radiate most of the gravitational energy through Comptonization of photons from the outer disc, the solution is that of a Shapiro–Lightman–Eardley (SLE (Shapiro *et al.* 1976)) flow, while, if the protons carry most of the accretion energy into the black hole, then this forms the advection dominated accretion flows (Narayan & Yi 1995). These two solutions are related, as in general both advection and radiative cooling are important (Chen *et al.* 1995; Zdziarski 1998). They are sketched as the grey line in Fig. 5.1(*a*). The SLE flows are unstable, although not dramatically so (Zdziarski 1998), while ADAFs are stable (Narayan & Yi 1995).

Importantly, both optically thin flows give typical electron temperatures of around 100 keV, as required to explain the low/hard spectra, and both can only exist (in fact they merge) at mass-accretion rates below a critical value of a few per cent of Eddington (Chen *et al.* 1995) as, when the flow becomes optically thick, the Coulomb collisions become efficient so that the flow collapses back into the one-temperature SS disc solution. Hence these flows could produce both the quiescent and LS spectra, and the collapse of such flows may give the physical mechanism for the hard–soft state transition (Esin *et al.* 1997).

X-ray emission in the high/soft states

The high-mass-accretion-rate spectra are dominated by the disc emission, but there is always a weak, power-law, X-ray tail (HS) and sometimes strong Comptonization of the disc emission (VHS). Magnetic reconnection above the disc is really the only known contender for producing the X-ray tail, but here the electrons must have a non-thermal spectrum (Gierlinski *et al.* 1999). The strong Comptonization of the disc may be connected to this same mechanism, with the electrons partially thermalizing to produce a thermal/non-thermal hybrid plasma (Poutanen & Coppi 1998). Alternatively, the Comptonization may be connected to the inner portion of the disc, either overheating in the standard disc equations (SS (Beloborodov 1998)), or as a result of trying to change its structure when it reaches the radiation pressure instability (Kubota *et al.* 2001).

Figure 5.4 sketches all these potential hard X-ray emission mechanisms from the accretion flow. However, it is known that there is also a radio/IR jet in these

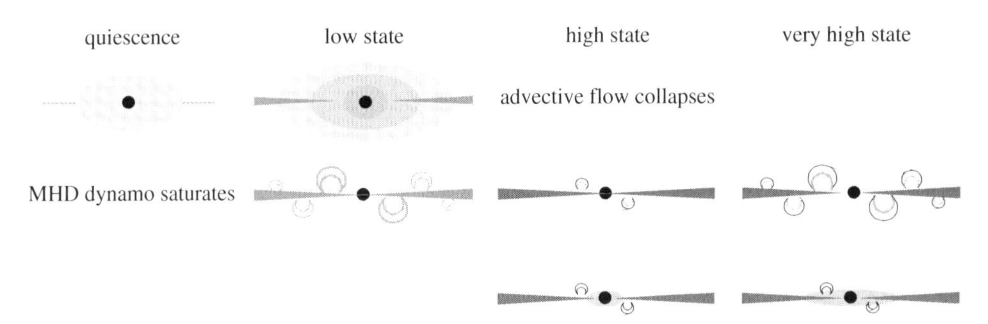

Figure 5.4. Potential X-ray emission mechanisms in the various spectral states. In quiescence the disc is *not* in a steady state (as indicated by the dotted line). Hydrogen is mostly neutral and the MHD dynamo probably cannot operate. In the low state the accretion flow is in (quasi-)steady state. Hydrogen is ionized so the MHD dynamo works and, if there is reconnection above the disc, then this could heat the electrons. Alternatively, the X-ray emission in both quiescence and the low state could be powered by an advective flow. In the high and very high states the only serious contender for the hard power-law tail is magnetic reconnection, leading to a non-thermal electron distribution (indicated by the black loops), while the thermal electrons could be part of the same mechanism (grey loops) or could be associated with the inner disc.

sources (see, for example, the review by Fender (2001)). This is poorly understood but could contribute to (or even dominate) the X-ray emission (Markoff *et al.* 2001) and/or change the accretion disc structure (Janiuk *et al.* 2002).

A unified description of accretion flows in X-ray binaries

A truncated disc/inner-hot-flow model can explain the evolution of the different classes of X-ray binaries on a colour–colour and colour–luminosity diagram. For all types of sources we assume that the evolution of the source can be explained if the main parameter driving the spectral evolution is the average mass-accretion rate, $\dot{m}$. This is *not* the same as the instantaneous mass-accretion rate, inferred from the X-ray luminosity. There is some much longer time-scale in the system, presumably tied to the response of the disc and/or inner flow (e.g. van der Klis 2000, 2001). While the truncation mechanism is not well understood, it seems likely that conductive heating of a cool disc by the hot plasma (especially if they are magnetically connected) could lead to evaporation of the disc (Meyer & Meyer-Hofmeister 1994; Rozanska & Czerny 2000), so that there is a smooth transition from an outer disc to inner hot flow.

A qualitative picture could be as follows, starting at low $\dot{m}$. For the black holes, as $\dot{m}$ increases, the truncation radius of the disc decreases, so it penetrates further into the hot flow. The changing geometry gives more soft photons to cool the hot flow and so leads to softer spectra. Both hard and soft colours soften together as the power law produced by Compton scattering in the inner flow is still the only component within the PCA bandpass. Then the inner flow reaches its maximum luminosity and collapses. As the inner disc replaces the hot flow, the disc is close enough to contribute to the spectrum above 3 keV, so the soft colour abruptly softens. The hard-X-ray tail is produced by the small fraction of magnetic reconnection which takes place outside of the optically thick disc (HS). At even higher $\dot{m}$, the disc structure is not well understood, but there seems to be a choice of two disc states, one in which the inner-disc emission is strongly Comptonized (VHS), characterized by moderate hard and soft colours or else is extremely disc dominated, as characterized by very small hard colours.

The atoll neutron stars show similar evolution, except that they also have a solid surface and so have a boundary layer. At low $\dot{m}$, the disc is truncated a long way from the neutron star, and the boundary layer is mostly optically thin, so it joins smoothly onto the emission from the inner accretion flow. Reprocessed photons from the X-ray-illuminated surface form the seed photons for Compton cooling of the inner flow. As $\dot{m}$ increases, the disc starts to move inward, but the cooling is dominated by seed photons from the neutron star rather than from the disc, so the geometry, and hence the high-energy spectral shape, does not change. The atolls keep constant hard colour, but the soft colour softens as the seed photon energy moves into the PCA band. The inner flow/boundary layer reaches its maximum luminosity when it becomes optically thick. This causes the hard colour to soften as the cooling is much more effective as the boundary layer thermalizes, so its temperature drops. The disc replaces the inner hot flow, so the disc temperature starts to contribute to the PCA bandpass so the soft colour softens abruptly. The track moves abruptly down and to the left during this transition. After this, increasing $\dot{m}$ increases the disc temperature, so the motion is to higher soft colour (Gierlinski & Done 2002a,b). The 'banana state' is then analogous to the high/soft state in the galactic black holes, but with additional luminosity from the boundary layer.

The Z sources can be similar to the atolls, but with the addition of a magnetic field (Hasinger & van der Klis 1989). In general they are stable to the disc instability because of the high mass-accretion rates, so they vary only within a factor of 2. The disc-evaporation efficiency decreases as a function of increasing mass-accretion rate, so this cannot truncate the disc in the Z sources. Instead the truncation is likely to be caused by stronger magnetic field, but here the increased mass-accretion rate means that the inner flow/boundary layer is already optically thick, and therefore cooler (Gierlinski & Done 2002a).

Models with a moving inner-disc radius at low mass-accretion rates can also qualitatively explain the variability power spectra of these sources. They show characteristic frequencies in the form of both breaks and quasi-periodic oscillations

(QPOs). These features are related ($f_{\rm break} \sim 10 f_{\rm QPO}$ for the low-frequency QPO), and they *move*, with the frequencies generally being higher (indicating smaller size scales) at higher $\dot{m}$ (see, for example, the review by van der Klis (2000)). Progress has concentrated on the similarity between the relationship between the QPO and break frequencies in black holes *and* neutron-star systems (see, for example, the review by van der Klis (2000)). If they truly are the same phenomenon, then the mechanism *must* be connected to the accretion-disc properties and not to the magnetosphere or surface of the neutron star. While the variability is not yet understood in detail, *all* QPO and break-frequency models use a sharp transition in the accretion disc in some form to pick out a preferred time-scale (van der Klis 2000), so by far the easiest way to change these frequencies is to change the inner-disc radius.

Observational tests of the low/hard-state geometry

While disc truncation with an inner X-ray-hot flow can form a coherent picture for the low-mass-accretion-rate X-ray binaries and Z sources, there is the alternative model in which the low state is produced by magnetic flares above an untruncated disc. This is a very different geometry to that of a truncated disc, so there should be some testable, observational signatures of the inner disc which can enable us to discriminate between these two models.

Direct disc emission

The most direct way to see whether there is an inner disc is look at the disc emission. The SS disc models make clear predictions about the temperature and luminosity of the disc, and these can easily be modified for the case where some fraction f of the energy is dissipated above and below the disc (Svensson & Zdziarski 1994). Disc temperature, luminosity and the fraction of energy emitted in the corona can all be *observed* with a broadband spectrum showing both disc and hard-X-ray emission, so the disc inner radius can be constrained directly from the data.

However, the expected temperatures for discs around black holes accreting at a few per cent of Eddington are in the extreme-ultraviolet (EUV)/soft-X-ray region, where interstellar absorption is important. Most black holes and neutron stars are in the galactic plane, so there is high obscuration. The Advanced Satellite for Cosmology and Astrophysics and SAX observations of Cyg X-1 in the low/hard state infer disc temperatures of *c.* 0.1 keV (di Salvo *et al.* 2001a), while the broadband SAX spectrum also shows its total bolometric luminosity was *c.* 2% of Eddington, with 70% of this emitted in the hard-X-ray spectrum (di Salvo *et al.* 2001b). Both energetics and disc temperature are consistent with a disc truncated at *c.* 50 Schwarzschild radii, but an untruncated disc with 70% of the power dissipated above the optically

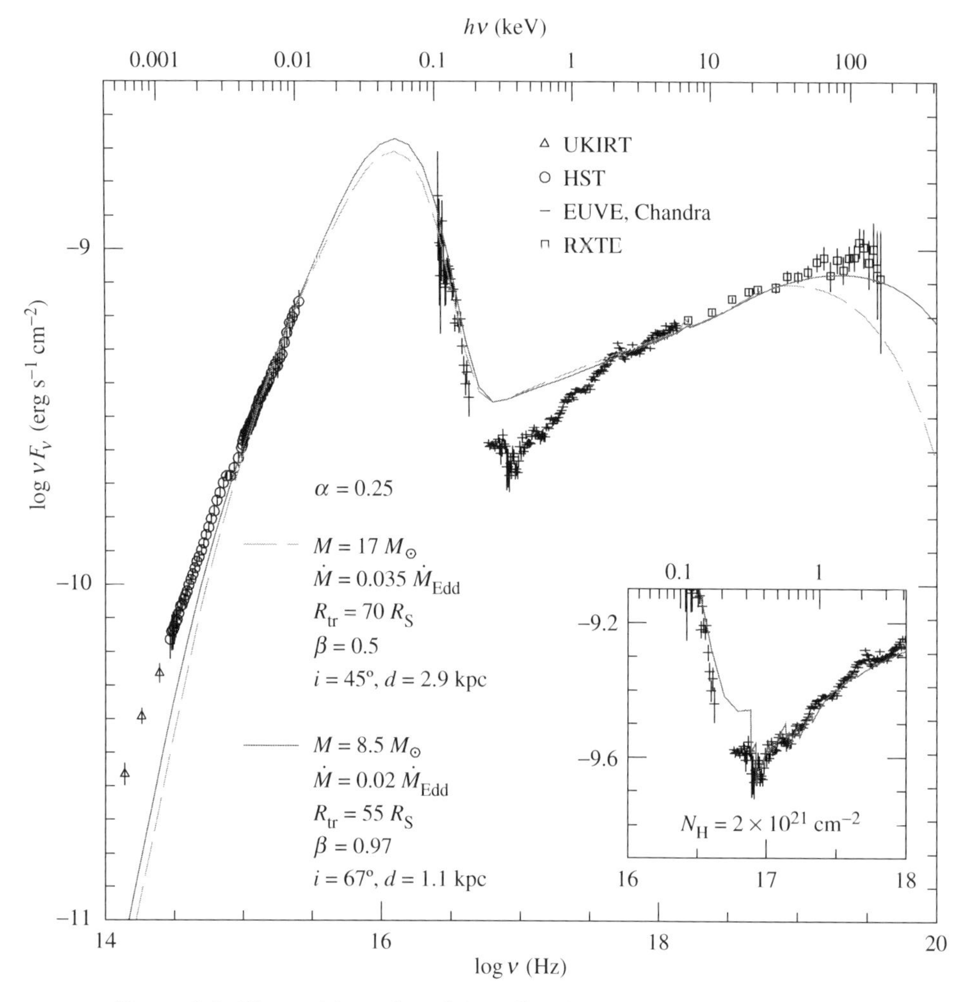

Figure 5.5. The multiwavelength broadband spectrum of the black-hole transient RXTE J1118 + 480, reproduced from Esin *et al.* (2001). The low galactic column to this source means the direct disc emission can easily be observed with the HST and EUVE.

thick disc material has a temperature of *c.* 0.2 keV. This is marginally inconsistent with that observed, but not dramatically so, and the absorption to Cyg X-1 is fairly high, so there are some systematic uncertainties.

The disc is much more clearly seen in the transient black-hole system RXTE J1118 + 480. This has extremely low galactic absorption, so the disc emission can be detected with the Hubble Space Telescope (HST) and Extreme Ultraviolet Explorer EUVE as well as in soft X-rays. The temperature is *c.* 0.02–0.04 keV, while the luminosity is *c.* 0.1% of Eddington, with *c.* 40% of the total emitted in hard X-rays. This low disc temperature is completely inconsistent with a disc extending down to the last stable orbit (McClintock *et al.* 2001; Frontera *et al.* 2001), even with 90% of the power dissipated in a hot corona. Esin *et al.* (2001) successfully fit the overall spectral shape with a truncated-disc/inner-advective-flow model (Fig. 5.5).

Reflection

An independent way to study the extent of the optically thick accretion flow is via reflection. Wherever hard X-rays illuminate optically thick material there is some probability that the X-rays can be reflected. The reflection probability is given by a trade-off between the importance of electron scattering and photoelectric absorption. Since the latter is energy dependent, the albedo is also energy dependent, with higher-energy photons being preferentially reflected due to the smaller photoelectric opacity of the material. This gives rise to a reflected spectrum that is harder than the intrinsic spectrum, with photoelectric edge features and associated fluorescent lines imprinted on it. Since iron is the highest atomic number element which is astrophysically abundant, the iron K edge is particularly prominent, at 7.1–9.3 keV depending on the ionization state of the reflecting material, together with its associated iron Kα fluorescence-line emission at 6.4–6.9 keV (see, for example, the review by Fabian *et al.* (2000)).

The reflected spectrum (line and continuum) is smeared by special and general relativistic effects of the motion of the disc in the deep gravitational potential well (Fabian *et al.* 1989, 2000). The fraction of the incident flux which is reflected gives a measure of the solid angle of the optically thick disc as observed from the hard X-ray source, while the amount of smearing shows how far the material extends into the gravitational potential of the black hole. This should give a clear test of whether the inner disc is present.

The galactic black holes in the low/hard state show overwhelmingly that the solid angle is significantly less than 2π, and that the smearing is less than expected for a disc extending down to the last stable orbit (Zycki *et al.* 1997, 1998, 1999; Gierlinski *et al.* 1997; Done & Zycki 1999; Zdziarski *et al.* 1999; Gilfanov *et al.* 1999; Revnivtsev *et al.* 2001). While this is clearly consistent with the idea that the disc is truncated in the low/hard state, an alternative explanation for the lack of reflection and smearing is that the inner disc or top layer of the inner disc is completely ionized. There are then no atomic features, and the disc reflection is unobservable in the 2–20 keV range, as it appears instead to be part of the power-law continuum (Ross & Fabian 1993; Ross *et al.* 1999; Nayakshin *et al.* 2000). Ionization could be especially important in GBH due to the high disc temperature predicted by SS untruncated disc models, and indeed in the HS and VHS the reflection signature from the disc in GBH is clearly ionized (Gierlinski *et al.* 1999; Wilson & Done 2001; Miller *et al.* 2001). However, for the low/hard state, the *observed* disc temperatures of less than 0.1 keV are not sufficient to strongly ionize iron. Collisional ionization does not strongly distort the GBH reflection, but photoionization can be very important, and models with complex photoionization structure can fit the observed reflection signature by an untruncated disc (Young *et al.* 2001; Done & Nayakshin 2001; Ballantyne *et al.* 2001). However, the inner disc in such models is not completely invisible as the illuminating photons at *c.* 100 keV cannot be reflected elastically The reflected spectrum decreases at high energies, so models fitting the 2–20 keV spectrum with large amounts of ionized reflection predict a smaller 100–200 keV flux than truncated disc models in which reflection is intrinsically weak. Broadband spectral observations clearly favour the truncated disc models, and show that the high-energy data can only be consistent with a large amount of ionized reflection if the continuum is complex (Maccarone & Coppi 2002; Barrio *et al.* 2003). Another prediction of the ionized disc models is that the Compton downscattering of high-energy photons heats the disc, so the hard X-rays are reprocessed and thermalized down to temperatures which are typically of the order of the temperature expected from an SS inner disc (Haardt & Maraschi 1993). While detailed models of the thermalized emission from complex ionization models have not yet been fitted to the data, at least 20% of the illuminating flux should be thermalized by the inner disc, leading to a higher temperature than observed in RXTE J1118 + 480.

Reflection and reprocessed thermal emission from the disc can be suppressed entirely if the magnetic flares are expanding relativistically away from the disc (Beloborodov 1999). However, the disc emission in RXTE J1118 + 480 and Cyg X-1 would then have to be the intrinsic disc emission, and the expected temperature of an untruncated disc is again higher than those observed.

Conclusions

A truncated disc at low mass-accretion rates is compatible with *all* the constraints on the extent of the accretion disc as measured by direct emission and reflection. It can also give a unified picture of the evolution of the broadband spectral shape in both accreting black holes and neutron stars, and qualitatively explain the variability power spectra if the disc extends further into the gravitational potential as the mass-accretion rate increases. The most convincing single measurement which challenges alternative, untruncated disc models for the low/hard state is the observed low-temperature disc in the black-hole transient RXTE J1118-480, with supporting evidence for a low disc temperature in Cyg X-1. Evaporation of the disc into a hot inner-accretion flow gives a plausible physical mechanism for the truncation, and the mass-accretion rate at which the inner flow becomes optically thick gives a mechanism for the spectral switch from hard to soft spectra. At higher mass-accretion rates the structure of the accretion disc is not well understood, and the data seem to show a variety of soft spectral shapes which may be associated with optically thick advection (slim discs) and/or the radiation-pressure instability and/or overheating of the inner disc.

We can scale these ideas up to AGN if the accretion flow is simply a function of radius in terms of Schwarzchild radii, and mass-accretion rate in terms of fraction of the Eddington rate. The disc-evaporation mechanism should still work in the same way around supermassive black holes (Rozanska & Czerny 2000). Thus, it seems likely that those AGN which accrete at a low fraction of the Eddington limit should also be similar to the low/hard state from galactic black holes. Conversely, AGN at high mass-accretion rates (narrow-line Seyfert-1s and MCG-6-30-15?) are probably the counterparts of the HS and VHS spectra seen in the GBH where the disc probably extends down to the last stable orbit. To test these ideas we again need some way to determine the extent of the inner disc, but as the disc temperature scales as $M^{-1/4}$, it is even harder to observe the direct disc emission in AGN than in GBH. Reflection is then probably the best current way to track the extent of the optically thick accretion disc, and the very broad line seen in MCG-6-30-15 (Tanaka *et al.* 1995; Wilms *et al.* 2001) points to a disc which extends down to at least the last stable orbit. However, results from other AGN are currently controversial. Lubinski & Zdziarski (2001) find that AGN with harder spectra and probably lower mass-accretion rates show less relativistic smearing and less reflection, consistent with the accretion picture outlined above for the galactic sources, while previous literature has stressed the similarity of the line profiles to that of MCG-6-30-15 (Nandra *et al.* 1997). Current challenges to observers are to disentangle the shape of the line in AGN, especially those with hard spectra, so as to determine the nature of the accretion flow in low mass-accretion-rate AGN, and to theoreticians are to develop a better understanding of the high mass-accretion rate discs and jets.

Much of this review includes ideas developed in collaborations with Marek Gierlinski, Aya Kubota and Piotr Zycki.

References

Abramowicz, M. A., Czerny, B., Lasota, J. P. & Szuszkiewicz, E. 1988 *Astrophys. J.* **332**, 646.
Balbus, S. A. & Hawley, J. 1991 *Astrophys. J.* **376**, 214.
Ballantyne, D. R., Ross, R. R. & Fabian, A. C. 2001 *Mon. Not. R. Astr. Soc.* **327**, 10.
Barret, D. & Vedrenne, G., 1994 *Astrophys. J. Supp.* **92**, 505.
Barret, D., Olive, J. F., Boirin, L., Done, C., Skinner, G. K. & Grindlay, J. E. 2000 *Astrophys. J.* **533**, 329.
Barrio, F. E., Done, C. & Nayakshin, S. 2003 *Mon. Not. R. Astr. Soc.* **342**, 557.
Beloborodov, A. M. 1998 *Mon. Not. R. Astr. Soc.* **297**, 739.
Beloborodov, A. M. 1999 *Astrophys. J.* **510**, L123.

Chen, X., Abramowicz, M. A., Lasota, J.-P., Narayan, R. & Yi, I. 1995 *Astrophys. J. Lett.* **443**, L61.
di Salvo, T., Done, C., Zycki, P. T., Burderi, L. & Robba, N. R. 2001a *Astrophys. J.* **547**, 102.
di Salvo, T., Robba, N. R., Iaria, R., Stella, L., Burderi, L., & Israel, G. L. 2001b *Astrophys. J.* **554**, 49.
Done, C. & Nayakshin, S. 2001 *Mon. Not. R. Astr. Soc.* **328**, 616.
Done, C. & Zycki, P. T. 1999 *Mon. Not. R. Astr. Soc.* **305**, 457.
Esin, A. A., McClintock, J. E. & Narayan, R. 1997 *Astrophys. J.* **489**, 865.
Esin, A. *et al.* 2001 *Astrophys. J.* **555**, 483.
Fabian, A. C., Rees, M. J., Stella, L. & White, N. E. 1989 *Mon. Not. R. Astr. Soc.* **238**, 729.
Fabian, A. C., Iwasawa, K., Reynolds, C. S. & Young, A. J. 2000 *Publ. Astr. Soc. Pac.* **112**, 1145.
Fender, R. 2001 *Astrophys. & Space Sci.* **276**, 69.
Frontera, F. *et al.* 2001 *Astrophys. J.* **546**, 1027.
Gammie, C. 1998 *Mon. Not. R. Astr. Soc.* **297**, 929.
Gierlinski, M. & Done, C. 2002a *Mon. Not. R. Astron. Soc.* **331**, L47.
Gierlinski, M. & Done, C. 2002b *Mon. Not. R. Astron. Soc.* **337**, 1373.
Gierlinski, M. *et al.* 1997 *Mon. Not. R. Astr. Soc.* **288**, 958.
Gierlinski, M., Zdziarski, A. A., Poutanen, J., Coppi, P. S., Ebisawa, K. & Johnson, W. N. 1999 *Mon. Not. R. Astr. Soc.* **309**, 496.
Gilfanov, M., Churazov, E. & Revnivtsev, M. 1999 *Astron. Astrophys.* **352**, 182.
Grove, J. E. *et al.* 1998 *Astrophys. J.* **500**, 899.
Haardt, F. & Maraschi, L. 1993 *Astrophys. J.* **413**, 507.
Hasinger, G. & van der Klis, M. 1989 *Astron. Astrophys.* **225**, 79.
Janiuk, A., Czerny, B. & Siemiginowska, A. 2002 *Astrophys. J.* **576**, 908.
King, A. R. & Ritter, H. 1998 *Mon. Not. R. Astr. Soc.* **293**, L42.
Kong, A. K. H., Kuulkers, E., Charles, P. A. & Homer, L. 2000 *Mon. Not. R. Astr. Soc.* **312**, L49.
Krolik, J. H. 1998 *Astrophys. J. Lett.* **498**, L13.
Kubota, A., Makashima, K. & Ebisawa, K. 2001 *Astrophys. J. Lett.* **560**, L147.
Lasota, J. P. 2001 *New Astron. Rev.* **45**, 449.
Laurent, P. & Titarchuk, L. 1999 *Astrophys. J.* **511**, 289.
Lubinski, P. & Zdziarski, A. A. 2001 *Mon. Not. R. Astr. Soc.* **323**, L37.
Maccarone, T. J. & Coppi, P. 2002 *Mon. Not. R. Astr. Soc.* **336**, 817.
Markoff, S., Falke, H. & Fender, R. 2001 *Astron. Astrophys* **372**, L25.
McClintock, J. E. *et al.* 2001 *Astrophys. J.* **555**, 477.
Meyer, F. & Meyer-Hofmeister, E. 1994 *Astron. Astrophys.* **288**, 175.
Miller, J. *et al.* 2001 *Astrophys. J.* **546**, 1055.
Miller, K. A. & Stone, J. M. 2000 *Astrophys. J.* **534**, 398.
Nandra, K., George, I. M., Mushotzky, R. F., Turner, T. J. & Yaqoob, T. 1997 *Astrophys. J.* **477**, 602.
Narayan, R. & Yi, I. 1995 *Astrophys. J.* **452**, 710.
Nayakshin, S., Kazanas, D. & Kallman, T. R. 2000 *Astrophys. J.* **537**, 833.
Nowak, M. A. 1995 *Publ. Astr. Soc. Pac.* **107**, 1207.
Popham, R. & Sunyaev, R. 2001 *Astrophys. J.* **547**, 355.
Poutanen, J. & Coppi, P. S. 1998 *Physica Scr.* **T77**, 57.
Revnivtsev, M., Gilfanov, M. & Churazov, E. 2001 *Astron. Astrophys.* **380**, 520.
Ross, R. R. & Fabian, A. C. 1993 *Mon. Not. R. Astr. Soc.* **261**, 74.
Ross, R. R., Fabian, A. C. & Young, A. J. 1999 *Mon. Not. R. Astr. Soc.* **306**, 461.

Rozanska, A. & Czerny, B. 2000 *Astron. Astrophys.* **360**, 1170.
Shakura, N. I. & Sunyaev, R. A. 1973 *Astron. Astrophys.* **24**, 337.
Shapiro, S. L., Lightman, A. P. & Eardley, D. M. 1976 *Astrophys. J.* **204**, 187.
Stella, L. & Rosner, R. 1984 *Astrophys. J.* **277**, 312.
Svensson, R. & Zdziarski, A. A. 1994 *Astrophys. J.* **436**, 599.
Tanaka, Y. & Lewin, W. H. G. 1995 In *X-ray Binaries* (eds. W. H. G. Lewin, J. van Paradijs & E. van den Heuvel), p. 126. Cambridge: Cambridge University Press.
Tanaka, Y. *et al.* 1995 *Nature* **375**, 659.
Turner, N., Stone, J. M. & Sano, T. 2002 *Astrophys. J.* **566**, 148.
van der Klis, M. 1995 In *Astronomy, Astrophysics and Gravitation.* Springer Lecture Notes in Physics, vol. 454, p. 321. Palo Alto Springer.
van der Klis, M. 2000 A. *Rev. Astron. Astrophys.* **38**, 717.
van der Klis, M. 2001 *Astrophys. J.* **561**, 943.
Watarai, K.-Y. & Mineshige, S. 2001 *Publ. Astr. Soc. Jpn* **53**, 915.
Wilms, J. *et al.* 2001 *Mon. Not. R. Astron. Soc.* **328**, L27.
Wilson, C. D. & Done, C. 2001 *Mon. Not. R. Astr. Soc.* **325**, 167.
Yoshida, K. *et al.* 1993 *Pub. Astron. Soc. Jpn* **45**, 605.
Young, A. J., Fabian, A. C., Ross, R. R. & Tanaka, Y. 2001 *Mon. Not. R. Astr. Soc.* **325**, 1045.
Zdziarski, A. A. 1998 *Mon. Not. R. Astr. Soc.* **296**, L51.
Zdziarski, A. A., Lubinski, P. & Smith, D. A. 1999 *Mon. Not. R. Astr. Soc.* **303**, 11.
Zycki, P. T., Done, C. & Smith, D. A. 1997 *Astrophys. J. Lett.* **488**, 113.
Zycki, P. T., Done, C. & Smith, D. A. 1998 *Astrophys. J. Lett.* **496**, 25.
Zycki, P. T., Done, C. & Smith, D. A. 1999 *Mon. Not. R. Astr. Soc.* **305**, 231.

6

Recent X-ray observations of supernova remnants

BY CLAUDE R. CANIZARES

Massachusetts Institute of Technology

Introduction

Supernova remnants (SNRs) have been important targets for X-ray astronomers since the earliest days, described in Chapter 1. The capabilities of the Chandra X-ray Observatory and XMM-Newton to provide both high-quality imaging and spectroscopy are particularly well suited to the study of these objects. There are already a great many new results, more than I can summarize here. Instead, I will illustrate with a few examples the kinds of things we are learning and the prospects for future progress in this field (see Holt & Hwang (2001) for background and context).

Supernovae, the cataclysmic explosions that end the lives of stars, shine brightly for only a few weeks or months. Most of the energy of the explosion, aside from that radiated in neutrinos, is carried off as kinetic energy. The material in the outer parts of the star is expelled at velocities of 5000–10 000 km s^{-1}. Over time-scales of years, decades and centuries, this supersonic outflow of stellar ejecta, which itself cools quickly after the explosion, sweeps up and shock heats the surrounding interstellar or circumstellar matter to X-ray temperatures (this is the primary shock). The swept-up material progressively retards the motion of the ejecta, causing another shock to work its way backwards (in the reference frame of the expanding ejecta), so it too is heated to X-ray temperatures. The result is an extended, diffuse region of X-ray-emitting plasma, most of which has temperatures above several million kelvins. By studying this debris, we can learn about the progenitor star, the explosion, and the interstellar and circumstellar matter. Because supernovae are thought to be the primary mechanism for chemical enrichment of the Universe, such studies have far-reaching implications.

SNRs come in a wide variety of sizes, structures and types, depending on what caused them, how old they are and how close they are to Earth. This means that the

Frontiers of X-Ray Astronomy, ed. A.C. Fabian, K.A. Pounds and R.D. Blandford. Published by Cambridge University Press.

complementary imaging and spectral capabilities of Chandra and XMM-Newton can be used to best advantage to span this range. Chandra's sub-arc-second resolution is best for the smaller remnants and for fine details, whereas XMM-Newton excels in the study of SNRs which are larger and have lower surface brightness. Both missions have the ability to perform simultaneous imaging with moderate resolution spectroscopy, which, as I will show, is absolutely essential in probing the detailed properties of SNRs. They also both cover a wide bandwidth, roughly 0.1–10 keV, which spans a correspondingly wide range of temperatures and ionization states, and can discriminate non-thermal- from thermal-emission processes. The high-resolution spectrometers are also complementary. The XMM-Newton reflection-grating spectrometer (RGS) is optimized for lower energies and for obtaining composite spectra of extended sources and spatially resolved spectra of larger SNRs. The Chandra transmission-grating spectrometers extend up to higher energies and give spatially resolved spectra for smaller SNRs or for bright filaments in the larger remnants.

Supernovae come in various types, depending on the properties of the star and its pre-explosion history. My examples will focus on three well-studied remnants of core-collapse supernovae of different ages. The explosion in these cases comes from the gravitational collapse of the innermost, iron-rich, core of a massive star, when the nuclear processes can no longer provide enough energy to counterbalance gravity. The explosions are classified as either type Ia or type II (based on the optical characteristics), with the differences caused by whether or not the outer envelope of the star has been ejected prior to the explosion. An important characteristic of the core-collapse supernovae is that they are thought to supply the bulk of the oxygen in the Universe.

The three remnants listed in Fig. 6.1 are Cassiopeia A, E0102-72 and N132D. They are among the brightest and best-studied SNRs and cover an interesting range

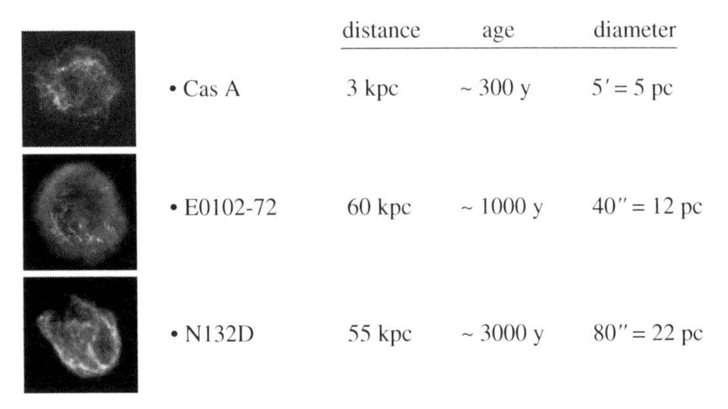

	distance	age	diameter
• Cas A	3 kpc	~ 300 y	$5' = 5$ pc
• E0102-72	60 kpc	~ 1000 y	$40'' = 12$ pc
• N132D	55 kpc	~ 3000 y	$80'' = 22$ pc

Figure 6.1. The three oxygen-rich supernova remnants. (Images courtesy of NASA/CXC/SAO.)

of ages, and therefore stages of evolution, from 300 to 3000 y after the supernova explosion. I will also mention some work on SN1987A, which is a mere 14-year-old SNR infant of a core-collapse supernova.

X-ray imaging

High-spatial-resolution images of SNRs show a rich structure of X-ray-emitting filaments, with size scales extending down to the arc-second scale, which corresponds to *c*. 0.015 parsec at the distance of Cas A (Hughes *et al.* 2000). Cas A was, in fact, the first astronomical target for Chandra. The high resolution of that image revealed the presence of a central point source thought to be the long-sought-for remnant of the collapsed core (e.g. Chakrabarti *et al.* (2001)). The image also shows the structural signature of the two shocks described above, one at the outermost edge of the SNR, propagating into the surrounding material, and another that has heated the denser and somewhat lower-temperature stellar ejecta in the interior of the remnant (Gotthelf *et al.* 2001). This 'textbook' picture of forward and reverse shocks is even more clearly seen in the image of E0102-72 (Gaetz *et al.* 2000). The oldest of our examples, N132D, does not show such a clear structure, presumably because the primary shocked material already dominates (as indicated by elemental abundances similar to those of the interstellar matter (ISM), see Hughes *et al.* (1998)) and also because this remnant is likely to have evolved in a low-density cavity in the ISM which could give it somewhat peculiar characteristics (Hughes 1987).

Proper motion studies of SNRs in the X-ray range are an especially exciting application of high-resolution imaging. Hughes *et al.* (2000) compared Chandra images of E0102-72 with earlier ones from the Röntgensatellit (ROSAT) and detected a *c*. 0.8 arcsec increase in diameter over 20 years. This corresponds to an age consistent with the earlier estimates of 1000 y and an expansion velocity of 6000 km s^{-1}. The latter would also be the primary shock velocity. Because the observed X-ray temperatures are far below those implied by this shock velocity (even accounting for slow equilibration of proton and electron energies), they could conclude that a significant fraction of the shock energy must be going into particle acceleration. More recently, Park *et al.* (2002) were even able to obtain marginal evidence for the expansion of the infant remnant of SN1987A, which is only 1 arcsec in diameter. By comparing images obtained between October 1999 and April 2001 they found an expansion of 0.08 karcsec in diameter, corresponding to a velocity of 5000 km s^{-1}.

X-ray spectroscopy

The spectra of SNRs are generally dominated by bright emission lines of highly ionized elements from C through to Fe or Ni (the exceptions are SNRs surrounding

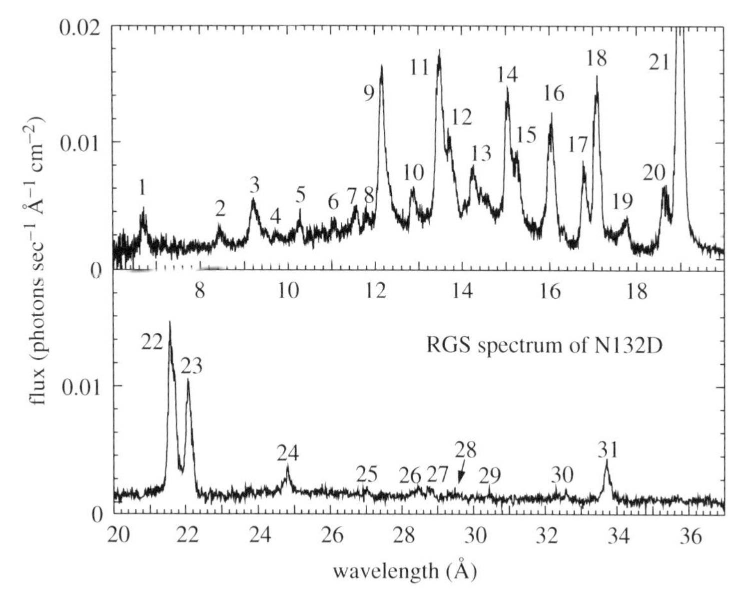

Figure 6.2. The composite RGS spectrum of N132D from Behar *et al.* (2001). Numbers refer to line identifications.

active pulsars, where the spectra are instead dominated by non-thermal continuum). With the RGS, Rasmussen *et al.* (2001) obtained a high signal-to-noise spectrum of E0102-72 that bristles with emission lines. A similar composite RGS spectrum of N132D, shown in Fig. 6.2, reveals over 30 emission lines (Behar *et al.* 2001).

E0102 appears to the RGS as a point source, so the spectrum is a composite of the entire remnant, in contrast to what we describe below from the high-energy transmission-grating spectrometer (HETGS) on Chandra. This is the complementarity referred to above: the RGS is untroubled by the spatial extent of moderately extended sources, so it measures the global properties. For example, Rasmussen *et al.* (2001) were able to detect weak Fe lines and to perform preliminary plasma diagnostics. Conversely, the HETGS can distinguish between different regions of the remnant and obtain monochromatic images in the light of the stronger emission lines. On larger remnants, the RGS also gives monochromatic images as mentioned below.

The RGS was also used to obtain a composite spectrum of N132D which shows 31 lines (Behar *et al.* 2001). This remnant is large enough to broaden the profiles considerably, but the effective resolution is still far better than that obtained with a charge-coupled device (CCD). Behar *et al.* found a wide range of ionization states of iron, from Fe XVII to Fe SSII. Their European photon imaging camera (EPIC) spectrum shows that Fe XXV is also present.

SN1987A is small enough to appear as nearly a point source to the HETGS, and Burrows *et al.* (2000) have obtained the first high-resolution spectrum of such a very young SNR. The spectrum is dominated by lines of He-like and H-like O, Ne, Mg and Si, with relatively little Fe. Their preliminary analysis shows consistency with an ionizing plasma at temperature $kT \sim 3$ keV.

Imaging spectroscopy

The ultimate power of Chandra and XMM-Newton for SNR studies is the capability of obtaining spectrally resolved images. Of course, observers using the Advanced Satellite for Cosmology and Astrophysics (ASCA) pioneered such studies on larger remnants (Hwang & Gotthelf 1997), but it could not easily resolve smaller SNRs such as those discussed here.

For Cas A, Hughes *et al.* (2000) and Hwang *et al.* (2000) obtained monochromatic images corresponding to the strong emission lines of Si, S, Ar, Ca and Fe (see Fig. 6.3). There are very clear spatial differences in the distribution of these elements. A surprising finding is that Fe, which should be interior to the Si and S zone in the progenitor star, appears to have overtaken and passed through the Si/S in the SE portion of the remnant. Clearly, this elemental segregation must be accounted for when one attempts to measure the abundances, temperatures and ionization time-scales.

Behar *et al.* (2001) used EPIC-MOS to obtain similar ionic distributions for N132D, whose larger extent matches the imaging resolution of XMM-Newton better. As would be expected for this older remnant with global abundances close to those of the host LMC galaxy (Hughes 1987), the different elements generally appear to have approximately the same distribution. However, there is a very prominent oxygen-rich region, which we also see in the Chandra data and which indicates that the ejecta are by no means completely dominated by or fully mixed with the swept up circumstellar material.

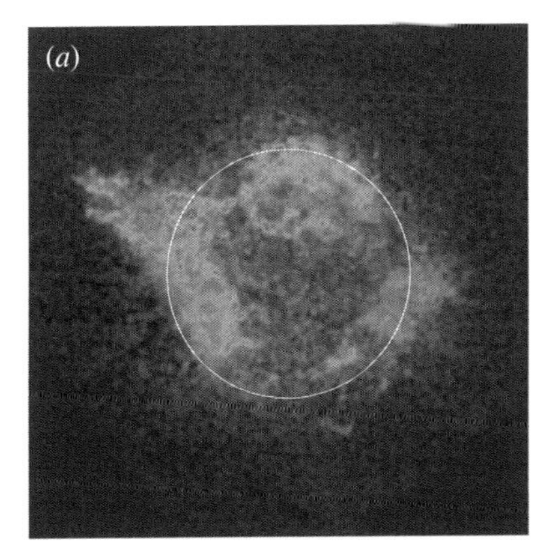

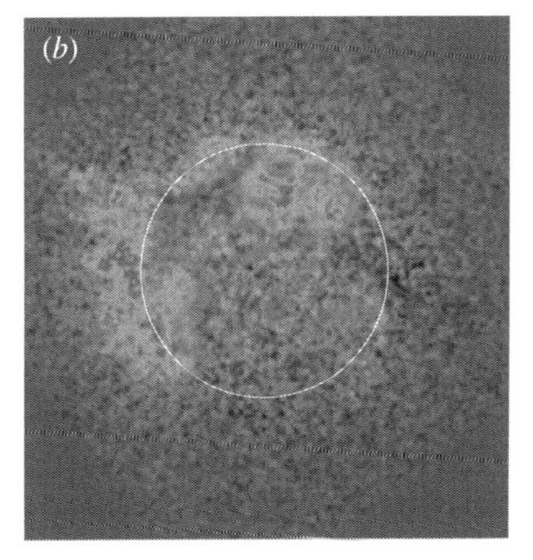

Figure 6.3. Monochromatic images of Cas A corresponding to emission from (*a*) Si and (*b*) Fe L. Reproduced from Hwang *et al.* (2000).

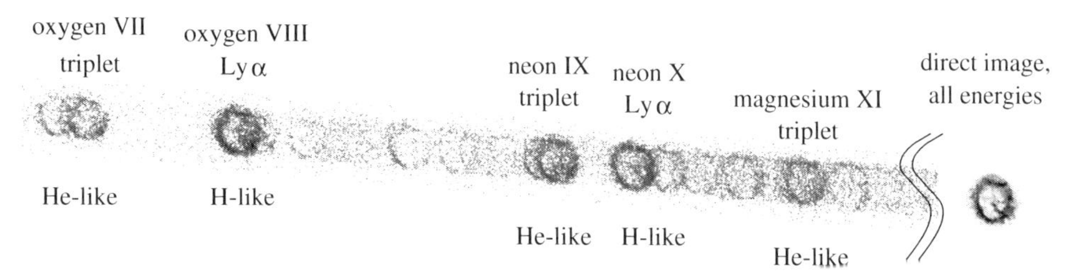

Figure 6.4. Dispersed high-resolution X-ray spectrum of E0102-72. Shown here is a portion of the dispersed spectrum formed by the medium-energy gratings. The zeroth order, which combines all energies in an undispersed image, is at the right-hand side. Images formed in the light of several strong X-ray-emission lines are labelled.

For moderately extended SNRs, such as E0102-72, the HETG produces a spectrum that resembles a spectroheliogram, with successive images of the remnant in the light of individual emission lines (Canizares *et al.* 2001). An example is shown in Fig. 6.4. This, of course, complicates the analysis of the composite spectrum, but it does allow us to perform spatially resolved spectroscopy at the much higher resolution that one can achieve with the CCD cameras, at least for the stronger, more isolated lines. The RGS provides the same capability for studying larger remnants such as Cas A (Bleeker *et al.* 2001). In contrast, for the HETG the spectral images of larger remnants like Cas A and even N132D have so many overlapping lines that it is difficult to obtain a complete monochromatic image except possibly in the strongest lines. However, many of these SNRs have individual bright filaments whose high-resolution spectra can be extracted (by using a narrow window in the cross-dispersion direction and using the zeroth-order image to correct for the effective-line-spread function of the extended feature). In N132D, we have studied spectra from several bright knots at different locations, finding clear evidence for radical variation in elemental abundance (Canizares *et al.* 2001), including the dominance of oxygen, as also noted by Behar *et al.* (2001) from EPIC-MOS images.

Ionization fronts

It has been known for many years that, for many SNRs, the time-scale for ionizing the plasma to equilibrium is comparable with or greater than the age of the remnant (or than the still shorter time since shock passage, since some material will only have very recently been shocked). The degree of ionization at a given temperature actually depends on the product of age and electron density, which is proportional to the number of potentially ionizing collisions suffered by a particular atom. Indeed, it has been customary to use non-equilibrium models to fit the composite spectra obtained from previous missions, most recently the ASCA (e.g. Hughes *et al.* 1998).

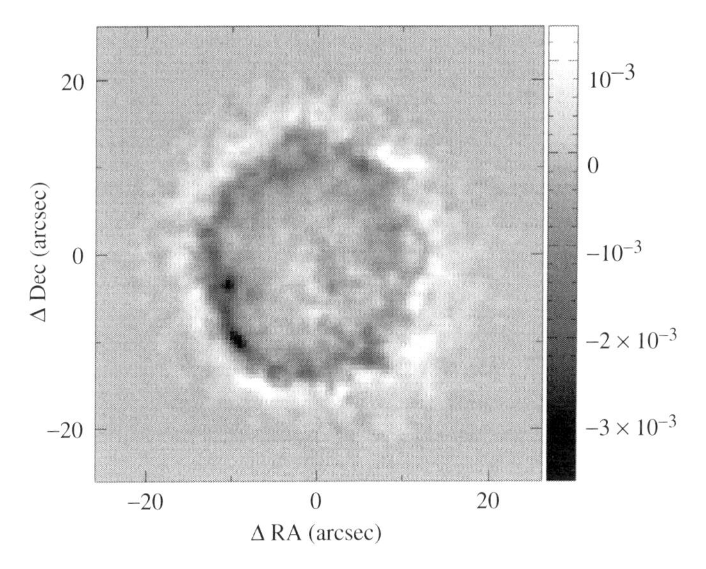

Figure 6.5. A difference image between bands centred at lines from O VIII (white) and O VII (black), taken from Gaetz *et al.* (2000). The inner regions are less ionized than the outer regions, reflecting the inward propagation of the reverse shock.

With Chandra and XMM-Newton we can now can obtain graphic images of the progressive ionization that occurs behind the shock front.

Gaetz *et al.* (2000) used Chandra's AXAF (Advanced X-ray Astrophysics Facility) CCD Imaging Spectrometer (ACIS) images of E0102-72 to show that H-like O VIII comes from a region with a larger diameter than that of He-like O VII (Fig. 6.5). The oxygen is in the expanding stellar ejecta, so it is heated and stripped by the reverse shock that propagates inwards towards the centre. The inner regions have been heated more recently and have not had time to strip off the extra electron to turn He-like into H-like ions. This shows up as well in the monochromatic images obtained with the HETG, not only for O but also for Ne, Mg and Si (Canizares *et al.* 2001; Flanagan *et al.* 2002). The relative radius of a given ion (measured in the cross-dispersion direction to avoid the spectral/spatial confusion) increases monotonically with its ionization time-scale. Bleeker *et al.* (2002) used the EPIC images of Cas A to construct maps of temperature and ionization time-scale across the remnant. These do not show such a clear progression of the ionization front, presumably partly because of projection effects (see below).

Doppler studies

The high spectral-resolution and spectral-imaging capabilities of Chandra and XMM-Newton are beginning to reveal interesting information about the kinematics of SNRs, through the measurement of Doppler shifts. The first such measurement was performed by Markert *et al.* (1983) on Cas A using the focal plane crystal spectrometer on the Einstein Observatory. They found a puzzling asymmetry across the remnant, with relative Doppler offsets of several thousand km s^{-1}, which was later confirmed by the ASCA (Holt *et al.* 1994). With data from Chandra's ACIS camera, Hwang *et al.* (2001) constructed a detailed Doppler map of Cas A, using the Si line, that shows clear differences in the line-of-sight velocities of the SE portion of the remnant relative to the NW. As noted above, for large remnants, the HETG is able to focus on individual bright knots, and we have detected Doppler velocities of 2000 km s^{-1} for knots in Cas A. Willingale *et al.* (2002) also mapped the Doppler velocities in Cas A for several lines and then, using the assumption that the flow (but not the matter) is spherically symmetric, constructed a three-dimensional distribution of the emitting material. Naturally, the asymmetry of the outflow translates into an asymmetric distribution of material, which appears to be concentrated to portions of an annular fraction of the spherical volume inclined to the line of sight (something already suggested by Markert *et al.* (1983)).

The monochromatic images of E0102-72 obtained with the HETG reveal similar asymmetric Doppler shifts. These are seen by comparing the two (plus- and minus-order) dispersed images from a strong line, such as Ne X Lyα, with the zeroth-order image constructed by cutting the ACIS pulse heights at the corresponding energy. The fact that we have both plus- and minus-order images breaks the spectral/spatial degeneracy that would otherwise confuse such an analysis. A dispersed image in the light of a single line, which is distorted due to intrinsic spatial variations, will look identical on either side of zeroth order, whereas a distortion due to a wavelength (Doppler) shift will appear with opposite offsets in the plus and minus orders (i.e. a shift to longer wavelength moves to the right in the plus-order image but to the left in the minus-order image, showing reflectional symmetry about the zeroth order). Furthermore, comparison of plus and minus orders also allows one to identify and avoid confusion from the overlapping images of nearby lines.

Canizares *et al.* (2002) used the strong Ne X Lyα line to construct a Doppler map of E0102-72 that shows even more clearly the asymmetries that emerged from their preliminary analysis (Canizares *et al.* 2001). Velocities are in the range ± 2000 km s^{-1} and show systematic variation around the bright ring of emitting ejecta. As in the case of Cas A, the bulk of the emitting material in E0102-72 appears to be confined to an expanding ring inclined to the line of sight. Hughes (1988) had suggested such a geometry already, based on the surface-brightness distribution measured with ROSAT. Although the RGS cannot image E0102-72, it does detect Doppler broadening in the line profiles (Rasmussen *et al.* 2001).

The pronounced asymmetries in Cas A and E0102-72 could conceivably reflect underlying asymmetries in the supernova explosion itself. But it seems more plausible to invoke asymmetries in the circumstellar material, allowing the ejecta to be more nearly spherically symmetric. Then the reverse shock would be asymmetric even though the ejecta are not. Blondin (2001) has begun to explore hydrodynamic models that suggest such a scenario might work. It is interesting that, on a much smaller scale, an analogue of this process is occurring in SN1987A. There the primary shock is just beginning to hit the innermost ring of circumsource material imaged with the Hubble telescope (see, for example, Burrows *et al.* (2000)), causing a ring-shaped emitting region in the X-ray.

Conclusion

The power of XMM-Newton and Chandra to provide a three-dimensional view of the kinematics, structure and physical conditions in SNR will add profoundly to our knowledge of these objects. I have presented only a small portion of the results that have emerged in the first few years; there is much more to come. I am particularly excited about the prospects for achieving three-dimensional imaging by combining direct images, spatially resolved Doppler measurements and, where possible, proper motion studies. One goal here is to gain new insights into the kinematics of the ejecta and the dynamics of their interaction with the surrounding medium. We also need to explain the apparent asymmetries in the structure of younger SNRs, about which we presently know very little. We are sure to learn much more about the shock phenomena by mapping the ionization stages of various ions and by refining techniques like those applied by Hughes *et al.* (2000) to estimate the energy channelled into cosmic rays. Elemental distributions will tell us about the degree of mixing in the explosion, and elemental abundances will constrain the nature of the progenitor and the models of nucleosynthesis. Altogether, the prospects for major advances in our knowledge of all these phenomena have never been better.

I am very grateful to K. Flanagan, D. Dewey, J. Houck, A. Fredricks, D. Davis and the other members of the MIT HETG and CXC groups who have contributed to the observations and analysis of SNR spectra. This work was supported in part by NASA under contracts NAS8-38249 and NAS8-01129 for the HETG, and via the Smithsonian Astrophysical Observatory contract SAO SV1-61010 for the CXC.

References

Behar, E. *et al.* 2001 *Astron. Astrophys.* **365**, L242–L247.
Bleeker, J. *et al.* 2001 *Astron. Astrophys.* **365**, L225–L230.
Bleeker, J., Vink, J., van der Heyden, K., Willingale, D., Kaastra, J. & Lamming, M. 2002 Preprint astro-ph/0202207. (In the press.)
Blondin, J. 2001 In *Young Supernova Remnants* (eds. S. Holt & U. Hwang), pp. 59–68. New York: American Institute of Physics.
Burrows, D. *et al.* 2000 *Astrophys. J.* **543**, L149–L152.

Canizares, C., Flanagan, K., Davis, D., Dewey, D., Houck, J. & Schattenburg, M. 2001 In *Young Supernova Remnants* (eds. S. Holt & U. Hwang), pp. 213–221. New York: American Institute of Physics.
Chakrabarti, D., Pivovaroff, M., Hernquist, L., Heyl, J. & Narayan, R. 2001 *Astrophys. J.* **548**, 800–810.
Flanagan *et al.* 2002 In preparation.
Gaetz, T. *et al.* 2000 *Astrophys. J.* **534**, L47–L50.
Gotthelf, E. V., Koralesky, B., Rudnick, L., Jones, T. W., Hwang, U. & Petre, R. 2001 *A. Astrophys. J.* **552**, L39–L43
Holt, S. & Hwang, U. (eds.) 2001 *Young Supernova Remnants*. New York: American Institute of Physics.
Holt, S., Gotthelf, E., Tsumnemi, H. & Negoro, H. 1994 *Publ. Astron. Soc. Jpn* **46**, L151–L155.
Hughes, J. 1987 *Astrophys. J.* **314**, 103–110.
Hughes, J. 1988 In *Supernova Remnants and the Interstellar Medium. Proceedings of IAU Colloquium 101*. (eds. R. S. Roger & T. L. Landecker), p. 125. Cambridge: Cambridge University Press.
Hughes, J., Hayashi, I. & Koyama, K. 1998 *Astrophys. J.* **505**, 732–748.
Hughes, J., Rakowski, C. & Decourchelle, A. 2000 *Astrophys. J.* **543**, L61–L65.
Hwang, U. & Gotthelf, E. V. 1997 *Astrophys. J.* **75**, 665.
Hwang, U., Holt, S. & Petre, R. 2000 *Astrophys. J.* **537**, L119–L122.
Hwang, U., Szymkowiak, A., Petre, R. & Holt, S. 2001 *Astrophys. J.* **560**, L175–L179.
Markert, T., Clark, G., Winkler, F. & Canizares, C. 1983 *Astrophys. J.* **268**, 134–144.
Park, S. *et al.* 2002 *Astrophys. J.* **567**, pp. 314–322.
Pounds, K. 2002 *Phil. Trans. R. Soc. Lond.* A **360**, 1905–1921.
Rasmussen, A., Behar, E., Kahn, S. M., den Herder, J. W. & van der Heyden, K. 2001 *Astron. Astrophys.* **365**, L231–L236.
Willingale, R., Bleeker, J., van der Heyden, K., Kaastra, J. & Vink, J. 2002 *Astron. Astrophys.* **381**, 1039–1048.

7

X-ray components in spiral and star-forming galaxies

BY MARTIN WARD

University of Leicester

Introduction

Early surveys of normal galaxies with the Einstein X-ray satellite revealed a number of distinct source populations: low-mass X-ray binaries associated with the old disc, the bulge and globular clusters; high-mass X-ray binaries associated with spiral arms and star-forming regions; plus a few supernova remnants. In addition to these discrete sources, an extended soft-X-ray component was discovered in the form of sometimes highly spectacular superwinds, most notably seen in galaxies undergoing energetic star formation. This supernova-driven wind is a mixture of ejecta and entrained interstellar medium. The interaction gives rise to high shock velocities of 500–800 km s^{-1}, which produce a complex multi-phase X-ray-emitting medium. This process is important in terms of chemical enrichment, galactic feedback and shock excitation physics. In this review I shall begin with a summary of some new results on the extended X-ray components. I will then concentrate on a subset of the discrete-source population within galaxies, now refered to as ultraluminous sources (ULXs). A relative term used because their luminosities are considerably more, and sometimes far in excess of that predicted for accretion of material onto a solar mass compact object. This topic is now receiving particularly close attention because it is exceptionally well suited to studies that exploit the excellent spatial resolution (better than an arc second) delivered by the Chandra observatory.

Extended/diffuse X-ray components

In this section I review some results on the extended X-ray components found in galaxies, excluding ellipticals (for discussion of some new results on early type

Frontiers of X-Ray Astronomy, ed. A.C. Fabian, K.A. Pounds and R.D. Blandford. Published by Cambridge University Press.

systems, see Irwin *et al.* (2002)). One of the consequences of high spatial resolution is that point sources can easily be identified and hence removed from the spectral analysis. This means that the diffuse emission can be studied free from 'contamination' by populations of point sources. Using a combination of spatial separation and spectral component fitting, a self consistent picture of the various X-ray components found in spiral and star-forming galaxies can be formulated (Persic and Rephaeli 2002).

Compared with emission from X-ray binary systems, the diffuse thermal plasma contributes mostly at energies below 1 keV, and relatively less at harder energies. For example in M82 75% of the total 2–10 keV emission comes from point sources (Griffiths *et al.* 2000). The origin of this diffuse thermal plasma is believed to be hot gas in the ISM extending out to the galactic halo, which is heated by the combined effect of multiple supernova explosions. In the case of NGC253, and probably for star-forming galaxies in general, this emission has been shown to arise from regions of interaction between the dense, cool ambient interstellar medium (ISM), and the fast moving hot supernova (SN) driven wind, and not from the ejected wind material itself (Strickland *et al.* 2000). The X-ray spectral signatures are those of a non-uniform, multi-temperature plasma, ranging from 0.13 keV up to 6 keV in the case of NGC 253 and 1782 (Pietsch *et al.* 2001; Stevens *et al.* 2003).

Size of extended components and cospatial Hα emission

It is easier to detect extended emission in edge-on galaxies, because of their favourable projection factor, which reduces contamination from other sources close to the centre of the galaxy in actual de-projected distance. Wang *et al.* (2001) detected a giant diffuse X-ray halo, around the edge-on disc galaxy NGC 4631. This component extends 8 kpc from the plane of the galaxy, and has a temperature of around 5 million degrees. There is a morphological similarity between the X-ray corona and the radio halo, suggesting a possible link between hot gas outflows, and magnetic field and cosmic ray distribution.

In the more energetic star-forming galaxies we often see so-called superwinds of X-ray emission, which in M82 and NGC253 extend out to about 11 kpc and 8 kpc from their respective galactic planes. A characteristic property of such winds appears to be cospatial emission in X-rays and ionized gas detected via Hα line emission. A clear explanation of this important relationship between hot tenuous, and warm dense gaseous components has not yet emerged. For example we do not know whether shocked disc gas has been carried out into the halo, or whether *in-situ* halo material was swept up. Alternatively, in some cases tidal debris left over from galaxy interactions may provide a working surface as the ejected wind material ploughs into it. As Strickland *et al.* (2002) point out, we must await better data on the detailed spatial distributions of the X-ray and Hα emitting gas to further constrain these various models. However, prospects for this are good, and in the case of the X-ray/Hα superbubble in NGC 3079, Cecil *et al.* (2002) have used Chandra observations to show that these filaments form a contact shock interface between the galaxy's ambient gas and the shocked wind.

High-resolution X-ray spectra

High-spatial-resolution images are a powerful tool to help us understand the nature of the extended X-ray component. Another perspective can be gained from the use of X-ray spectral diagnostics.

Active galaxies (Seyferts) are outside the scope of this review. However, since many star-forming galaxies and some Seyferts exhibit clear evidence of extended X-ray emission, it is instructive to make a brief comparison of the spectral properties of this component for both classes. Extended X-ray components are seen in some active galaxies, notably in Seyfert 2s e.g. NGC1068 (Brinkman *et al.* 2002), Circinus (Rita *et al.* 2001) and Markarian 3 (Sako *et al.* 2000). Unlike the situation in spirals and star-forming galaxies extended X-ray emission from Seyferts is not consistent with hot plasma models, which would require extremely low metal abundances. For the best studied example, NGC1068, the extended X-ray emission is best fit by models of photionization and fluorescence due to continuum radiation from the hidden (from our direct view) Seyfert nucleus (Young *et al.* 2001). High-resolution X-ray spectroscopy of NGC1068 was obtained using the reflection grating spectrometer (RGS) aboard XMM/Newton by Kinkhabwala *et al.* (2002), and using the LETGS on Chandra by Brinkman *et al.* (2002). These spectra display strong narrow radiative recombination continua, suggesting that most of the soft X-ray emission arises in a plasma with a temperature of only a few eV.

In contrast to the Seyfert 2s, high-resolution RGS X-ray spectra of M82 are consistent with hot gas in collisional ionization equilibrium, with evidence of high metal abundance for several species (Read and Stevens, 2002). Comparison with an RGS spectrum for another star-forming galaxy NGC253 (Pietsch *et al.* 2001) shows similar spectral features but the gas in M82 is more highly ionized, perhaps because the SN wind/ISM interaction in M82 is more energetic. The RGS spectra of both these star-forming galaxies are similar to that of an intermediate age SNR, supporting an interpretation in terms of collisionally ionized gas. It seems therefore that although extended X-ray components may be present in AGN as well as star-forming galaxies, in the former case it is the AGN which exerts the dominant influence on the physical processes responsible for the X-ray emission, even when its continuum cannot be directly detected at optical/ultraviolet wavelengths.

In summary, a better understanding of the origin and evolution of these extended X-ray components, and in particular superwinds in relatively nearby galaxies, will have broad and far reaching implications. It is apparent that such SN driven winds also exist in high-redshift galaxies, and these may heat and enrich the intergalactic medium with metals (Pettini *et al.* 2001). A local example of gas destined to enrich the IGM is inferred from a detailed Chandra study of the dwarf starburst galaxy NGC1569, (Martin *et al.* 2002). Winds may also influence the growth of galactic bulges, and affect the evolution of isolated dwarf galaxies and the satellites of large galaxies.

A history of ULX studies

Even the earliest imaging X-ray surveys identified a number of intermediate to high luminosity (10^{39}–10^{40} erg s^{-1}) X-ray sources (see Fabbiano (1989)). This result immediately attracted interest, since unobscured active galactic nuclei (AGN) in nearby spirals are typically orders of magnitude more luminous, whereas black-hole X-ray binaries (BHXB) known in our Galaxy are at least an order of magnitude less luminous. Subsequent work showed that these X-ray-luminous extra-nuclear sources are common (Colbert & Mushotzky 1999). Furthermore, a ROSAT High-Resolution Imagery (HRI) survey of 83 bright, nearby galaxies (Roberts & Warwick 2000) revealed 28 off-nuclear X-ray sources with $L_X > 10^{39}$ erg s^{-1}, with *c.* one in five galaxies hosting at least one such source, leading to the prediction that they are likely to be found in the nuclear regions of *c.* 50% of normal galaxies.

In the following dicussion we should bear in mind that the luminosity of 10^{39} erg s^{-1} is an arbitrary one. It is simply a convenient filter imposed to remove well-known 'normal' populations of low-mass X-ray binaries (LMXBs), high-mass X-ray binaries (HMXBs) and supernova remnants (SNRs). In particular, if one assumes a particular model, i.e. symmetric accretion onto a 1.4 $M_\odot$ compact object, then this luminosity implies either a substantially super-Eddington accretion rate or that the compact object is a black hole of many solar masses. To help dispel any confusion that may result from the use of multiple nomenclatures in the literature, I recall that, depending on the authors of the papers, these sources are referred to as IXOs (intermediate X-ray objects), SESs (super-Eddington sources), SLSs (superluminous sources, which unfortunately sounds very similar to, but is physically totally distinct from, superluminal sources!) and ULXs (ultraluminous sources). As yet there is no general consensus on which acronym to adopt. I prefer to use the last term ULXs, because it does not imply any particular model, it simply states the observational fact that these sources have X-ray luminosities that are considerably above those typical for known discrete-source populations in galaxies (although as

I discuss later, a complication may be introduced by the possibility of a luminosity dependence on system orientation).

The existence of ULXs has been known of observationally since the early 1980s. A brief historical perspective reveals that they were interpreted in terms of accreting systems as early as 1983 (Long & Van Speybroeck 1983). The earliest journal reference known to this author appeared in a paper on M51 by Palumbo *et al.* (1985), in which they suggest that the brightest extra-nuclear sources have compact object masses of *c.* 10 $M_\odot$. However, it was much later and following the discovery of more extreme versions of ULXs that suggestions their mass range might extend up to *c.* 100 $M_\odot$ were made, for example, for the ULX in NGC 5408 by Fabian & Ward (1993). Now, with Chandra's arc-second imaging we can separate multiple compact sources very close to each other and against a diffuse background component, and so the study of ULXs has received a strong impetus. The pinning down of a variable extra-nuclear ULX in M82, which reached a luminosity of more than 10^{41} erg s^{-1} (Kaaret *et al.* 2001; Matsumoto *et al.* 2001), has resulted in a plethora of new ideas being put forward to explain their properties and we are now entering an interesting phase of beginning to constrain these models.

In the following sections I first briefly describe some case studies of ULXs discovered in a range of host galaxies. I then describe a number of the currently favoured models that seek to explain and predict their properties, followed by discussion of their X-ray spectra and variability. Next I consider attempts to identify ULXs with counterparts at other wavelengths, and also to study their local environments. Finally, I suggest routes for future studies.

Some individual case studies

Given the large and increasing number of ULXs known, it is not feasible or productive to list the properties of each in turn. However, by way of illustration I will briefly review examples of four galaxies which harbour one or more examples of ULXs.

M82

This is a nearby irregular star-forming galaxy, which, because of its spectacular out-flowing wind visible both in Hα and soft X-rays, has been a favourite target for X-ray observations since the era of the Einstein Observatory (Watson *et al.* 1984). Collura *et al.* (1994) drew attention to a very bright variable X-ray source in M82, and suggested that it might be an extreme example of an X-ray binary with a mass of 20 $M_\odot$. The problem at that time was one of confusion in a complex region containing many sources and also the presence of a strong soft extended X-ray component.

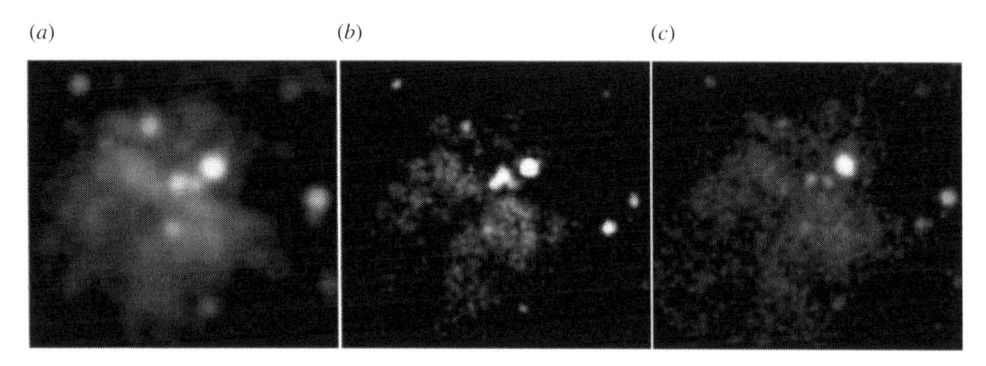

Figure 7.1. Chandra images of the central kpc region of M82: (*a*) September 1999, (*b*) October 1999 and (*c*) January 2000. The bright source (up and to the right of centre) is the ULX. A transient source is clearly seen in the October 1999 image, situated just to the left of the ULX. Also, another highly variable source is seen near the right-hand edge of the image. The diffuse emission is constant, but appears more obvious in some images because of the different integration times.

Although based on ROSAT and ASCA observations, several authors proposed the possible existence of a weak AGN in M82 (Matsumoto & Tsuru 1999; Ptak & Griffiths 1999), it was not until the exact spatial location of the brightest source was revealed by Chandra observations that the AGN versus ULX interpretation could be resolved (Matsumoto *et al.* 2001; Kaaret *et al.* 2001). The Chandra snapshots of the centre of M82 at arc-second spatial resolution unambiguously revealed a very bright and highly variable X-ray source located in projection *c*. 200 pc from the dynamical centre (see Fig.7.1). The source had increased in luminosity by about a factor of 10 in the three months between two of these observations. There are no obvious optical or mid-infrared counterparts to the ULX. Somewhat surprisingly, in view of the lack of a mid-infrared source, it is claimed that near-infrared data reveal a star cluster at the position of the X-ray source (Ebisuzaki *et al.* 2001). There is currently no radio source present, although, intriguingly, a moderately bright radio source was detected in 1981, which rapidly faded below the threshold of detection. Kronberg *et al.* (2000) claimed that this radio source was related to a supernova, but its very rapid decline suggests that it may be an example of the less-common type of radio-transient supernova discussed by Muxlow *et al.* (1994). At the time of writing this review, the ULX in M82 has the distinction of being the most X-ray luminous, at *c*. 10^{41} erg s^{-1}.

Given that there are currently a number of fundamentally different models that seek to explain ULXs, it remains to be seen whether this particular source will help us to understand this class of object or whether, as has often been the case in other areas of astronomy, the most extreme case may require rather special circumstances in order to fit within a generalized interpretation.

M81 X-9

This source lies in a dwarf companion galaxy to the east of M81, called Ho IX (therefore it is also known as Ho IX X-1). It is located within a nebula of diameter 250 pc, suggested to be the result of a supernova. However, an archival study of the X-ray emission from M81 X-9 by LaParola *et al.* (2001) showed it to be variable by a factor of 4, and hence the ULX itself is unlikely to arise from a supernova. Wang (2002) reported the 3-sigma detection of a 1 mJy radio source at the position of M81 X-9. If the radio emission were to be associated with the ULX, then it would be substantially higher than that expected from galactic micro-quasars (see discussion of the models in the pervious section). This ULX is worthy of note, as it appears outside the main body of M81, and because of its feeble host galaxy (Ho IX). Its existence proves that ULXs may be found in faint low-mass galaxies and that without follow-up investigations some may have been mis-identified as distant background AGN.

The Antennae system

At the time of writing, the Antennae (NGC 4038/39) is the best-studied star-forming galaxy observed by Chandra. In a series of papers, Fabbiano *et al.* (2001) and Zezas *et al.* (2002, and references therein) have described the X-ray source-luminosity function, and analysed the X-ray spectral properties of the sources, together with multi-wavelength associations of the sources within this interacting system. Their principal results are that ULXs are very numerous in the Antennae system. In fact there are 18 sources with absorption corrected L_X in the band 0.1–10 keV of greater than 10^{39} erg s^{-1}! But it should be noted that this assumes a distance calculated using H_0 of 50 km s^{-1} Mpc^{-1}. If the now more generally accepted value of H_0 equal to 75 km s^{-1} Mpc^{-1} is used, this reduces the luminosities by a factor of 2.25. A study of X-ray variability shows that 7/9 ULXs are variable, confirming they are probably compact binaries (Fabbiano *et al.* 2003). They divide the sources into three luminosity bins, low, medium and high, and then produce an average coadded X-ray spectrum for each sample. The results are consistent with a mixture of binaries and SNRs, for the low and medium bins, with some of the latter population having radio-source counterparts. The bin containing the ULXs is believed to be dominated by binaries: an interpretation supported by, albeit limited, information on their variability, and the fact that the spectral parameters of the Antennae ULXs are similar to those of galactic micro-quasars. This point is returned to in the next section. Finally, small observed offsets between the optical positions of star associations and some X-rays sources, including ULXs, led Zezas *et al.* (2002) to suggest a model invoking runaway binaries ejected from a star cluster following supernova explosions. It is clear that the

Antennae system is a rich laboratory for the study of ULXs as well as other X-ray source populations.

M101

This face-on spiral galaxy at *c*. 7.2 Mpc has been very well studied at most frequencies. The wide-field Hα-image (Fig. 7.2(*a*)) shows the hundreds of ionized hydrogen H II regions and SNR nebulae that populate the whole area of the galaxy. The Chandra image shown in Fig 7.2(*b*) contains more than 100 sources and is the subject of a detailed paper by Pence *et al.* (2001). Comparison of these two images underlines the major task required to first associate the X-ray sources with counterparts at other wavelengths and then to undertake the follow-up spectroscopic observations required by defining the ionization, metallicity and stellar populations within their close environments. Originally it was suggested by Wang (1999) that several X-ray sources were associated with so-called hypernovae, based on radio-source positions. However, subsequent refined astrometry and X-ray-variability studies now rule out most if not all of these suggested hypernovae (Snowden *et al.* 2001). Despite the high number of X-ray sources, only one fits our luminosity definition of a ULX. However, it does display unusual short-term X-ray variability, as well as having an atypical X-ray spectrum for a ULX (Pence *et al.* 2001). The X-ray luminosity functions of M101 and other grand design spirals, combined with high-quality multi-frequency data, are needed to provide the essential context within which to understand the more spectacular examples of star-forming galaxies, like the Antennae, which contain an order of magnitude more ULXs. Such a survey has already begun and preliminary results are reported in Kilgard *et al.* (2002).

The above cases are all galaxies in the local Universe. It is also interesting to note a possible link between these examples and their more distant counterparts. Using Chandra data from deep-field surveys, Alexander *et al.* (2002) have discussed the X-ray emission arising from distant emission-line and starburst galaxies (see also the review by Brandt *et al.* (2002)). A population of ULXs may dominate the X-ray emission from some non-AGN detected in these deep surveys, although overall they will probably not contribute more than a few per cent of the integrated background.

What are ULXs?

Intermediate-mass black holes

As the generic explanation of ULXs, the intermediate-mass BH model has been touted for many years. However, it was following the confirmed existence of variable ULXs with luminosities above 10^{40} erg s^{-1} that interest became sharpened. If they are powered by accretion, and emitting isotropically at close to their

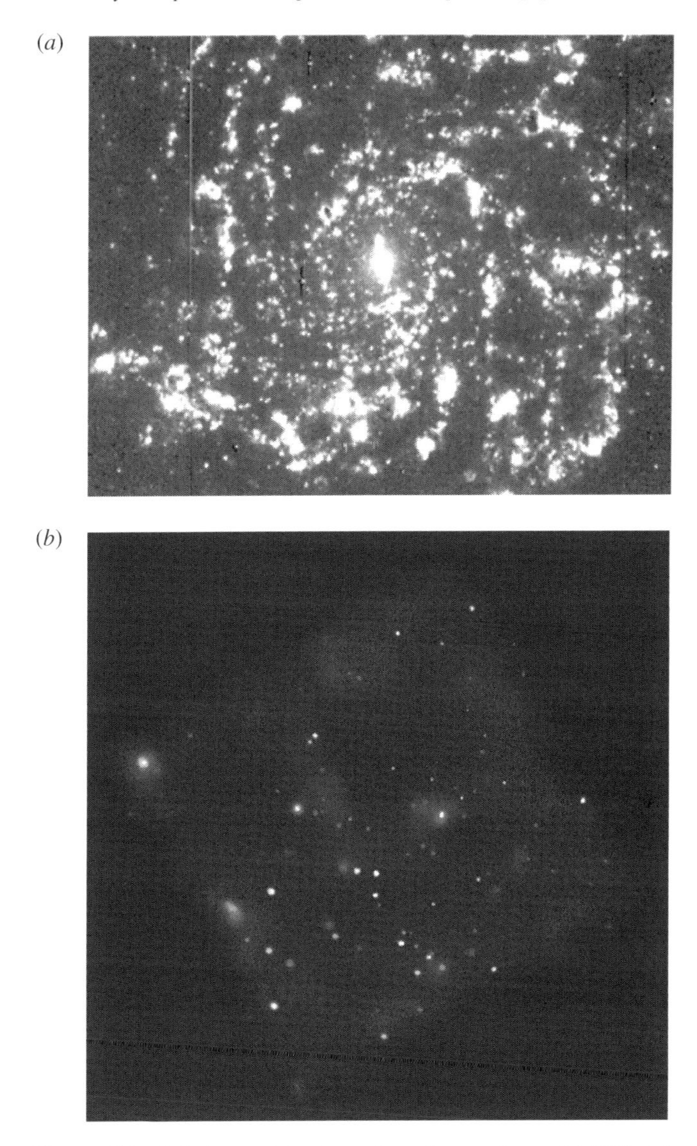

Figure 7.2. (*a*) An archival service observation of M101 taken through an Hα filter, using the wide-field camera on the INT on La Palma. (*b*) An X-ray image of M101, extracted from the Chandra data archive (by Roy Kilgard). North is up, and the Chandra data square is *c*. 8 arcmin across. The Hα image has a similar scale to the X-ray image.

Eddington limits, then these systems have BHs of *c*. 100 $M_{\odot}$. Understanding how such intermediate-mass BHs are formed is a challenge for theory, since BHs with maximum mass of only *c*. 20 $M_{\odot}$ arise from supernovae and massive star-core collapse. A minor industry has developed with the aim of proposing plausible explanations for the existence of intermediate-mass BHs located away from the nuclei of galaxies. Two interesting ideas are that they may be formed directly via the evolution of Population III stars (Madau & Rees 2001) or they may start out within globular clusters, growing more massive over the lifetime of the cluster, and then merge with their host galaxy to eventually accrete in a new dense environment within molecular clouds (Miller & Hamilton 2002). These novel ideas have yet to face crucial observational tests. However, if even a subset of ULXs is associated with such processes, then their locations and frequency of occurrence may provide constraints on models of galaxy formation.

First let us consider the constraints that can be set on the upper-mass range. Many ULXs are offset from the optical nucleus and, when known, from the dynamical centre of their host galaxy (Colbert & Mushotzky 1999). Such displacements are not expected if the object is a supermassive BH with mass *c.* $10^6\ M_\odot$, since that is often comparable with the dynamical mass within a radius of *c.* 1 kpc. The offset distances from the nucleus place typical upper limits on the mass of 10^5–$10^6\ M_\odot$, otherwise dynamical friction would cause the object to sink into the nucleus in 10^9–10^{10} y (see Tremaine *et al.* (1975)).

Within these mass constraints, ULX luminosities may be explained via accretion in several ways. Either the accretion rate may be quite low, *c.* 1% Eddington, with a BH mass close to the dynamical upper limit 10^5–$10^6\ M_\odot$. This is a version of the so-called ADAF (advection-dominated accretion flow) model, in which the material is advected into the BH without radiating significantly (Narayan & Yi 1994). Although still a possibility for some ULXs, this model appears unlikely for bright ULXs in dwarf star-forming galaxies, where the ADAF BH mass would be a significant fraction of the total mass of the system. In principle the ADAF model could be rejected or supported by means of multi-frequency observations, particularly in the near-infrared, where the predicted emission should be easily detectable using current instrumentation on 4–8 m class telescopes.

Alternatively, ULX BH masses may be *c.* 100 $M_\odot$, accreting near to their Eddington limits. In addition to X-ray spectroscopy/variability, the search for counterparts, possibly via detection of the accretion disc (see Chaty *et al.* (1996)), and the properties of the local ISM all offer good prospects for investigating this hypothesis. See Miller and Colbert (2003) for a review of the evidence for intermediate mass BHs. A third possibility is that the X-rays are beamed. In this case the true bolometric luminosities are reduced, and consequently the masses required are moved into a range similar to that found for known BHXBs, i.e. *c.* 10 $M_\odot$.

Models involving anisotropic X-ray emission

There are two classes of models that seek to explains ULXs in terms of anisotropic emission. In the first case a link has been suggested between ULXs and galactic superluminal sources (somctimes referred to as galactic micro-quasars or GMQs) such as GRS1915+105. This concept was first discussed in detail for the case of

Dwingeloo X-1 (Reynolds *et al.* 1997), and has been expanded upon by Körding *et al.* (2002). The X-ray spectral properties of ULXs are similar on average to the GMQs, although there are exceptions. In the GMQs that do exhibit relativistic jets at radio frequencies, it is not thought that their X-ray emission is relativistically beamed because their X-ray spectra are well fitted by accretion-disc models. Zezas *et al.* (2002) suggested that an X-ray Doppler-boosted component may be associated with the power-law component required to fit some but not all ULXs. The strongest evidence in favour of this class of model is the similarity in the X-ray spectra of many ULXs and GMQs, which will be discussed in the next section. The recent detection of radio emission from the ULX in NGC 5408 is consistent with jet emission from a stellar mass BH (Kaaret *et al.* 2003).

The second class of anisotropic model invokes funnelled (not intrinsic) collimation via a thick accretion disc (King *et al.* 2001). They associate ULXs with close-separation X-ray binaries undergoing mass transfer on a thermal time-scale. Super-Eddington accretion can proceed as the donor overflows its Roche lobe. One constraint of the model is that the required donor star must be of early type, to avoid quenching of the X-rays in the resulting common envelope. The model naturally explains why ULXs are common in active star-forming regions. This model has the advantage that the mass range required for the compact object is reduced by a factor of a few, but potential difficulties remain to account for the most luminous ULXs; the severity of these problems depends on their duty cycle between the high and low states.

Association of ULXs with SNRs

The range in X-ray luminosity of SNRs is extremely broad, depending on their progenitor, local environment and age. Although a typical X-ray luminosity for an SNR is *c*. 10^{36}–10^{37} erg s^{-1}, it is known that in some cases they can reach those of ULXs, and Roberts *et al.* (2002) claim that between 10 and 20% of these sources can be identified with young SNRs.

An example is SN1987K in NGC 1313 (Schlegel *et al.* 1999) (source number 11 in Schlegel *et al.* (2000)), which has an L_X of 4×10^{39} erg s^{-1} in the soft ROSAT band, and is roughly constant over the course of ROSAT observations from 1992–1998 (corresponding to 12–16 y after the outburst). Another X-ray luminous SNR lies in NGC 6946 and has an L_X of 3×10^{39} erg s^{-1} in the 0.2–3 keV band (Blair *et al.* 2001). A less luminous SNR is SN1979C in M100, which also has roughly constant L_X at *c*. 10^{39} erg s^{-1}, over the time-frame 16–20 y after the outburst (Kaaret 2001). An interesting, although very faint, probable supernova was detected by ROSAT at the position of SN1988Z in the galaxy MCG+03-28-022, which has a redshift of 0.022 (Fabian & Terlevich 1996). Although it was too faint to be studied spectroscopically, if it is assumed to have a bremsstrahlung X-ray spectrum with temperatures between 1 and 5 keV, then its total source luminosity is between 1 and 2×10^{41} erg s^{-1}. SN1988Z was observed in X-rays 6.5 y after the explosion, and therefore it may have been considerably more luminous shortly after the explosion (see Aretxaga *et al.* (1999)). We do not know the frequency of occurrence of such dramatically X-ray-luminous SNRs, but clearly energetically they are, in principle, capable of producing even the most luminous ULXs. In order to distinguish between binary accretion and a supernova exploding in dense environments, we need to use X-ray monitoring and spectroscopy and also optical and near-infrared spectroscopy, as well as sensitive radio observations. A supernova shell expanding into dense ISM can certainly provide X-ray variability, but the long-term trend in the light curve over many years must be downward. Also, general arguments about the working surface of interaction between the expanding shell and the ISM preclude very high amplitude variations. Optical and infrared observations could reveal the presence of a nebula and/or star cluster associated with the SNR, although this will depend strongly on age. Progress can also be made using statistics by searching for coincidences between radio and X-ray source positions (see the remarks about the Antennae and M82 in the previous section).

The X-ray spectra of ULXs

The X-ray spectra of many ULX systems can be modelled by fitting black-body BH emission from an optically thick standard accretion disc around a BH (Colbert & Mushotzky 1999; Makishima *et al.* 2000). Since the integrated spectrum is the result of adding together regions of different temperature (producing corresponding thermal 'colours'), this fit is referred to as the multi-colour disc (MCD) model. However, there are several problems with this model when the fit parameters are compared with the assumed BH masses. The temperature of material at the inner accretion radius implied using MCD models is generally too high for the BH mass. One escape route for the MCD model is to postulate that the BH is spinning, and hence the inner radius is closer to the BH compared with a non-rotating BH of the same mass (see Mizuno *et al.* (2001)).

In addition, it is observed that not all X-ray spectra of ULXs are well fitted using the MCD model, counterexamples being the ULXs in Holmberg II (Miyaji *et al.* 2001), IC342 (Kubota *et al.* 2001) and NGC 3628 (Strickland *et al.* 2001), which require the presence of a hard spectral component with photon index 1–2. In cases for which the spectral fitting requires two components, the soft component may originate from an optically thick accretion disc, and the hard component from Comptonization of soft photons by hot electrons.

While the available X-ray spectroscopy of ULXs strongly favours binary accretion, some models of supernovae evolving in high-density environments also predict power-law spectra with photon indices 1.6–2.0 (see, for example, Plewa (1995)). However, further supporting evidence for binary systems comes from observations of a spectral transition between a high state (with a soft-X-ray spectrum) and low state (with a hard-X-ray spectrum) for two ULXs in the galaxy IC342 (Kubota *et al.* 2001) and also for a ULX in NGC 1313. It is not known whether such transitions between hard and soft states are a fundamental property of ULXs, but it strengthens the argument for continued monitoring of a sample of them.

X-ray variability

Statements concerning the variability of ULXs must always be viewed in the context of the duration of a single observation and the number of epochs used for comparison. It is a statistical fact that ULXs will tend to be discovered preferentially while in their high state. Multi-epoch observations will be essential if we are to understand transitions between high and low states and to measure their duty cycles.

At the time of writing there are two claims of possible periodic behaviour in ULXs, and one confirmed periodicity for a borderline case in terms of its luminosity. For the source Circinus CGX-1, Bauer *et al.* (2001) claimed periodic variations of 7.5 h, and argued against it being a foreground galactic binary because of the absence of an optical counterpart to $m_V < 25.3$ magnitudes, and on general statistical grounds. However, one galactic AM-Her system is known with a period similar to CGX-1, and if it were in fact a galactic system more distant than 1.3 kpc, then the data are consistent with an AM-Her-type system. Clearly, further study of Circinus CGX-1 is of major interest, since, if it is a ULX, then the X-ray period will give unique information on its nature. The other possibly periodic ULX is in the galaxy IC342 (Sugiho *et al.* 2001) and has a possible period of 31 or 41 h (both periods are consistent with the current data). Both periods are also consistent with the orbital period of a semidetached binary consisting of a BH and a main-sequence star of several tens of solar masses.

Finally, M33 X-8 had a reported 106-day periodicity (Dubus *et al.* 1997). However, subsequent *Beppo*SAX observations do not support the continued presence of this period. Its X-ray spectrum was fitted by Parmar *et al.* (2001) using an MCD model plus a harder power law, suggesting that it is a close binary system containing a BH of *c.* 10 $M_{\odot}$. M33 X-8 is a borderline ULX in terms of its luminosity, which was 1.3×10^{39} erg s^{-1} over the 1–10 keV band in July 2000 when observed by *Beppo*SAX and assuming a distance of 795 pc. It is worth noting that, prior to the detailed X-ray spectroscopy, this source was believed to be an example of a low-luminosity AGN, because it lies so close to the centre of M33. One may speculate on how many other ULXs residing very close to the centres of their parent galaxy might be misidentified with the presence of a low-luminosity AGN, such as those found in the sample studied by Ho *et al.* (2001). This is more than a merely semantic distinction, since the formation processes of BHs outside of binary systems and those within are presumably quite different and, hence, any meaningful census of AGN and ULXs must be able to distinguish between them.

A search for multi-wavelength counterparts of ULXs and their local environment

Identification of multi-wavelength counterparts to ULX can provide important clues to their nature. Such studies are now possible based on the sub-arc-second X-ray astrometry provided by Chandra. This has allowed us to identify the first possible stellar counterpart to a ULX, with the discovery of a blue optical-continuum source ($m_V = 19.7$) coincident with NGC 5204 X-1 (Roberts *et al.* 2001). Crucially, the optical spectrum of this source is consistent with one or more O stars. Indeed, inspection of HST archive data from the wide-field planetary camera resolves the ground-based optical source into two closely separated point sources, providing support for the King *et al.* (2001) model of ULX as high-mass X-ray binary systems. In addition, a very significant optical study has been reported by Pakull & Mirioni (2002). Their study includes Holmberg II, which is a dwarf irregular galaxy in the M81 group, whose X-ray emission is dominated by a single compact X-ray source that varies on both yearly and daily time-scales and reaches an unabsorbed X-ray luminosity of 10^{40} erg s^{-1} (Zezas *et al.* 1999; Miyaji *et al.* 2001). Pakull & Mirioni found strong evidence for an X-ray-ionized nebula around the ULX, based on the presence of intense He II 4686Å and strong [O I] 6300Å emission. Such observations may hold the key to estimating the bolometric luminosity of ULXs, since a simple photon-counting argument implies that the ionized gas sees a continuum in the range $(3 - 13) \times 10^{39}$ erg s^{-1}, thereby ruling out significant beaming in this case. Interestingly, Pakull & Mirioni noted that this ULX is *c*. 30 times more luminous in both X-rays and the He II than the correspondingly values for the BH candidate LMC X-1.

Several other ULX (e.g. two in NGC 1313 and one in NGC 4559) appear to be located at the centre of ionized nebula ‘rings’, 200–400 pc in diameter, which may be related either to their formation or to the presence of stellar winds or jets. As both Pakull & Mironi (2002) and Roberts *et al.* (2002) pointed out, the evidence for compact and nebular optical counterparts strongly favours the association between ULXs and young stellar regions containing SNRs and high-mass X-ray binaries.

Summary and future observations

Largely as a result of high-spatial-resolution observations using Chandra, which have much reduced the effect of source confusion, it is now becoming clear that

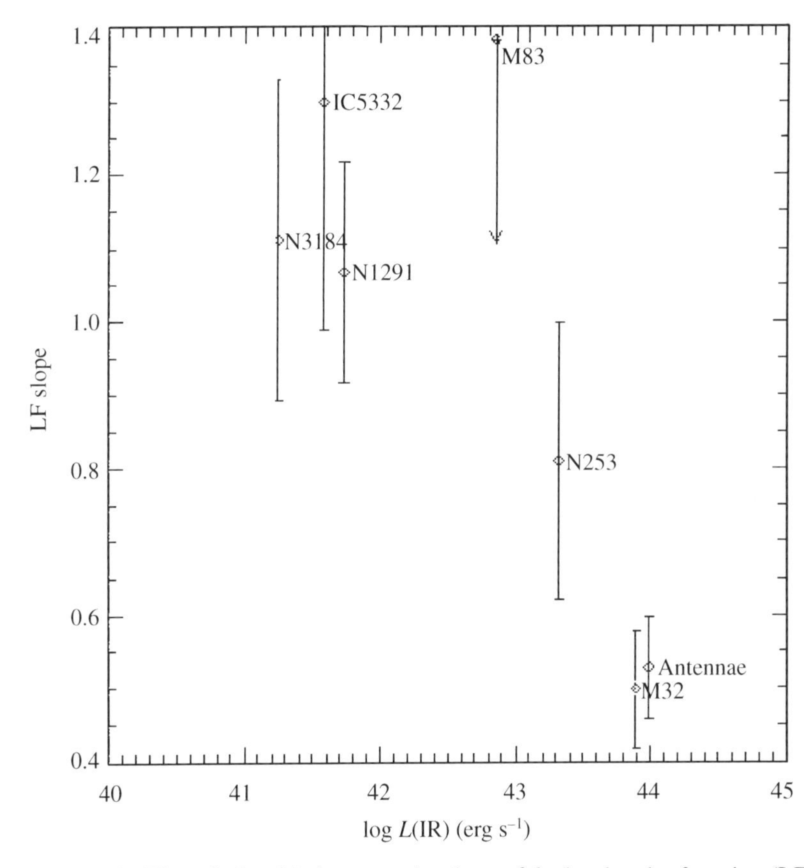

Figure 7.3. The relationship between the slope of the luminosity function (LF) and the star-formation rate determined using the integrated far-infrared luminosity, for a sample of seven galaxies (Kilgard *et al.* 2002).

ULXs are common in spiral and star-forming galaxies. The present ULX definition is crude and based only on their X-ray luminosity. This almost certainly results in a mixed bag of objects, including bright supernovae and accreting systems. The dual approach of using temporal X-ray spectroscopy and multi-wavelength investigations to search for their actual counterparts, their associated ionized nebulae, or both, is beginning to yield results that constrain the models. But we are currently only at an early stage in this process, and in particular we still do not know whether the accreting versions of ULXs represent exotic objects, e.g. intermediate-mass BHs possibly formed at very early epochs, or, more prosaically, whether they are just manifestations of known classes of binary, at a special stage in their evolution or beamed, like the galactic micro-quasars.

There is also the bigger picture to consider. We need to understand why it is that some galaxies are exceptionally rich in ULXs, containing tens of them, while other large spirals contain only one or none. Here the X-ray source-luminosity functions (XLFs) may provide the answer. Kilgard *et al.* (2002) reported preliminary results for a number of galaxies and noted that there is an anticorrelation between the slope of the XLFs and the star-formation rate, as inferred from the mid-infrared luminosity (see Fig 7.3). A flat XLF implies many ULXs and is associated with a high star-formation rate. Deviations from a simple power-law fit to the XLFs are apparent. However, a representative sample of galaxies is needed in order to separate the effects on the XLF of population ageing, mass range of the accretor, and the influence of the beaming scenarios described earlier. Although I have focused exclusively on ULXs found within spiral galaxies, it is worth noting that they also exist in early type and elliptical galaxies. But, intriguingly a survey of 15 early-type galaxies with Chandra (Irwin *et al.* 2003) showed that discrete sources more luminous than 2×10^{39} erg s^{-1} are rare in such galaxies, with their observed fluxes and number statistics consistent with them being foreground or background sources, unrelated to the nearby galaxy. Two of the exceptions were found to lie in globular clusters of NGC 1399. They could be intermediate mass BHs with sub-Eddington accretion rates. These new results support the general picture in which ULXs are associated either directly or indirectly with star-forming activity.

Whichever model(s) turn out to be correct, it is likely that the study of sources found at the extreme end of the X-ray luminosity function will provide new insights into accretion processes and, in a few cases, the supernova phenomenon.

It is a pleasure to thank all of my colleagues with whom I have had useful discussions during the preparation of this review. In particular I thank Andrea Prestwich, Roy Kilgard, Tim Roberts, Andrew King, Aya Kubota and Felix Mirabel.

References

Alexander, D. *et al.* 2002 *Astrophys. J.* **568**, L85.
Aretxaga, I. *et al.* 1999 *Mon. Not. R. Astr. Soc.* **309**, 343.
Bauer, F. E. *et al.* 2001 *Astron. J.* **122**, 182.
Blair, W. P. *et al.* 2001 *Astron. J.* **121**, 1497.
Brandt, W. N., Alexander, D. M., Bauer, F. E. & Hornschemeier, A. E. 2002 *Phil. Trans. R. Soc. Lond.* A **360**, 2057–2075.
Brinkman, A. *et al.* 2002 *Astron. Astrophys.* **396**, 761.
Cecil, G. *et al.* 2001 *Astrophys. J.* **576**, 745.
Chaty, S. *et al.* 1996 *Astron. Astrophys.* **310**, 825.
Colbert, E. J. M. & Mushotzky, R. F. 1999 *Astrophys. J.* **519**, 89.
Collura, A. *et al.* 1994 *Astrophys. J.* **420**, L63.
Dubus, G. *et al.* 1997 *Astrophys. J.* **490**, L47.
Ebisuzaki, T. *et al.* 2001 *Astrophys. J.* **562**, L19.
Fabbiano, G. 1989 *A. Rev. Astron. Astrophys.* **27**, 87.
Fabbiano, G., Zezas, A. & Murray, S. S. 2001 *Astrophys. J.* **554**, 1035.
Fabbiano, G., *et al.* 2003 *Astrophys. J.* **584**, L5.

Fabian, A. C. & Terlevich, R. 1996 *Mon. Not. R. Astr. Soc.* **280**, L5.
Fabian, A. C. & Ward, M. J. 1993 *Mon. Not. R. Astr. Soc.* **263**, L51.
Griffiths, R. *et al.* 2000 *Science* **290**, 1325.
Ho, L. C. *et al.* 2001 *Astrophys. J.* **549**, 51.
Irwin, J.A. *et al.* 2002 In *X-rays at Sharp Focus – Chandra Science Symposium, ASP Conference Series* **262**, p157. eds. Eric Schlegel and Saega Dil Vrtilek. San Francisco, ASP.
Irwin, J.A., Athey, A.R. & Bregman, J. N. 2003 *Astrophys. J.* **582**, 356.
Kaaret, P. 2001 *Astrophys. J.* **560**, 715.
Kaaret, P. *et al.* 2003 *Science* **299**, 365.
Kaaret, P. *et al.* 2001 *Mon. Not. R. Astr. Soc.* **321**, L29.
Kilgard *et al.* 2002 *Astrophys. J.* **573**, 138.
King, A. R. *et al.* 2001 *Astrophys. J.* **552**, L109.
Kinkhabwala, A. *et al.* 2002 *Astrophys. J.* **575**, 732.
Körding, E., Falcke, H. & Markoff, S. 2002 *Astron. Astrophys.* **382**, L13.
Kronberg, P. P. *et al.* 2000 *Astrophys. J.* **535**, 706.
Kubota, A. *et al.* 2001 *Astrophys. J.* **547**, L119.
LaParola, V. *et al.* 2001 *Astrophys. J.* **556**, 47.
Long, K. S. & Van Speybroeck, L. P. 1983 In *Accretion Driven X-ray Sources*, p. 141. Cambridge: Cambridge University Press.
Madau, P. & Rees, M. J. 2001 *Astrophys. J.* **551**, L27.
Makishima, K. *et al.* 2000 *Astrophys. J.* **535**, 632.
Martin, C., Kobulnicky, H. A. & Heckman, T. M. 2002 *Astrophys. J.* **574**, 663.
Matsumoto, H. & Tsuru, T. G. 1999 *Publ. Astr. Soc. Jpn* **51**, 321.
Matsumoto, H. *et al.* 2001 *Astrophys. J.* **547**, L25.
Miller, M. C. & Colbert, E. M. 2003 (astro-ph/0308402).
Miller, M. C. & Hamilton, D. P. 2002 *Mon. Not. R. Astr. Soc.* **330**, 232.
Miyaji, T., Lehmann, I. & Hasinger, G. 2001 *Astron. J.* **121**, 3041.
Mizuno, T. *et al.* 2001 *Astrophys. J.* **554**, 1282.
Muxlow, T. W. B. *et al.* 1994 *Mon. Not. R. Astr. Soc.* **266**, 455.
Narayan, R. & Yi, I. 1994 *Astrophys. J.* **428**, L13.
Pakull, M. W. & Mirioni, L. 2002 In *Proc. New Visions of the X-ray Universe in the XMM-Newton and Chandra Era, 26–30 November 2001, ESTEC, The Netherlands.* (Preprint astro-ph/0202488.) (In the press.)
Palumbo, G. *et al.* 1985 *Astrophys. J.* **298**, 259.
Parmar, A. N. *et al.* 2001 *Astron. Astrophys.* **368**, 420.
Pence, W. D. *et al.* 2001 *Astrophys. J.* **561**, 189.
Persic, M. & Rephaeli, Y. 2002 *Astron. Astrophys.* **382**, 843.
Pettini, M. *et al.* 2001 *Astrophys. J.* **554**, 981.
Pietsch, W. *et al.* 2001 *Astron. Astrophys.* **365**, L174.
Plewa, T. 1995 *Mon. Not. R. Astr. Soc.* **275**, 143.
Ptak, A. & Griffiths, R. 1999 *Astrophys. J.* **517**, L85.
Read, A.M. & Stevens, I. R. 2002 *Mon. Not. R. Astr. Soc.* **335**, L36
Reynolds, C. S. *et al.* 1997 *Mon. Not. R. Astr. Soc.* **286**, 349.
Rita, M. *et al.* 2001 *Astrophys. J.* **546**, L9.
Roberts, T. P. & Warwick, R. S. 2000 *Mon. Not. R. Astr. Soc.* **315**, 98.
Roberts, T. P. *et al.* 2001 *Mon. Not. R. Astr. Soc.*, **325**, L7.
Roberts, T. P. *et al.* 2002 In *Proc. New Visions of the X-ray Universe in the XMM-Newton and Chandra Era, 26–30 November 2001, ESTEC, The Netherlands.* (Preprint astro-ph/0202017.) (In the press.)

Sako, M. *et al.* 2000 *Astrophys. J.* **543**, L115.
Schlegel, E. M. *et al.* 1999 *Astron. J.* **118**, 2689.
Schlegel, E. M. *et al.* 2000 *Astron. J.* **120**, 791.
Snowden, S. L. *et al.* 2001 *Astron. J.* **121**, 3001.
Stevens, I., Read, A. M. & Bravo-Guerrero, J. 2003 *Mon. Not. R. Astr. Soc* **343**, L47.
Strickland, D. *et al.* 2000 *Astron. J.* **120**, 2965.
Strickland, D. *et al.* 2001 *Astrophys. J.* **560**, 707.
Strickland, D. *et al.* 2002 *Astrophys. J.* **568**, 689.
Sugiho, M. *et al.* 2001 *Astrophys. J.* **561**, L73.
Tremaine, S., Ostriker, J. P. & Spitzer Jr, L. 1975 *Astrophys. J.* **196**, 407.
Wang, Q. D. 1999 *Astrophys. J.* **517**, L27.
Wang, Q. D. 2002 *Mon. Not. R. Astr. Soc.* **332**, 764.
Wang, Q. D. *et al.* 2001 *Astrophys. J.* **555**, L99.
Watson, M. G., Stranger, V. & Griffiths, R. E. 1984 *Astrophys. J.* **286**, 144.
Young, A. J., Wilson, A. S., & Shopell, P. L. 2001 *Astrophys. J.* **556**, 6.
Zezas, A. L. *et al.* 1999 *Mon. Not. R. Astr. Soc.* **308**, 302.
Zezas, A. L. *et al.* 2002 *Astrophys. J.* **577**, 710.

8

Cosmological constraints from Chandra observations of galaxy clusters

BY STEVEN W. ALLEN

University of Cambridge

Introduction

Accurate measurements of the masses of clusters of galaxies are of profound importance to cosmological studies. Clusters are the largest gravitationally bound objects known and their observed spatial distribution and redshift evolution provide sensitive probes of the underlying cosmology. The first measurements of cluster masses used the motions of individual galaxies to trace the gravitational potentials of clusters. Although such studies were prone to systematic uncertainties due to velocity anisotropies, substructure and projection effects (van Haarlem *et al.* 1997), work, based on large galaxy samples and employing careful selection techniques, has made important progress (e.g. Carlberg *et al.* 1996).

Two further techniques for measuring the masses of clusters, using X-ray observations and studies of gravitational lensing by clusters have now been developed. X-ray mass measurements are based on the assumption that the hot (*c.* 10^8 K), X-ray luminous gas which pervades clusters is in hydrostatic equilibrium; the total mass distribution is determined once the radial distributions of the X-ray gas density and temperature are known (Sarazin 1988). Since the X-ray emissivity is proportional to the square of the gas density, and the relaxation time-scale for the X-ray emitting gas is relatively short (of the order of a few sound crossing times), the X-ray method is relatively free from the projection and substructure effects that hamper the optical dynamical studies. However, although X-ray gas density profiles can be obtained easily from X-ray images of clusters, measurements of the temperature profiles require detailed spatially resolved spectroscopy which has only become available with the launch in 1999 of the Chandra X-ray Observatory and XMM-Newton.

In contrast to the aforementioned X-ray and optical dynamical techniques, gravitational lensing offers a method for measuring the projected masses through clusters

Frontiers of X-Ray Astronomy, ed. A.C. Fabian, K.A. Pounds and R.D. Blandford. Published by Cambridge University Press.

that is essentially free from assumptions about the dynamical state of the gravitating matter (see, for example, Mellier (1999) for a review). The primary observational challenges of requiring deep exposures, excellent viewing conditions, wide-field imaging and accurate point spread function models have now been largely overcome with improved instrumentation, although the recovery of the three-dimensional mass distributions in clusters can be complicated by projection effects and uncertainties in the redshift distributions of the lensed sources.

Clearly, the best approach when attempting to obtain precise results on the masses of clusters is to make use of the best available data from all three methods. Here, we present results from Chandra X-ray and gravitational lensing studies (supplemented,

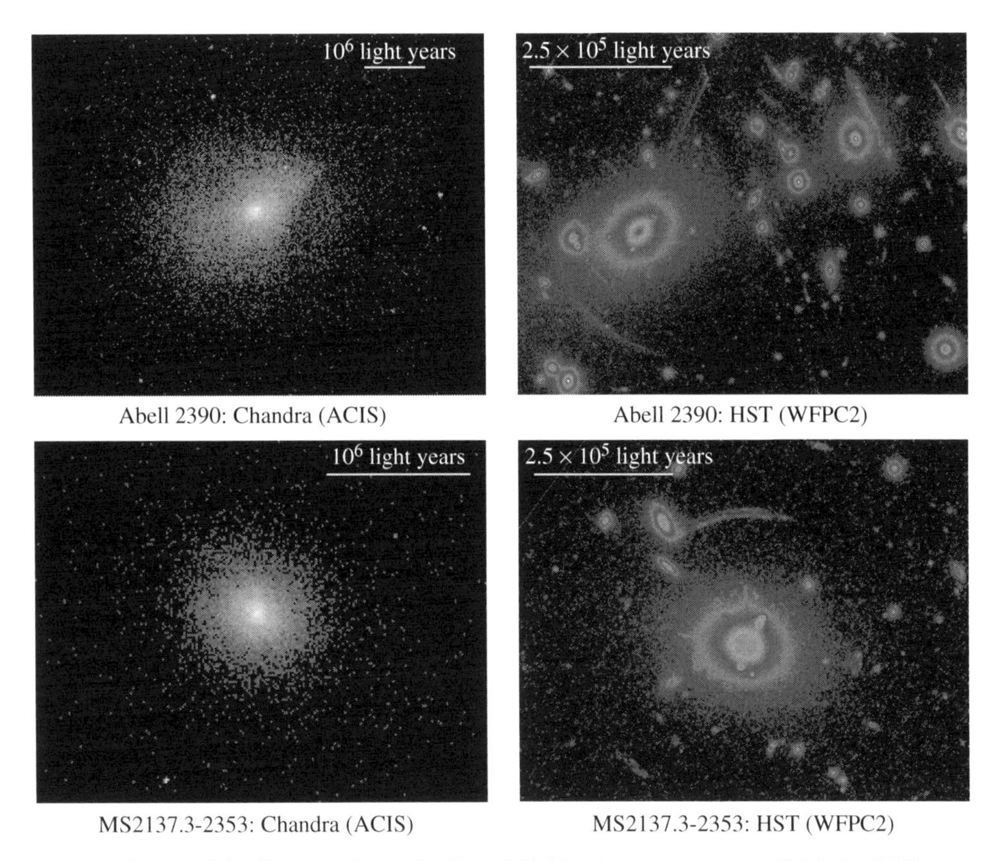

Figure 8.1. Chandra X-ray (left) and Hubble Space Telescope Wide Field Planetary Camera 2 optical (right) images of two of the dynamically relaxed, X-ray luminous lensing clusters discussed here. The clusters shown are Abell 2390 ($z = 0.230$) and MS2137.3-2353 ($z = 0.313$). The scale bars indicating distances of 10^6 light years correspond to angular sizes of 83 and 67 arcsec for Abell 2390 and MS2137.3-2353, respectively. (A standard ΛCDM cosmology with $h = H_0/100\ \mathrm{km\,s^{-1}\,Mpc^{-1}} = 0.7$, $\Omega_m = 0.3$ and $\Omega_\Lambda = 0.7$ is assumed.) Note the clear gravitational arcs in the HST images.

where available, by high-quality optical dynamical data) of a sample of six of the most luminous, dynamically relaxed clusters of galaxies known: PKS0745-191 ($z = 0.103$), Abell 2390 ($z = 0.230$), Abell 1835 ($z = 0.252$), MS2137.3-2353 ($z = 0.313$), RXJ1347.5-1145 ($z = 0.451$) and 3C 295 ($z = 0.461$). The high degree of dynamical relaxation in these clusters, which is evidenced by their regular optical and X-ray morphologies (e.g. Fig. 8.1), minimizes the systematic uncertainties in the mass measurements. The fact that these clusters are amongst the most luminous and, by implication, most massive clusters known (as identified from the ROSAT All-Sky Survey and follow-up observations) also makes them, individually, the most powerful probes for cosmological studies.

X-ray and lensing mass measurements

The methods employed in the X-ray and gravitational lensing analyses are described by Allen *et al.* (2001a,b) and Schmidt *et al.* (2001). The Chandra observations were made using the Advanced CCD Imaging Spectrometer (ACIS). The weak lensing results are drawn from the literature and use ground-based optical observations. The strong lensing analyses have been carried out using our own codes and are based primarily on data from the Hubble Space Telescope (HST).

Figure 8.2 shows a comparison of the Chandra X-ray and weak lensing results on the (total) mass profiles (luminous plus dark matter) for (*a*) Abell 2390

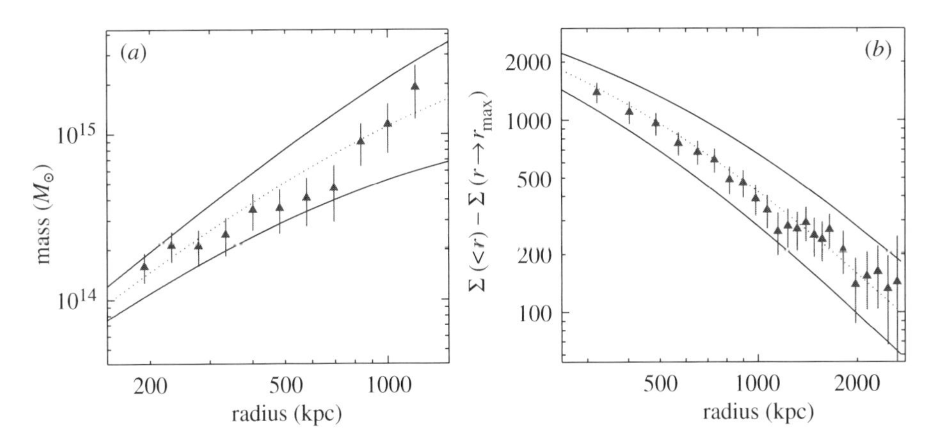

Figure 8.2. A comparison of the Chandra X-ray (curves) and weak lensing (triangles) mass measurements for (*a*) Abell 2390 (Allen *et al.* 2001b) and (*b*) RXJ1357.5-1145 (Allen *et al.* 2002a). For Abell 2390, the weak lensing results are from Squires *et al.* (1996). For RXJ1357.5-1145, the lensing data are from Fischer & Tyson (1997). For RXJ1357.5-1145, the surface mass density contrast (in units of $M_\odot$ pc^{-2}) rather than the projected mass has been plotted. An Einstein–de Sitter cosmology with $h = 0.5$ is assumed.

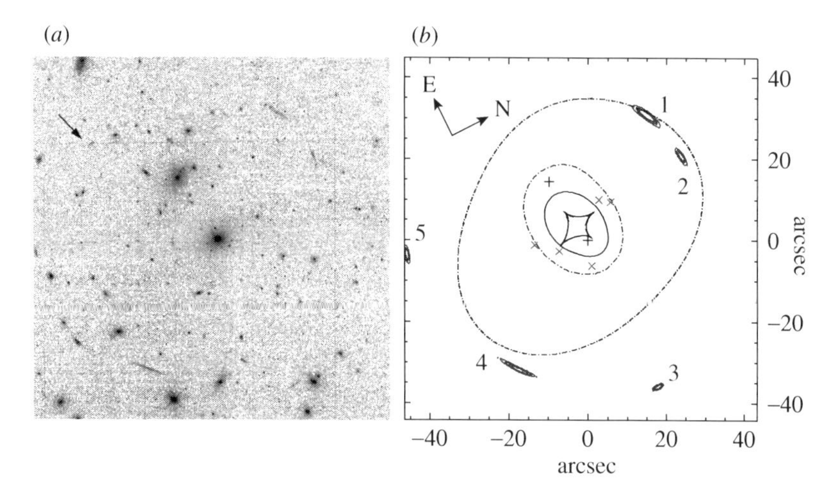

Figure 8.3. (*a*) The observed and (*b*) the predicted gravitational arc geometry in RXJ1347.5-1145 (Allen *et al.* 2002a). The image in (*a*) has been compiled from archival HST Space Telescope Imaging Spectrometer (STIS) data. (*b*) The arc geometry predicted by our best-fit mass model. The arcs have been simulated using circular Gaussian surface brightness distributions. The 10%, 30% and 50% brightness contours are plotted. The true source positions are denoted by crosses. The central positions for the two dominant mass clumps in the cluster are marked with plus signs. Also shown are the critical curves (dash-dotted) and caustic lines for a source at a redshift $z = 1.0$.

and (*b*) RXJ1347.5-1145. We find that for both clusters the mass profiles can be parametrized using the 'universal' form determined from simulations by Navarro *et al.* (1997; hereafter NFW), with concentration parameters and scale radii within the ranges expected for clusters of these masses. The agreement between the independent lensing and X-ray mass measurements is good in both cases, confirming the validity of the hydrostatic assumption used in the X-ray analysis and suggesting that the mass profiles in the clusters have been robustly determined.

Figure 8.3 shows (*a*) the observed and (*b*) the predicted gravitational arc geometry in RXJ1347.5-1145 (Allen *et al.* 2002*a*). We find that a simple two-component mass model, with ellipticities and orientations for the mass components matching those of the dominant cluster galaxies, provides a reasonable description for the overall arc geometry in the cluster. Such a model is consistent with the Chandra X-ray and weak lensing results. Similar results for Abell 1835 are presented by Schmidt *et al.* (2001).

Overall, for the five or six clusters in our sample with excellent X-ray and gravitational lensing data, we find good agreement between the X-ray and lensing mass measurements. For Abell 2390, high-quality optical dynamical data are also available (Borgani *et al.* 1999) and give consistent results.

The virial relations for relaxed galaxy clusters

The space density $n(M, z)$ of clusters predicted by analytical models (Press & Schechter 1974) and numerical simulations (Jenkins *et al.* 2001) can be related to (more easily) observable properties such as the X-ray temperatures and luminosities of clusters via simple scaling relations. Assuming that the X-ray gas in clusters is virialized and in hydrostatic equilibrium, the mass, M_Δ, within radius r_Δ (inside which the mean mass density is Δ times the critical density, $\rho_c(z)$, at that epoch) is related to the mean mass-weighted temperature within that radius, T_Δ, by $E(z)M_\Delta \propto T_\Delta^{3/2}$. Here,

$$E(z) = \frac{H(z)}{H_0} = (1+z)\sqrt{1 + z\Omega_m + \frac{\Omega_\Lambda}{(1+z)^2} - \Omega_\Lambda},$$

where $H(z)$ is the redshift-dependent Hubble constant (e.g. Bryan & Norman 1998). Since the X-rays from rich clusters are primarily bremsstrahlung emission, one can also show that $L_\Delta/E(z) \propto T_\Delta^2$, where L_Δ is the bolometric luminosity from within radius r_Δ. The validity of these simple scaling relations is supported by numerical simulations (Evrard *et al.* 1996; Mathiesen & Evrard 2001; Thomas *et al.* 2002), which also make firm predictions for the normalization of the mass–temperature relation. The normalization of the luminosity–temperature relation is more difficult to predict due to the complex physics of the X-ray gas in the innermost regions of clusters from where the bulk of the X-ray luminosity arises.

Pre-Chandra studies of the mass–temperature relation, using Advanced Satellite for Cosmology and Astrophysics (ASCA) and ROSAT data for relatively hot ($kT \gtrsim$3–4 keV) clusters (Horner *et al.* 1999; Nevalainen *et al.* 2000; Finoguenov *et al.* 2001) have recovered slopes consistent with the simple scaling-law predictions, and noted that the observed normalizations are typically *c.* 40% lower than predicted by standard (adiabatic) simulations. Given the limitations of the pre-Chandra X-ray data, however, these offsets were often ascribed to systematic problems. Pre-Chandra studies of the luminosity–temperature relation (Allen & Fabian 1998; Markevitch 1998; Arnaud & Evrard 1999) generally found $L_{\mathrm{Bol}} \propto T^3$, whereas theory predicts $L_{\mathrm{Bol}} \propto T^2$. This was taken as evidence for significant preheating and/or cooling in cluster cores (see, for example, Cavaliere *et al.* (1997)). Allen & Fabian (1998) noted that for hot ($kT \gtrsim 5$ keV), relaxed clusters $L_{\mathrm{Bol}} \underset{\sim}{\propto} T^2$ is recovered once the effects of cool, central components in the clusters are accounted for in the X-ray analysis.

A major goal of studies with the new generation of X-ray missions which permit the first detailed, spatially resolved X-ray spectroscopy of hot, distant clusters, is the verification and accurate calibration of the virial relations for galaxy clusters. In particular, studies of systems for which precise mass measurements

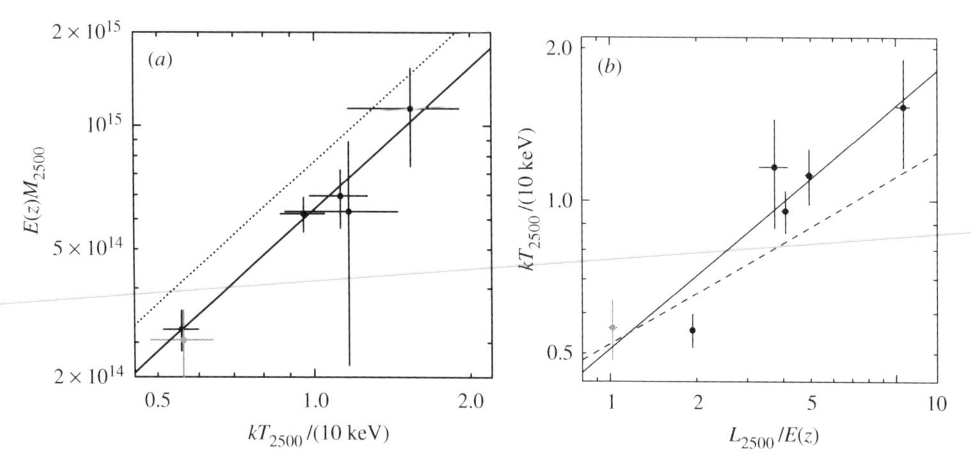

Figure 8.4. (*a*) The observed mass–temperature relation, assuming a ΛCDM cosmology ($h = 0.7$) with M_{2500} in $M_{\odot}$ and kT_{2500} in keV. The solid line is the best-fitting power-law model of the form $E(z)M_{2500} = A(kT_{2500}/10)^{\alpha}$, with $A = 5.38 \pm 0.74 \times 10^{14}$ and $\alpha = 1.52 \pm 0.36$. The dotted curve is the predicted result from the hydrodynamical simulations of Mathiesen & Evrard (2001). (*b*) The temperature–luminosity relation, with kT_{2500} in keV and L_{2500} in units of 10^{45} erg s^{-1}. The solid line is the best-fitting power-law model of the form $kT_{2500}/10 = B[L_{2500}/10^{45}E(z)]^{\beta}$, with $B = 0.51 \pm 0.05$ and $\beta = 0.48 \pm 0.06$. The dashed line shows the best-fit curve with $\beta = 0.33$ ($L \propto T^3$) fixed, which provides a poor description of the data. The data for 3C295, for which the X-ray mass measurement has not yet been confirmed by independent lensing data, are shown in a lighter shading.

are also available from other, independent methods (e.g. gravitational lensing) are required.

Figure 8.4 shows the results on the mass–temperature and temperature–luminosity relations determined from the Chandra data for the sample of clusters described in the first section. In determining the results on the virial properties, we adopt $\Delta = 2500$, since r_{2500} is well matched to the outermost radii at which reliable temperature measurements can be made from the Chandra data. kT_{2500} is the mean gas mass-weighted temperature within r_{2500}; note that the determination of gas mass rather than emission-weighted temperatures permits a direct comparison with the results from simulations. L_{2500} is the bolometric luminosity from within r_{2500}.

Fitting only the data for those clusters for which independent confirmation of the X-ray mass results is available from lensing studies (i.e. excluding 3C295 from the present sample) using a power-law model of the form

$$E(z)\left(\frac{M_{2500}}{1M_{\odot}}\right) = A\left(\frac{kT_{2500}}{10\ \mathrm{keV}}\right)^{\alpha} \tag{8.1}$$

and a χ^2 estimator, we obtain $A = 5.38 \pm 0.52 \times 10^{14}$ and $\alpha = 1.51 \pm 0.27$ (ΛCDM: $h = 0.7$). Using the modified least-squares estimator of Fasano & Vio (1988), which accounts for errors in both axes, we obtain values $A = 5.38 \pm 0.74 \times 10^{14}, \alpha = 1.52 \pm 0.36$. The measured slope is therefore consistent with the expected value of $\alpha = 1.5$. The dotted curve in Fig. 8.4(*a*) shows the predicted (zero-redshift) relation for a ΛCDM cosmology from the hydrodynamical (adiabatic) simulations of Mathiesen & Evrard (2001). We see that the predicted normalization lies *c*. 40% above the observed value.

Figure 8.4(*b*) shows the kT_{2500}–L_{2500} relation for a ΛCDM ($h = 0.7$) cosmology. Fitting the kT_{2500}–L_{2500} data for all six clusters using a power-law model of the form

$$\left(\frac{kT_{2500}}{10\ \mathrm{keV}}\right) = B\left(\frac{L_{2500}}{10^{45}\ \mathrm{erg\ s^{-1}}\,E(z)}\right)^{\beta} \tag{8.2}$$

and a χ^2 estimator, we obtain $B = 0.42 \pm 0.05$ and $\beta = 0.56 \pm 0.10$. Using the BCES($X_2 \mid X_1$) estimator of Akritas & Bershady (1996), which accounts for errors in both axes and the presence of possible intrinsic scatter, we obtain $B = 0.51 \pm 0.05$ and $\beta = 0.48 \pm 0.06$. We conclude that the slope of the temperature–luminosity relation for the present sample of hot, relaxed clusters is consistent with the predicted value of $\beta = 0.5$. Note that fixing $\beta = 0.33$ ($L \propto T^3$) results in a poor fit: $\chi^2 = 12.5$ for five degrees of freedom, as opposed to $\chi^2 = 6.7$ with $\beta = 0.5$.

The offset between the observed and simulated mass–temperature curves cannot be explained by invoking an earlier-formation redshift for the observed clusters (we assume that the clusters form at the redshifts they are observed) since, for the measured NFW mass distributions, $M_{2500}(z)$ drops as fast or faster than $E(z)$ rises as the formation redshift is increased. Our results suggest that, on the spatial scales studied here, important physics may be missing from the reference simulations. The most likely candidates are radiative cooling of the X-ray gas, which the Chandra data show to be significant within $r \sim 0.2r_{2500}$ (Allen *et al.* 2001*a*, *b*, 2002*a*; David *et al.* 2001; McNamara *et al.* 2001; Schmidt *et al.* 2001) and preheating. Pearce *et al.* (2000) show that the introduction of radiative cooling into hydrodynamical simulations can lead to significant increases in the mass-weighted temperatures within $r \sim r_{2500}$ (as cool, low-entropy gas is deposited, warmer, high-entropy material flows inwards and is compressed). This work has been extended by Thomas *et al.* (2002), who determine a M_{2500}–kT_{2500} relation from simulations, including the effects of cooling and preheating, in good agreement with the observations in Fig. 8.4(*a*).

The new virial relations for clusters have important implications for cosmological studies that we are just beginning to explore. By combining the Chandra mass–temperature and mass–luminosity relations with the space densities of clusters, $n(M, z)$, predicted by simulations for various cosmological parameters (Evrard *et al.* 2002), and comparing the results with the observed luminosity, $n(L, z)$, and temperature, $n(T, z)$, functions of clusters, we can obtain tight constraints on Ω_m and σ_8 (the root-mean-square variation of the linearly evolved density field, smoothed by a top-hat window function of size $8h^{-1}$ Mpc). Seljak (2002) has shown that the application of a mass–temperature relation consistent with that presented here to the observed temperature function for nearby clusters leads to a downwards revision of the best-fit value of σ_8 by *c*. 30% (well beyond the statistical uncertainties quoted in most previous studies). Analysis, combining the local luminosity function of clusters with the Chandra mass–luminosity relation, gives similar results ($\sigma_8 \sim 0.7$ (Allen *et al.* 2003)).

Cosmological constraints from the X-ray gas mass fraction

The matter content of rich clusters of galaxies is thought to provide a fair sample of the matter content of the Universe as a whole (White *et al.* 1993). The observed ratio of the baryonic to total mass in clusters should therefore closely match the ratio of the cosmological parameters Ω_b/Ω_m, where Ω_b and Ω_m are the mean baryon and total mass densities of the Universe, in units of the critical density. The combination of robust measurements of the baryonic mass fraction in clusters with accurate determinations of Ω_b from cosmic nucleosynthesis calculations (constrained by the observed abundances of light elements at high redshifts) can therefore be used to determine Ω_m.

This method for measuring Ω_m, which is particularly simple in terms of its underlying assumptions, was first highlighted by White & Frenk (1991) and subsequently employed by many groups (David *et al.* 1995; White & Fabian 1995; Evrard 1997; Ettori & Fabian 1999). The baryonic mass content of rich clusters of galaxies is dominated by the X-ray emitting intracluster gas, the mass of which exceeds the mass of optically luminous material by a factor of *c*. 6 (White *et al.* 1993; Fukugita *et al.* 1998). Since the X-ray emissivity of the X-ray gas is proportional to the square of its density, the gas mass profile can be precisely determined from the X-ray data. With the advent of accurate measurements of Ω_b (O'Meara *et al.* 2001, and references therein) and a precise determination of the Hubble constant, H_0 (Freedman *et al.* 2001), the dominant uncertainty in determining Ω_m from the baryonic mass fraction in clusters had lain in the measurements of the total (luminous plus dark) matter distributions in the clusters. However, the launch of Chandra and development of gravitational lensing studies have now provided the first precise mass measurements for clusters of galaxies. The agreement between the independent X-ray and gravitational lensing masses discussed earlier limits the systematic uncertainties in the baryonic mass fraction measurements to *c*. 10%, an accuracy comparable with the current Ω_b and H_0 results. Here, we explore the constraints that the new Chandra data can provide on Ω_m and the cosmological constant, Ω_Λ.

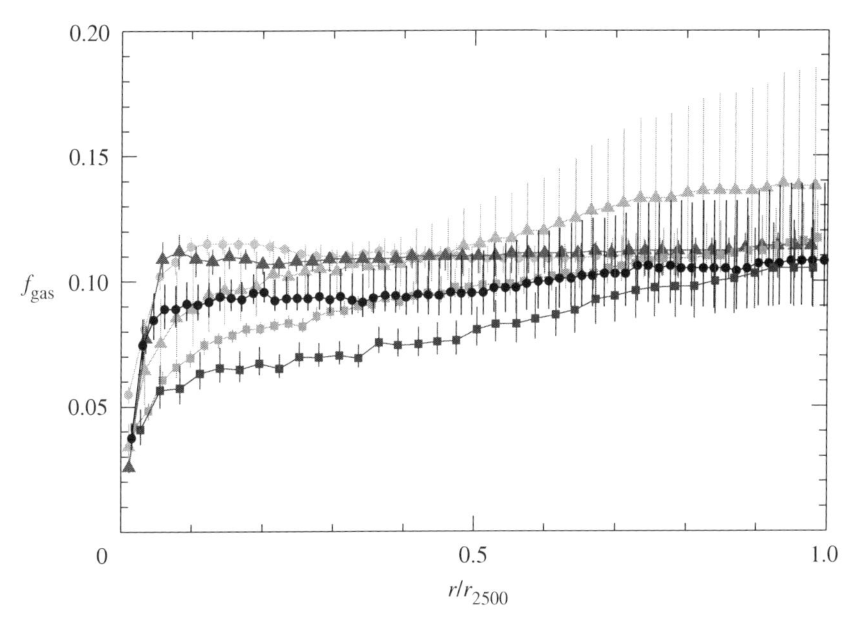

Figure 8.5. The observed X-ray gas mass fraction profiles in the clusters with the radial axis scaled in units of r_{2500} (Allen *et al.* 2002*b*). PKS0745-191 (light circles), Abell 2390 (light triangles), Abell 1835 (dark triangles), MS2137-2353 (light squares), RXJ1347-1145 (dark circles), 3C295 (dark squares). An $h = 0.7$, ΛCDM cosmology is assumed. $f_{gas}(r)$ is an integrated quantity and the error bars on neighbouring points in a profile are correlated.

As with the virial studies discussed in the previous section, in determining the results on the X-ray gas mass fraction, f_{gas}, we have adopted a canonical radius r_{2500}, within which the mean mass density is 2500 times the critical density of the Universe at the redshift of the cluster. Figure 8.5 shows the observed $f_{gas}(r)$ profiles. Although some variation is present from cluster to cluster, the profiles tend towards a similar value at r_{2500}. Taking the weighted mean of the f_{gas} results at r_{2500}, we obtain $\bar{f}_{gas} = 0.113 \pm 0.005$ (for a standard ΛCDM cosmology with $h = 0.7$).

Taking the optically luminous baryonic mass in galaxies as $0.19h^{0.5}$ times the X-ray gas mass (White *et al.* 1993; Fukugita *et al.* 1998) and ignoring other sources of baryonic matter (which are expected to make only very small contributions to the total mass), we can write $\Omega_m = \Omega_b/f_{gas}(1 + 0.19h^{0.5})$. For $\Omega_b h^2 = 0.0205 \pm 0.0018$ (O'Meara *et al.* 2001) and $h = 0.7$, the f_{gas} results in Fig. 8.5 give $\Omega_m = 0.319 \pm 0.032$

In addition to the simple calculation above, the Chandra data can be used to obtain more rigorous constraints on cosmological parameters using the apparent variation of the observed f_{gas} values with redshift. (Such methods were first proposed by Sasaki (1996) and Pen (1997).) The key point is that when measuring f_{gas} from the X-ray data, one needs to adopt a reference cosmology. The measured f_{gas} value for

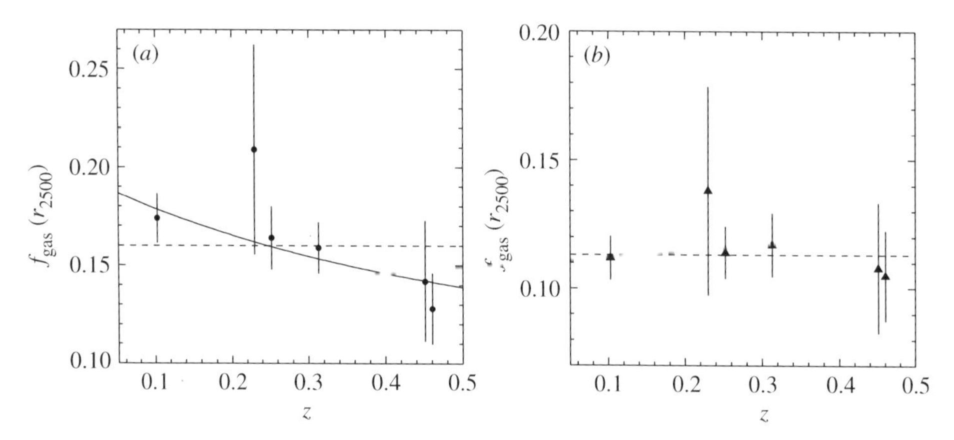

Figure 8.6. The apparent variation of the observed X-ray gas mass fraction (with root-mean-square 1σ errors) as a function of redshift for the default (*a*) SCDM ($\Omega_m = 1.0$, $\Omega_\Lambda = 0.0$, $h = 0.5$) and (*b*) ΛCDM ($\Omega_m = 0.3$, $\Omega_\Lambda = 0.7$, $h = 0.7$) cosmologies. The dashed lines show the results of fitting a constant value to the data in each case. The solid line in (*a*) shows the predicted curve for the best-fit cosmology with $\Omega_m = 0.30$ and $\Omega_\Lambda = 0.95$.

each cluster will depend upon the assumed angular diameter distance to the source ($f_{gas} \propto D_A^{1.5}$). Thus, although we expect the observed f_{gas} values to be invariant with redshift, this will only appear to be the case when the assumed cosmology matches the true, underlying cosmology.

Figures 8.6 shows the observed f_{gas} values as a function of redshift for an assumed Einstein–de Sitter (SCDM: $\Omega_m = 1.0$, $\Omega_\Lambda = 0.0$) and ΛCDM ($\Omega_m = 0.3$, $\Omega_\Lambda = 0.7$) cosmology. We see that whereas the results for ΛCDM are consistent with a constant f_{gas} value, the results for SCDM indicate an apparent drop in f_{gas} as the redshift increases. Thus, inspection of Fig. 8.6 favours the ΛCDM over the SCDM cosmology.

In order to constrain the relevant cosmological parameters, we have fitted the data in Fig. 8.6(*a*) with a model which accounts for the expected apparent variation in the $f_{gas}(z)$ values, measured assuming an SCDM cosmology (with $h = 0.5$), for different underlying cosmologies. The 'true' cosmology should be the cosmology that provides the best fit to the measurements. (Note that the $f_{gas}(r)$ profiles exhibit only small variations around r_{2500}, and so the effects of changes in r_{2500} as the cosmology is varied can be ignored.) The model function fitted to the data is

$$f_{gas}^{mod}(z) = \frac{\Omega_b}{(1 + 0.19\sqrt{h})\Omega_m} \left[\frac{h}{0.5} \frac{D_A^{\Omega_m=1,\Omega_\Lambda=0}(z)}{D_A^{\Omega_m,\Omega_\Lambda}(z)} \right]^{1.5}, \tag{8.3}$$

which depends on Ω_m, Ω_Λ, Ω_b and h. The ratio $(h/0.5)^{1.5}$ accounts for the change in the Hubble constant between the considered model and default SCDM cosmology

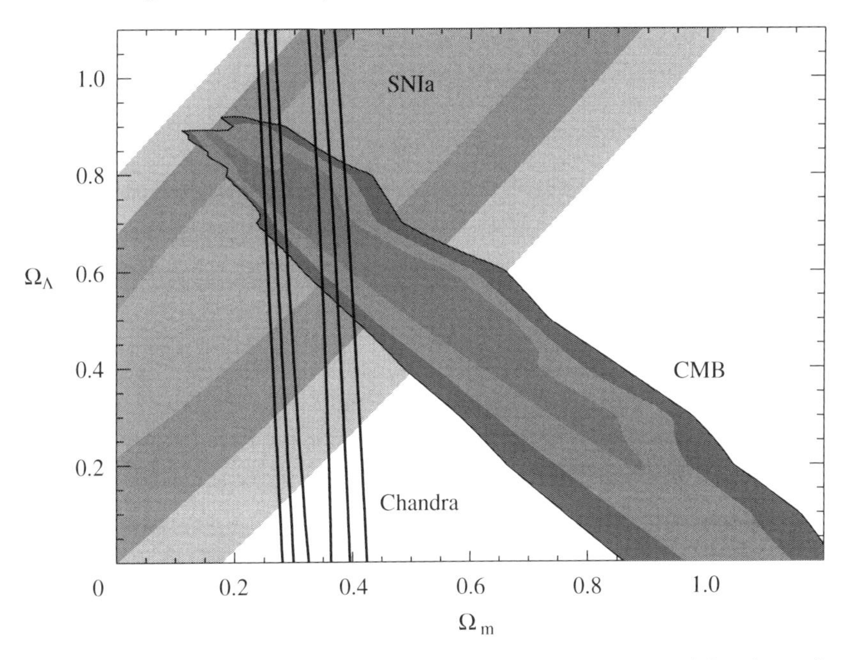

Figure 8.7. The joint 1, 2 and 3σ confidence contours on Ω_m and Ω_Λ determined from the Chandra $f_{gas}(z)$ data (bold contours), and independent analyses of cosmic microwave background (CMB) anisotropies and the properties of distant supernovae (from Allen *et al.* (2002*b*)).

($h = 0.5$), and the ratio of the angular diameter distances accounts for deviations in the geometry of the Universe from the Einstein–de Sitter case. We apply Gaussian priors of $\Omega_b h^2 = 0.0205 \pm 0.0018$ (O'Meara *et al.* 2001) and $h = 0.72 \pm 0.08$, the final result from the Hubble Key Project reported by Freedman *et al.* (2001).

We have examined a grid of cosmologies covering the plane $0.0 < \Omega_m < 1.0$ and $0.0 < \Omega_\Lambda < 1.5$. The joint 1, 2 and 3$\sigma$ confidence contours on Ω_m and Ω_Λ are shown in Fig. 8.7. The best-fit parameters and marginalized 1σ error bars are $\Omega_m = 0.30^{+0.04}_{-0.03}$ and $\Omega_\Lambda = 0.95^{+0.48}_{-0.72}$, with $\chi^2_{min} = 1.7$ for four degrees of freedom, indicating that the model provides an acceptable description of the data. The best-fit cosmological parameters are similar to the values that were assumed in Fig. 8.6(*b*), as can be expected given the approximately constant $f_{gas}(z)$ results shown in that figure.

Also shown in Fig. 8.7 is a comparison of the constraints on Ω_m and Ω_Λ determined from the Chandra $f_{gas}(z)$ data, with the results of Jaffe *et al.* (2001) from studies of cosmic microwave background (CMB) anisotropies (incorporating the COBE Differential Microwave Radiometer, BOOMERANG-98 and MAXIMA-1 data of Bennett *et al.* (1996), de Bernardis *et al.* (2000) and Hanany *et al.* (2000), respectively) and the properties of distant supernovae (incorporating the data of

Riess *et al.* (1998) and Perlmutter *et al.* (1999)). The agreement between the results obtained from the independent methods is striking: all three datasets are consistent, at the 1σ confidence level, with a cosmological model with $\Omega_m = 0.3$ and $\Omega_\Lambda = 0.7$–0.8. These results are also consistent with the findings of Efstathiou *et al.* (2002) from a combined analysis of the 2dF Galaxy Redshift Survey and CMB data.

The constraints on Ω_m, Ω_Λ and σ_8 are expected to improve as further Chandra, XMM-Newton and high-quality gravitational lensing data are gathered for more relaxed clusters, especially at high redshifts.

I am grateful to my collaborators Robert Schmidt, Andy Fabian and Stefano Ettori. I thank Gordon Squires and Phillipe Fischer for communicating their weak lensing results shown in Fig. 8.2, and Andrew Jaffe for providing the CMB and supernovae results shown in Fig. 8.7. This work was supported by a Royal Society University Research Fellowship.

References

Akritas, M. G. & Bershady, M. A. 1996 *Astrophys. J.* **470**, 706–714.
Allen, S. W. & Fabian, A. C. 1998 *Mon. Not. R. Astr. Soc.* **297**, L57–L62.
Allen, S. W. *et al.* 2001a *Mon. Not. R. Astr. Soc.* **324**, 842–858.
Allen, S. W., Ettori, S. & Fabian, A. C. 2001b *Mon. Not. R. Astr. Soc.* **324**, 877–890.
Allen, S. W., Schmidt, R. W. & Fabian, A. C. 2001c *Mon. Not. R. Astr. Soc.* **328**, L37–L41.
Allen, S. W., Schmidt, R. W. & Fabian, A. C. 2002a *Mon. Not. R. Astr. Soc.* **334**, L11.
Allen, S. W., Schmidt, R. W. & Fabian, A. C. 2002b *Mon. Not. R. Astr. Soc.* **335**, 256.
Allen, S. W., Fabian, A. C., Schmidt, R. W. & Ebeling, H. 2003 *Mon. Not. R. Astr. Soc.* **342**, 287.
Arnaud, M. & Evrard, A. E. 1999 *Mon. Not. R. Astr. Soc.* **305**, 631–640.
Bennett, C. *et al.* 1996 *Astrophys. J.* **464**, L1–L4.
Borgani, S., Girardi, M., Carlberg, R. G., Yee, H. K. C. & Ellingson, E. 1999. *Astrophys. J.* **527**, 561–572.
Bryan, G. L. & Norman, M. L. 1998 *Astrophys. J.* **495**, 80–99.
Carlberg, *et al.* 1996 *Astrophys. J.* **462**, 32–49.
Cavaliere, A., Menci, N. & Tozzi, P. 1997 *Astrophys. J.* **484**, L21–L24.
David, L. P., Jones, C. & Forman, W. 1995 *Astrophys. J.* **445**, 578–590.
David, L. P. *et al.* 2001 *Astrophys. J.* **557**, 546–559.
de Bernardis, P. *et al.* 2000 *Nature* **404**, 955–959.
Efstathiou, G. *et al.* 2002 *Mon. Not. R. Astr. Soc.* **330**, L29–L35.
Ettori, S. & Fabian, A. C. 1999 *Mon. Not. R. Astr. Soc.* **305**, 834–848.
Evrard, A. E. 1997. *Mon. Not. R. Astr. Soc.* **292**, 289–297.
Evrard, A. E., Metzler, C. A. & Navarro, J. F. 1996 *Astrophys. J.* **469**, 494–507.
Evrard, A. E. *et al.* 2002 *Astrophys. J.* **573**, 7–36.
Fasano, G. & Vio, R. 1988 *Bull. Inf. Cnt. Données Stellaires* **35**, 191–196.
Finoguenov, A., Reiprich, T. H. & Böhringer, H. 2001. *Astron. Astrophys.* **368**, 749–759.
Fischer, P. & Tyson, J. A. 1997 *Astron. J.* **114**, 14–25.
Freedman, W. *et al.* 2001 *Astrophys. J.* **553**, 47–72.
Fukugita, M., Hogan, C. J. & Peebles, P. J. E. 1998 *Astrophys. J.* **503**, 518–530.
Hanany, S. *et al.* 2000 *Astrophys. J.* **545**, L5–L9.
Horner, D. J., Mushotzky, R. F. & Scharf, C. A. 1999 *Astrophys. J.* **520**, 78–86.

Jaffe, A. H. *et al.* 2001 *Phys. Rev. Lett.* **86**, 3475–3479.
Jenkins, A. *et al.* 2001 *Mon. Not. R. Astr. Soc.* **321**, 372–384.
McNamara, B. *et al.* 2001 *Astrophys. J.* **562**, L149–L152.
Markevitch, M. 1998 *Astrophys. J.* **504**, 27–34.
Mathiesen, B. F. & Evrard, A. E. 2001 *Mon. Not. R. Astr. Soc.* **546**, 100–116.
Mellier, Y. 1999 *A. Rev. Astr. Astrophys.* **37**, 127–189.
Navarro, J. F., Frenk, C. S. & White, S. D. M. 1997 *Astrophys. J.* **490**, 493–508.
Nevalainen, J., Markevitch, M. & Forman, W. 2000 *Astrophys. J.* **532**, 694–699.
O'Meara, J. M. *et al.* 2001 *Astrophys. J.* **552**, 718–730.
Pearce, F. R., Thomas, P. A., Couchman, H. M. P. & Edge, A. C. 2000 *Mon. Not. R. Astr. Soc.* **317**, 1029–1040.
Pen, U. 1997 *New Astron.* **2**, 309–317.
Perlmutter, S. *et al.* 1999 *Astrophys. J.* **517**, 565–586.
Press, W. H. & Schechter, P. 1974 *Astrophys. J.* **187**, 425–438.
Riess, A. G. *et al.* 1998 *Astron. J.* **116**, 1009–1038.
Sarazin, C. L. 1988 *X-ray Emission from Clusters of Galaxies.* Cambridge: Cambridge University Press.
Sasaki, S. 1996 *Publ. Astr. Soc. Jpn* **48**, L119–L122.
Schmidt, R. W., Allen, S. W. & Fabian, A. C. 2001 *Mon. Not. R. Astr. Soc.* **327**, 1057–1070.
Seljak, U. 2002 *Mon. Not. R. Astr. Soc.* **337**, 769.
Squires, G. *et al.* 1996 *Astrophys. J.* **469**, 73–77.
Thomas, P. A., Muanwong, O., Kay, S. T. & Liddle, A. R. 2002 *Mon. Not. R. Astr. Soc.* **330**, L48–L52.
van Haarlem, M. P., Frenk, C. S. & White, S. D. M. 1997 *Mon. Not. R. Astr. Soc.* **287**, 817–832.
White, D. A. & Fabian, A. C. 1995 *Mon. Not. R. Astr. Soc.* **273**, 72–84.
White, S. D. M. & Frenk, C. S. 1991 *Astrophys. J.* **379**, 52–79.
White, S. D. M., Navarro, J. F., Evrard, A. E. & Frenk, C. S., 1993 *Nature* **366**, 429–433.

9

Clusters of galaxies: a cosmological probe

BY RICHARD MUSHOTZKY

NASA Goddard Space Flight Center

Introduction

Clusters and groups of galaxies represent the largest virialized structures in the Universe, with masses between $10^{13}\ M_{\odot}$ and $5 \times 10^{15}\ M_{\odot}$. Since their dynamical time-scales are not much shorter than the age of the Universe, studies of their evolution in mass, number density and other characteristics can place strong constraints on all theories of large-scale-structure formation and, in principle, determine precise values for many of the cosmological parameters (such as Ω, Λ (the value of a cosmological constant), σ_8 (the normalization of the density fluctuations), etc.). Clusters probably retain all the enriched material created inside their potential well, unlike galaxies, star clusters, globular clusters and H II regions. The elemental abundances in clusters and their evolution with redshift strongly constrain the origin and evolution of the elements. The distribution of the elements in the cluster gas reveals how the metals were removed from the stellar systems in which they were created and delivered into the intergalactic medium.

Theoretical studies indicate that clusters should be 'fair samples' of the Universe. Therefore, their bulk properties, like their baryonic fraction, chemical composition and entropy distribution, sample the 'gross' properties of the Universe as a whole. While clusters represent initially high overdensities in the matter distribution, the processes that proceed within them are representative of the Universe as a whole and thus clusters represent a rather efficient way of studying the bulk properties of the Universe.

Clusters are relatively 'simple' objects, and cluster formation and evolution are amenable to detailed numerical and analytical studies, in contrast to galaxies, whose formation and evolution are very complex. Clusters are truly 'X-ray' objects, since

Frontiers of X-Ray Astronomy, ed. A.C. Fabian, K.A. Pounds and R.D. Blandford. Published by Cambridge University Press.

c. 80% of the baryons and the metals are in the hot X-ray-emitting gas (Allen 2002), which radiates primarily in the X-ray band.

Cluster evolution and cosmology

The distribution of clusters in space and mass and the evolution of these quantities with redshift are a strong function of the cosmological model (Borgani & Guzzo 2001; Bahcall 2000). The discriminating power grows and the uncertainty in the free parameters (e.g. Λ, Ω) shrinks as the redshift, size and mass range of the sample increases. X-ray observations are crucial because: of the ease and robustness with which clusters can be identified (they are the only bright, hard, extended X-ray sources at high latitude); they are bright enough to detect with Chandra and XMM at $z > 1$ (Hashimoto *et al.* 2002; Stanford *et al.* 2001); it is relatively easy to obtain large samples (Böhringer *et al.* 2000); and there are very few selection effects in constructing complete mass-limited samples (Donahue *et al.* 2002). There is (Allen 2002) a direct connection between the bulk X-ray properties such as luminosity and temperature and the theoretically desired quantities: mass, mass evolution and mass distribution. The combination of these properties allows X-ray surveys to truly probe the mass spectrum of the Universe over the redshifts, physical scales and mass range of interest.

The ROSAT All-Sky Survey (RASS) and deep pointed surveys, as well as Advanced Satellite for Cosmology and Astrophysics (ASCA) spectral observations, have already constrained the cosmological parameters. The cluster-number evolution against redshift (e.g. Bahcall *et al.* (1997)), the power spectrum of clusters (Schuecker *et al.* 2001), the local temperature function (Ikebe *et al.* 2002), the evolution of the temperature function with redshift (Henry 2000), the evolution of the luminosity function with redshift (Borgani *et al.* 2001) and the two-point correlation function (Moscardini *et al.* 2001) all provide significant measurements of Ω and σ_8. Taken together, these data require $0.2 < \Omega < 0.4$ and $0.6 < \sigma_8 < 1$. The formal errors are small and strongly rule out $\Omega_m \sim 1$.

These results led me and my colleagues to develop a medium-class-explorers proposal, the Dark Universe Exploration Telescope (DUET), designed to derive precise estimates of the cosmological parameters from an X-ray cluster survey. Using the spare XMM telescope to survey the 10 000 deg^2 of the Sloan Digital Sky Survey (SDSS) (giving optical counterparts for much of the survey) over a three-year period, along with a deep field of *c*. 150 deg^2 in the south, which overlaps with very large telescope (VLT) surveys, one obtains a complete sample of massive clusters to $z \sim 1$. This survey will be *c*. 50 times more sensitive than the RASS in the Sloan region and 500 times deeper in the VLT region. The southern field overlaps with a deep Sunyaev–Zeldovich effect (SZ effect) survey. The XMM telescope

PSF is sharp enough to resolve clusters at $z > 1$ (Hashimoto *et al.* 2002) and our large-format charge-coupled device (CCD) and observing strategy produce uniform sky coverage and a large contiguous solid angle to measure the power spectrum and to minimize systematic errors caused by large-scale structure. XMM cannot perform such a survey because of its smaller field of view and the mission operations constraints. DUET will determine the distribution of dark matter by accumulating a catalogue of positions, redshifts and masses of over 20 000 clusters of galaxies. Redshifts will be obtained from the SDSS, VLT surveys or dedicated follow-up observation. Cluster masses will be estimated from X-ray luminosity and, for about 500 clusters, X-ray temperature. The overlap with the SDSS and South Pole SZ effect surveys will statistically test the mass estimation using weak lensing and SZ measurements.

The power of such a survey is immense (Fig. 9.1) with formal errors in Ω of ± 0.02 and σ_8 of ± 0.02 from the redshift distribution and the evolution of the temperature function. The total baryonic mass is derived from the acoustic oscillations in the power spectrum and these data also test models of the growth of large-scale structure via gravitational instabilities. The precision estimates of the power spectrum of clusters to $z \sim 1$ on scales from 10 to 1000 Mpc measure the dark energy (Λ), and strongly bound the mass density of neutrinos and the equation of state of the dark energy (w). The correlation between cosmological parameters is different in the DUET dataset from many other experiments (such as Planck and SNAP), allowing detailed estimates of systematic errors. The survey will detect more than 100 000 serendipitous active galaxies and will be the finding chart for future X-ray missions like Constellation-X and XEUS. The sensitivity of DUET is an excellent match to the sensitivity limits of the Sloan, PRIME and other surveys.

The abundances of the elements

Background

The light elements (H, He, Li) were produced in the 'Big Bang', but all the other abundant elements are produced in the cores of stars. Massive stars with $M > 8\ M_{\odot}$ produce virtually all of the α-burning elements (O, Ne, Mg, Si, S) and some of the Fe in type-II supernova (SN) explosions. Type-I supernova produce copious amounts of Fe and Ni. The relative yields of the elements depend on the mass of the star and the distribution of stellar masses (the interplanetary magnetic field (IMF)). There are still major uncertainties about the exact mechanisms that produce type-II supernovae explosions and the astrophysical origin of type Is (Thielemann *et al.* 2002) and thus many of the yields are uncertain.

Because the α-element yields increase with increasing SN II progenitor mass (e.g. Woosley & Weaver 1995), the [α/Fe] ratio is sensitive to the IMF. The abundances

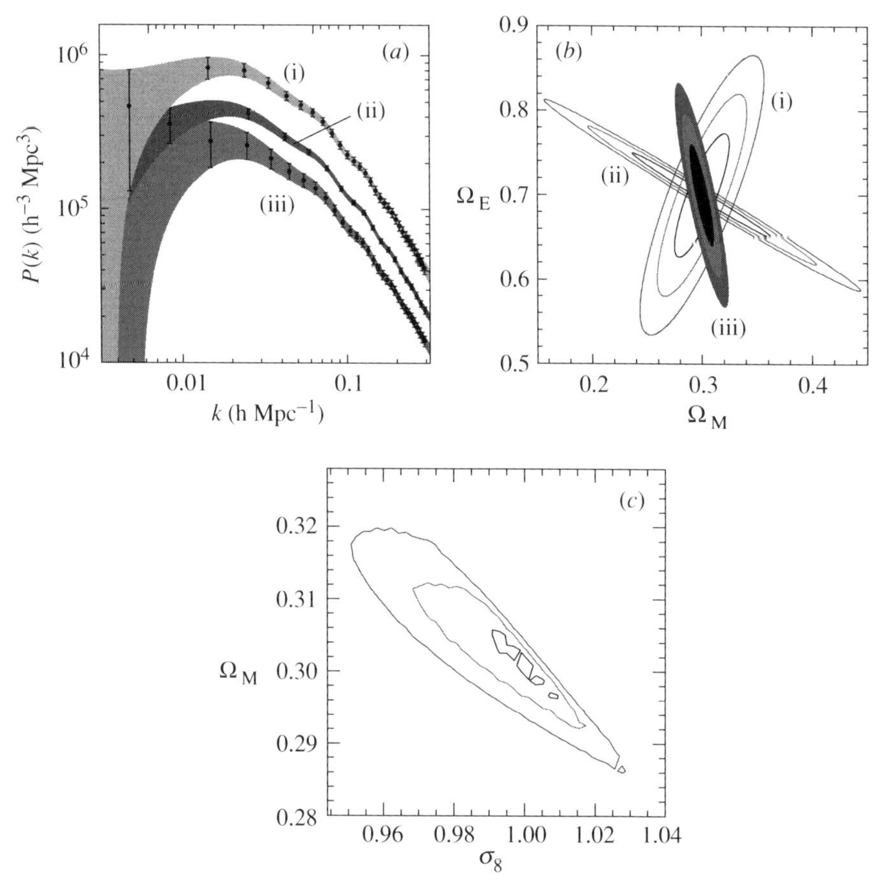

Figure 9.1. Uncertainties in the cosmological parameters that could be obtained by DUET. (*a*) Power spectrum in three redshift shells out to $z \sim 1$ (*c.* 6000 clusters per z-bin): (i) $0.45 < z < 1$; (ii) $0.25 < z < 0.45$; (iii) $0 < z < 0.25$. (*b*) Errors in Ω_E (often called Λ) and Ω_M from (i) SNAP (2366 SNe), (ii) Planck (polarization) and (iii) DUET Wide (18 000 clusters). (*c*) Error in σ_8 and Ω_M from DUET (north + south, *c.* 20 000 clusters, $H_0 = 72 \pm 7$ km s^{-1} Mpc^{-1}). Notice the very small errors in all the quantities.

are due to a sum over the initial mass function and over cosmic time and thus there is not a simple one-to-one relationship between the relative abundances of the elements, the absolute abundances and the IMF. The most detailed studies of the creation of the elements have been based primarily on the stars and the interstellar matter (ISM) in the Milky Way (MW) (Wheeler *et al.* 1989; Henry & Worthey 1999), the gas in the Lyman-limit objects (Prochaska & Wolfe 2002) and the H II regions and the ISM of nearby galaxies (Sembach *et al.* 2000). The systematic errors due to complex radiative transfer in stars and the effects of selective depletion onto dust in gas dominate over the statistical errors.

X-ray astronomy and abundances

The hot cluster gas contains most of the baryons and the heavy elements, and X-ray spectroscopic measurements are necessary to derive its chemical composition. Theoretical calculations (Davé *et al.* 2001) indicate that most of the baryons in the Universe are hot and potentially only observable via X-ray imaging and spectroscopy.

The hot gas contains a record of nucleosynthesis over all cosmic time. Galaxies are open systems and much of the material produced inside them is lost via a variety of physical process, such as galactic winds and ram-pressure stripping (Heckman 2002). Accurate stellar abundances in the galaxies in the clusters are hard to derive because of degeneracies between the age and metallicity of stellar systems and the absence of sharp abundance diagnostics in the summed spectra of stars. Despite the very high signal-to-noise spectra obtained for elliptical galaxies, accurate abundances for any element other than Fe are still controversial (Trager *et al.* 2000).

The derivation of abundances from X-ray spectra is simple and straightforward (Kahn *et al.* 2002). The gas is in collisional equilibrium and the strength of the lines is a direct measure of abundance. The atomic physics is simple, since the strongest lines are from the H- and He-like ions in hot systems. The electron temperature is obtained from the bremmstrahlung continuum and is precise and accurate. Only in the X-ray band is the electron temperature for a collisionally ionized gas measured directly. The high temperatures and moderate densities destroy the dust, and depletion corrections are not needed. Almost all the lines are optically thin and thus radiative-transfer effects are small.

The abundances of the elements with the highest equivalent lines in the temperature range from 2×10^6 to 1×10^8 K (C, N, O, Ne, Mg, Si, S, Ca, Ar, Fe and Ni) can be determined over a wide range of redshifts and cluster masses, often from one broadband X-ray spectrum. The less abundant elements, like Cr, Zn, Na and Al, are beyond our present capabilities but should be accessible to the next generation of missions.

The most accurate data are for Fe, since it has the highest equivalent-width lines, with the next most accurate abundances being obtained for O, Si, S and Ni. The ability of X-ray CCDs to derive abundances depends on the temperature of the object, since the equivalent widths of the lines vary strongly with temperature and, when the equivalent width of the lines is much below the spectral resolution of the instrument, the derived abundances are not robust. For ASCA and Chandra, but not XMM, calibration problems at low energies have prevented the determination of accurate oxygen abundances. Ne and Mg are very difficult to measure properly with CCD spectrometers because the H- and He-like lines of these elements are blended with strong Fe L lines. The equivalent widths of Ca and Ar are low and thus somewhat suspect with ASCA data. C and N are not in the ASCA CCD bandwidth and even the best CCDs have low spectral resolution at the energies of their strongest lines.

In objects with multi-temperature structure there is often not a unique solution for CCD-quality data, particularly for objects with $kT < 1.5$ keV, where the Fe L complex is extremely strong. The difficulty of interpreting these data has resulted in various workers finding rather different abundances. With the higher spatial resolution of Chandra and XMM and the better spectral resolution of the XMM RGS these regions are often resolved spatially and spectrally and accurate abundances can be derived.

Status of the field

There are hundreds of high-quality X-ray images and *c.* 150 high-quality CCD X-ray spectra. These data allow the reliable determination of the abundances of Ni, Si, S and Fe in 20–30 individual clusters and a similar number of groups.† Reasonable samples of clusters (*c.* 20–30 objects) at $z > 0.2$ are now available with images, temperatures and Fe and Si abundances. Groups at $z > 0.2$ are being found in deep ROSAT/XMM and Chandra fields (e.g. Bauer *et al.* 2002) and spectra and images for *c.* 40 low-z groups have been published.

Large solid-angle surveys with ROSAT find numerous clusters in a reliable uniform fashion out to $z \sim 0.3$ (Böhringer *et al.* 2000) and smaller solid-angle surveys find objects with $z > 1$ (Borgani *et al.* 2001).

XMM RGS data have been obtained for over 20 clusters and are now being analysed in detail. These observations should give precise and robust abundances (Peterson *et al.* 2002), particularly for low-temperature objects (Xu *et al.* 2002).

New results

The ASCA cluster project

(i) Overall metallicity and correlations

Over 270 clusters of galaxies were observed by ASCA during its seven-year lifetime: the largest sample of cluster temperatures and abundances. Our group (Horner *et al.* 2002; Baumgartner *et al.* 2002) has analysed the global spectra of all of these systems and has obtained precise average properties for them.

Fe abundances The Fe abundance is not constant from cluster to cluster (Figs. 9.2 and 9.3), but shows a variance of ± 0.13 about the mean. This variation

† I define groups as objects less massive than $2 \times 10^{14}\ M_{\odot}$ and clusters to be more massive than this limit; the division is subjective but roughly corresponds to the least massive clusters included by G. Abell in his classic 1958 cluster-survey paper (Abell 1958).

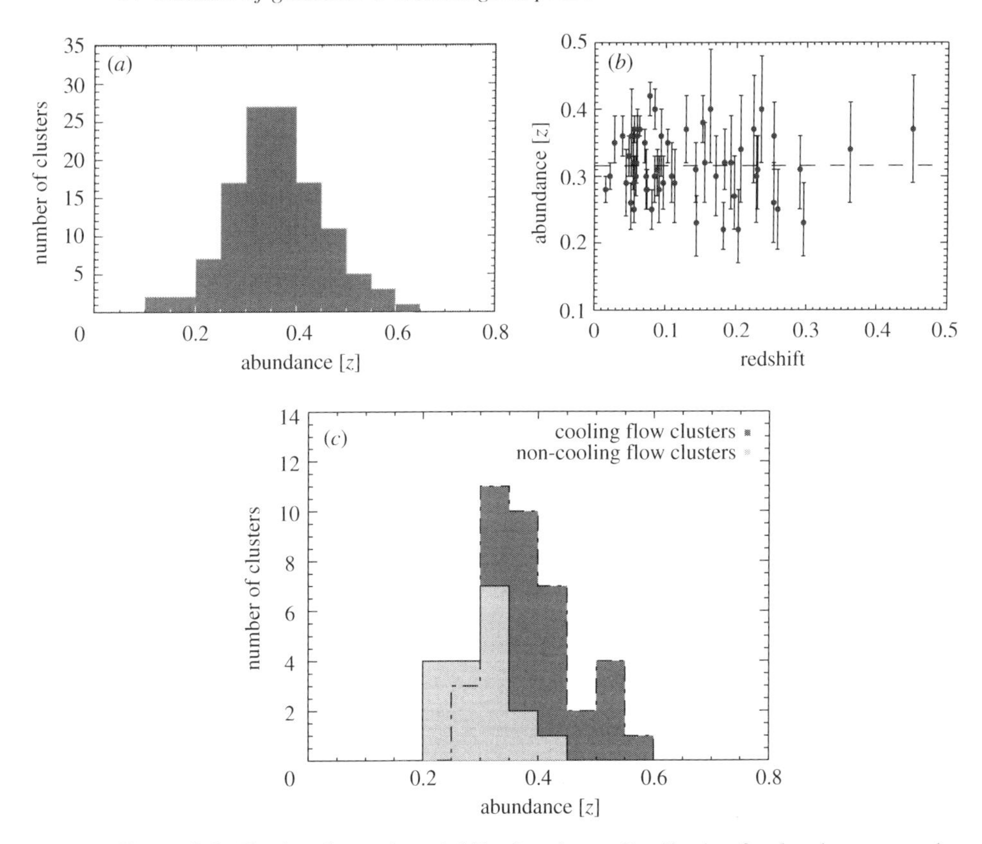

Figure 9.2. Fe abundance data: (*a*) Fe abundance distribution for the cluster sample and (*b*) lack of evolution in the metal abundance with redshift. (*c*) The abundance distribution for cooling flow clusters in dark shading is compared with that of non-cooling flow clusters in light shading.

in abundance is correlated with the cluster temperature, the cluster central density and the existence of cooling flows (Figs. 9.2 and 9.3).

In clusters more massive than *c.* $5 \times 10^{14}\ M_{\odot}$, corresponding to $T \sim 5$ keV, the Fe abundance is roughly constant. However, it increases at lower temperatures, reaching a peak at $T \sim 3$ keV, and then drops again for lower-temperature systems. The effect is subtle, requiring a large ASCA sample with small error bars to discern it, but real. Similarly, the correlation between cooling flow rate, cluster central gas density and Fe abundance (Fabian *et al.* 1994) has been confirmed and enhanced. We believe that the fundamental correlations are with the cluster central gas density (Fig 9.3) and temperature rather than with the cooling rate (Horner 2001).

In the simplest concepts all massive clusters are coeval structures which retain within them all the nucleosynthesis products and have identical ratios of stellar to gas mass. Given these assumptions, the abundances should be the same from cluster to cluster. That is, the yield of the heavy elements, due to stars normalized by the total

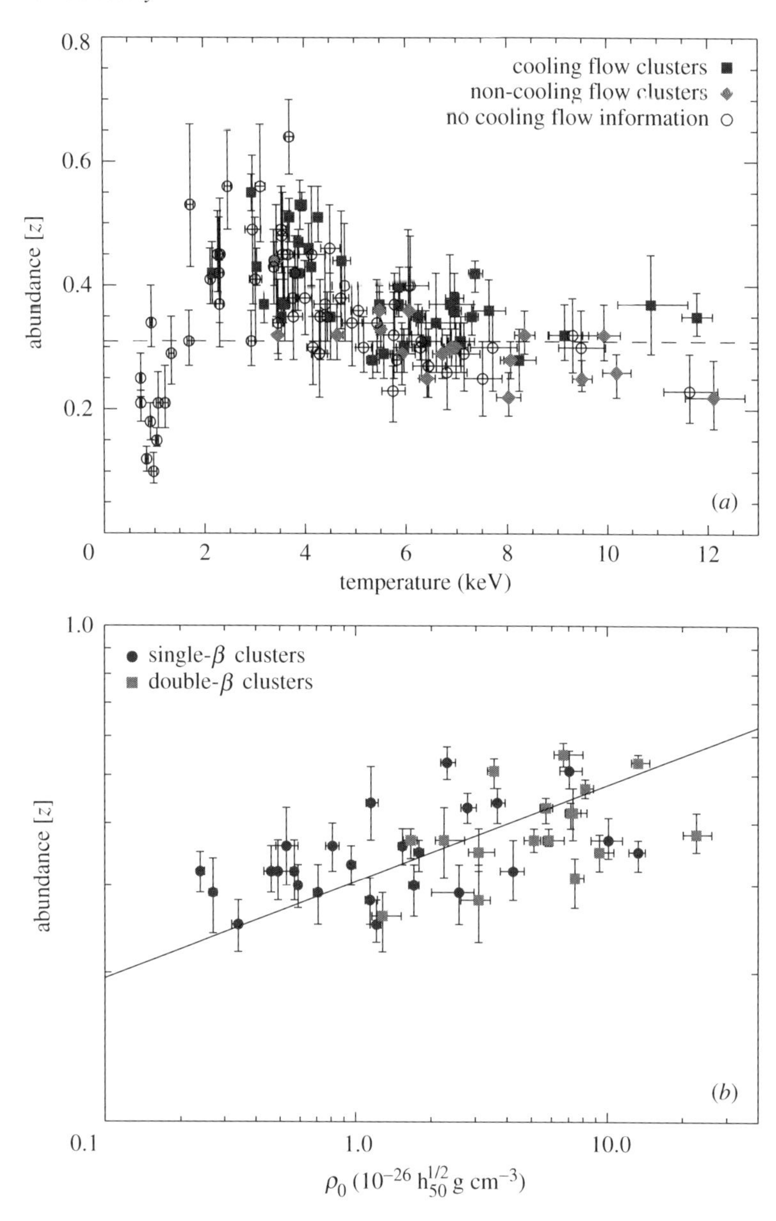

Figure 9.3. (*a*) Fe abundance against temperature on an object-by-object basis. (*b*) Correlation of the Fe abundance with the cluster central density as derived by Mohr *et al.* (1999).

baryonic mass, most of which is in the hot X-ray-emitting gas, should be the same in each cluster. Clearly, there is something wrong with this scenario. One possibility is that not all clusters are closed systems and in the low-mass clusters and groups there is enough energy produced by nucleosynthesis or active galaxies to drive cluster winds which expel the metal-enriched gas (Davis *et al.* 1998). This might explain the lower abundances at the lower mass scales and the reduced abundances at the lower gas densities (the gas becomes 'puffed up' when extra energy is injected). This idea is supported by the observations of extra entropy in cluster cores (Ponman *et al.* 1999). However, an additional mechanism is needed to account for the reduced abundances at larger masses. One idea is that there is a relationship between the age of a cluster and its central density (Scharf & Mushotzky 1997). Since higher overdensities collapse first, and the rate of star formation was much higher in the past, one can speculate that those systems that formed early captured more of the metals in the proto-cluster potential well and thus have higher relative abundances. Another possibility is that the ratio of gas mass to stellar mass is increasing and the stellar mass fraction is decreasing as the cluster mass is increasing (Bryan 2000). This would produce the trends in the abundance data; more gas and fewer stars decrease the Fe abundance but would need to be truncated at the highest masses. However, this solution is controversial (Balogh *et al.* 2001).

The cluster Fe abundance does not evolve with redshift (Fig. 9.2), indicating that most of the elemental production occurred at early times. The absence of copious star formation (most of the element production occurs in massive stars) in high-redshift clusters at $z \sim 0.8$ (and even larger (Nakata *et al.* 2001)) indicates that the epoch of most of the metal creation occurred at least 2 Gy (the lifetime of A stars on the main sequence) earlier, giving a minimum redshift of *c.* 1.2 for the epoch of cluster metal formation. While clusters form early in ΛCDM models, it is surprising that they have finished all of their metal creation at high redshifts.

Si, S *and* Ni *abundances* Si and S are produced primarily in type-II supernovae and Ni comes almost exclusively from type-I supernovae, while Fe can come from both types. Thus, measurement of these elements provides a means of deciding how they are produced. The absolute abundance of these elements, compared with Fe, gives an estimate of the total yield of heavy elements from the stars (Loewenstein & Mushotzky 1996).

The ASCA data for all but the brightest individual clusters do not have sufficient signal-to-noise to obtain precise measurements for elements other than iron and perhaps Si (Mushotzky *et al.* 1996; but see also Finoguenov *et al.* 2000). To improve the signal-to-noise the spectra of many clusters in different temperature bins were summed, producing 'stacked' spectra with 10–30 clusters per temperature bin.

Figure 9.4 shows the Si, S and Ni data for the ASCA sample. The Si abundance rises with temperature up to $kT \sim 8$ keV (Fukazawa *et al.* 2000) and then levels off, while the Si/Fe ratio rises monotonically up to the highest temperatures. The S values are roughly constant at $4T > 4$ keV and drop from 1–4 keV, giving S/Fe values roughly constant at $T > 4$. Since Si and S come from almost exactly the

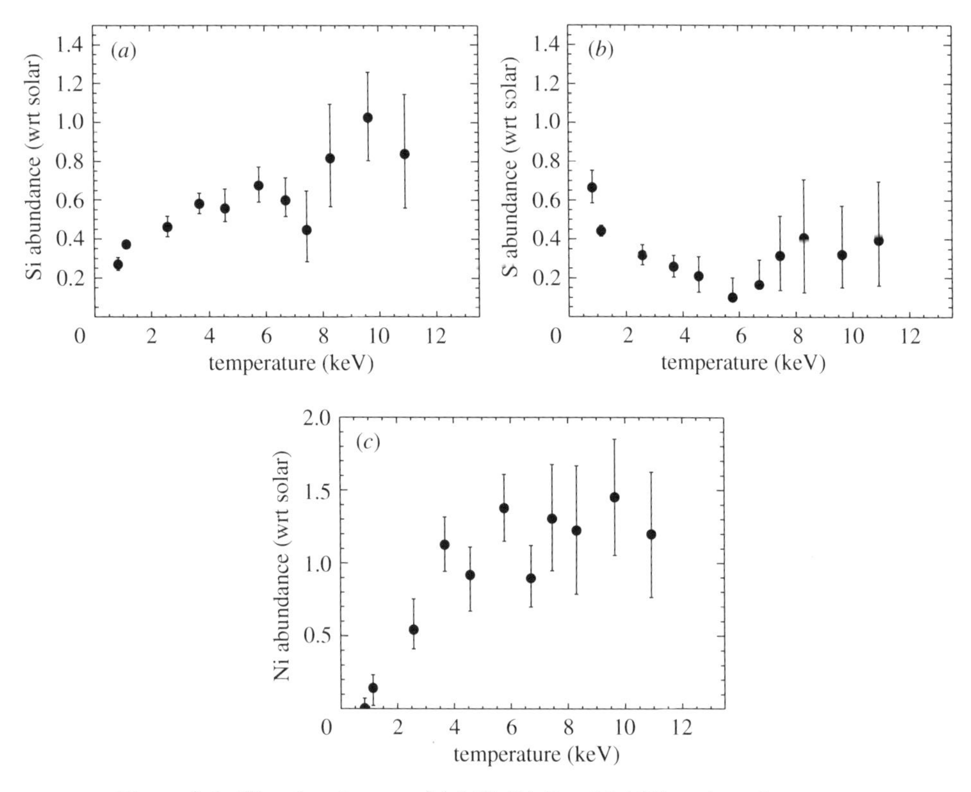

Figure 9.4. The abundances of (*a*) Si, (*b*) S and (*c*) Ni against cluster temperature. Notice the systematic rise of Si with temperature and the rough constancy of Ni and S.

same process (explosive oxygen burning (Woosley & Weaver 1995)), this is difficult to understand. However, the theoretical calculations for S do not agree well with the stars in the MW lying below the observed S/Fe ratios and are just barely consistent with the theoretical uncertainty. Ni, which is produced primarily by type-I supernovae, is overabundant with respect to Fe, indicating that some of the Fe is produced in type-I supernovae.

(ii) Inferences

Relative importance of type-I and type-II supernovae The drop in both Si/Fe and S/Fe with Fe abundance and their rise with temperature shows that there is a range in the relative importance of type-I and type-II supernovae as a function of cluster mass (Gibson *et al.* 1997). At high Fe abundances, type-I supernovae are relatively more important, while type IIs are relatively more important at high mass. However, the relative constancy of Ni/Fe against both temperature and Fe abundance suggests that all the Fe comes from type-Ia supernovae, inconsistent with the value of the Ni/Fe ratio and the Si/Fe trend. The disagreement in the type-II/type-I supernovae

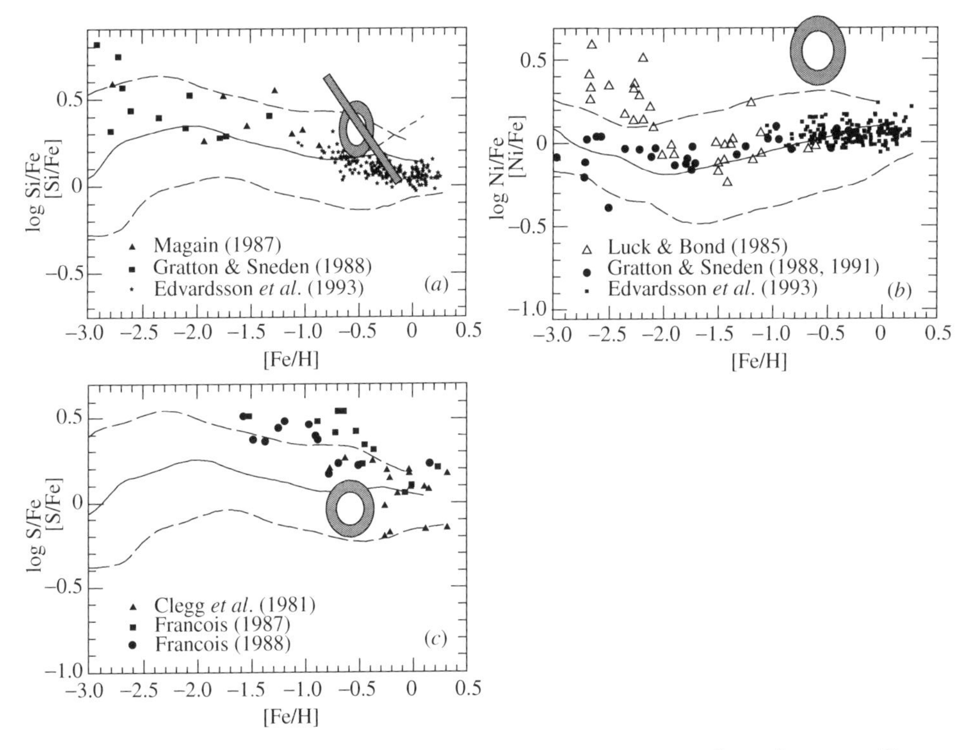

Figure 9.5. Comparison of the cluster abundance pattern, shown in green ellipses, with that seen in the stars in the MW (Timmes *et al.* 1995, where the references listed in the panels will be found.): (*a*) Si/Fe; (*b*) Ni/Fe; (*c*) S/Fe. The shaded bar in (*a*) represents trend of the element ratio.

ratio using the Si, S and Ni diagnostics led Finoguenov *et al.* (2002) to reach rather unusual conclusions, deriving high Ni yields from a particular type of type-I supernovae and adjusting the yields of S from type-II supernovae, and led Loewenstein (2001) to suggest an origin in hypernovae from Pop-III stars. The high Ni/Fe values have been confirmed by XMM in at least one cluster. XMM results (Tamura *et al.* 2001; Peterson *et al.* 2002) indicate that the O/Fe values, which should be very sensitive to the type-II/type-I supernovae ratio, are solar to subsolar, even at high masses, and are inconsistent with the Si and S data, leading to other difficulties.

Comparison with other data We compare the stellar data for our galaxy (Timmes *et al.* 1995) and the Lyman-limit data (Prochaska & Wolfe 2002) in Fig 9.5. The cluster data and the MW star data show some startling differences. At the values of Fe found in clusters there are no stars with similar Si/Fe or Ni/Fe values, while the stellar values of S/Fe are somewhat higher than those in clusters. The high Ni/Fe values are not seen in stars at any metallicity, and there are no stars with the value of Si/Fe and S/Fe seen in clusters. I believe the results indicate that, contrary to the usual assumptions (Wyse 1997), the IMF producing the metals in clusters is different from the IMF that produced the stars in the MW.

None of the Lyman-limit systems has the high Ni/Fe and low S/Fe values seen in clusters, but their Si/Fe values agree with the cluster value (Prochaska & Wolfe 2002). The damped-Lyman-limit systems' abundances are different from those of the stars in the MW, indicating some cosmic variation in the production of heavy elements.

Comparison with H II region abundances is difficult because the elements best determined in these objects (C, N, O) are not determined by the ASCA cluster database and because the effects of dust on elements like Fe and Si are very large (Peimbert *et al.* 2001).

Conclusions

These results challenge our understanding of the origin of the elements. There are several possible modifications to our standard assumptions.

(i) The average abundances could be misleading if there are strong differential abundance gradients. However, this does not seem to be true for clusters in general (see the next subsection).

(ii) The theoretical yields of the elements may have to be adjusted (see the discussion in Gibson *et al.* (1997) and Finoguenov *et al.* (2002)). Since the data for the stars in the MW and their evolution are reproduced by the standard theoretical yields (Timmes *et al.* 1995), this might be difficult. Alternatively, the normalization of the yields by the MW stars could be wrong. It is assumed that the stellar abundances are a fair sample of the heavy-element yield. This has been challenged by observations of rapidly forming star regions (Kobulnicky & Skillman 1998), in which the expected high abundances in the gas or young stars are not seen, and of the X-ray winds in rapidly forming star galaxies (Martin *et al.* 2000), which contain most of the metals produced by the large number of supernovae in these objects. I speculate that the abundances and abundance patterns in the cluster gas are fundamental and that whichever process lets some of the metal-enriched gas stay in the galaxies and form stars also chemically fractionates the gas.

(iii) The differences between the MW and Lyman-limit abundances and the clusters (in particular the high Ni/Fe ratio in the cluster gas) indicates that these systems could not have been the sources of the elements in the cluster gas. While not unexpected, this result means that element production in the Universe has not been from a universal IMF. Since most of the stellar mass in the Universe is in bulges and elliptical galaxies, whose chemical compositions are poorly known (Henry & Worthey 1999; but see McWilliam 1998), one must be very careful in extrapolating from the well-observed but relatively unimportant stars and gas in spiral galaxies to the nature of nucleosynthesis in the Universe as a whole.

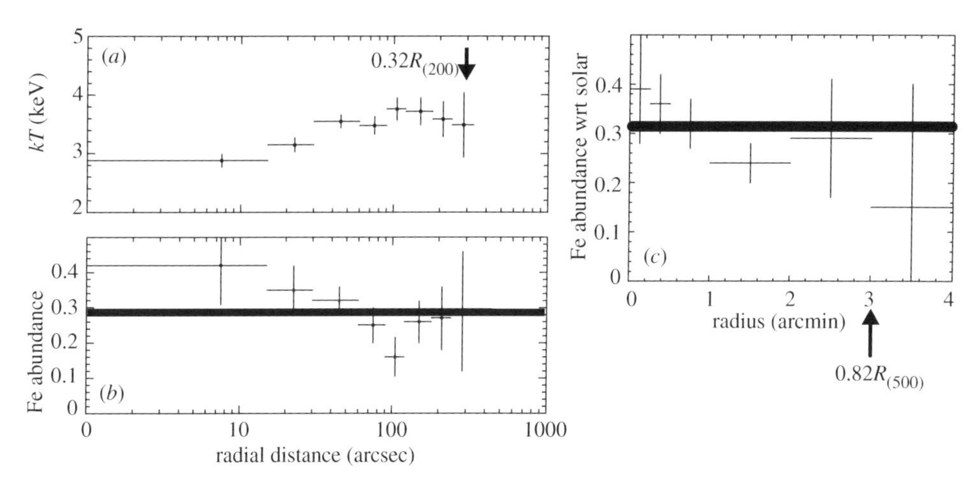

Figure 9.6. XMM abundance profiles for (*a*), (*b*) Abell 2597 and (*c*) ZW3146. (*a*) $100'' = 0.247$ Mpc, (*b*) viral radius = 2.2 Mpc (*c*. 15′), (*c*) $1' = 500$ kpc. The average cluster abundance from the ASCA sample Fe/H ~ 0.3 solar is indicated.

Cluster and group abundance gradients

I have assumed that the average cluster abundances measured by ASCA are representative of the IGM. If there are strong abundance gradients, and if they differ from element to element, then, because the X-ray data are emission-measure-weighted, there will be a bias towards the central regions of the cluster. The existence of strong abundance gradients in clusters is rather controversial, with some authors (Ezawa *et al.* 1997; DeGrandi & Molendi 2001; Irwin & Bregman 2001) finding relatively strong gradients in most clusters, while other authors (e.g. White (2000)) find weak or absent gradients from the same ASCA and SAX data.

XMM and Chandra data have much better spatial resolution and signal-to-noise and have taken a large step in resolving the issue (Fig. 9.5). A sample of 14 clusters observed by XMM (Kaastra *et al.* 2001) indicates that many clusters show a relatively sharp abundance gradient in their central regions and shallow or no gradients in the outer regions, out to very large radii. The length-scale of this enhanced abundance is small ($r < 100$ kpc) but the enhancement can be large, reaching almost solar values for Fe. At large radii (Fig. 9.6), many but not all (Kaastra *et al.* 2001) clusters seem to be isochemical. DeGrandi & Molendi (2001) showed that the 'excess' metallicity seen in the central regions of the cooling-flow clusters is consistent, in most cases, with the metals produced by the central luminous cD galaxy. The different chemical composition of the clusters' central regions compared with the outer parts (Makishima *et al.* 2001) also indicates a different enrichment history for the gas in the central regions. However, the vast bulk of the elements lies in the outer parts of the cluster and integration of the XMM abundance profiles shows that the total cluster abundances are not strongly affected by the observed gradients. Thus, the ASCA cluster averages are meaningful. The almost-flat abundance profiles for rich clusters at large radii indicate that the gas is well mixed, which has strong implications for the origin of the gas.

Groups and elliptical galaxies

So far, only two groups with good XMM or Chandra data (NGC 4325 and NGC 2563) have been presented, and analysis of the ASCA data for a much larger sample is controversial. Outside the central regions, these two objects are both isochemical and isothermal. However, the outer radius is only *c.* $\frac{1}{4}$–$\frac{1}{3}$ of the virial radius.

The ISM of elliptical galaxies is hot ($T \sim 0.3 \times 10^7$–1×10^7 K) and their ISM is the repository of the history of stellar mass loss and supernova explosions. The gas should have supersolar abundances based on stellar spectra and the observed type-Ia supernova rate. At $kT \sim 10^7$ K the X-ray emission is dominated by the Fe L complex (Fe XVI–XXIV) and the He-like lines of Ne, Mg, Si and S and the H-like lines of O and N.

The early ASCA result (Awaki *et al.* 1994) indicated a very low abundance, inconsistent with that inferred for the stars. Detailed follow-on studies by Buote *et al.* (1999) and Matsushita *et al.* (1997) showed that there were other possible abundances allowed by the ASCA data and, in fact, that for plasmas with $kT < 1.5$ keV, which includes all groups and clusters, CCD data could be degenerate if the plasma was multi-phase. This situation has been rectified with the much higher resolution XMM RGS data (Xu *et al.* 2002; Peterson *et al.* 2002) which shows subsolar abundances for all the elements except N. The three systems analysed to date (NGC 4636 (Xu *et al.* 2002); M87 (Sakelliou *et al.* 2002); and N533 (Peterson *et al.* 2002)) have a small range in abundances with O $\sim$ 0.45, Fe $\sim$ 0.7, Mg $\sim$ 0.7 and N $\sim$ 1, with small statistical errors. The gas in NGC 4636 shows a very small range in temperature, while M87 shows a sharp temperature gradient in the central regions.

These abundances and abundance ratios conflict with the high α/Fe ratio indicated by stellar spectra (Trager *et al.* 2000) and a high Fe value as required by the observed type-I supernova rate. However, the observed abundance ratios in the gas are almost exactly what are seen in the MW stars at the same Fe abundance values!

Conclusions

The new XMM-Newton and Chandra data, combined with an analysis of the very large ASCA dataset, have led to fundamental new results on the chemical abundances in clusters, groups and elliptical galaxies. The statistical errors are small and the number of well-determined elements has increased substantially. We compare the cluster abundances with theoretical models, the stars in the M and the high-redshift Universe. The observed abundances do not resemble any of these, leading to rather unusual new requirements on supernova yields and rates. The new level of precision allowed by high-resolution data, as it is extended to clusters by Astro-E, will strongly test these conclusions. However, it is already clear that the RGS data for elliptical galaxies are a major challenge for theoretical modelling.

Many of these results have been obtained by D. Horner for his PhD thesis and by W. Baumgartner. Extensive discussions with M. Loewenstein through the years have always been enlightening. I thank J. Peterson and J. Kaastra for communicating their results before publication.

References

Abell, G. O. 1958 *Astrophys. J. Suppl.* **3**, 211A.
Allen, S. W. 2002 *Phil. Trans. R. Soc. Lond.* A **360**, 2005–2017.
Awaki, H. *et al.* 1994 *Publ. Astr. Soc. Jpn* **46L**, 65A.
Bahcall, N. 2000 *Phys. Rep.* **333**, 233.
Bahcall, N., Neta, A., Fan, X. & Cen, R. 1997 *Astrophys. J.* **485L**, 53B.
Balogh, M. L., Pearce, F. R., Bower, R. G. & Kay, S. T. 2001 *Mon. Not. R. Astr. Soc.* **326**, 1228B.
Bauer, F. E. *et al.* 2002 *Astr. J.* **123**, 1163.
Baumgartner, W. H., Horner, D. J. & Mushotzky, R. F. 2002 In *Matter and Energy in Clusters of Galaxies, Chung-Li, Taiwan, 23–27 April.* (Preprint astro-ph/0207179.)
Böhringer, H. *et al.* 2000 *Astrophys. J. Suppl.* **129**, 435B.
Borgani, S. & Guzzo, L. 2001 *Nature* **409**, 39.
Borgani, S. *et al.* 2001 *Astrophys. J.* **561**, 13B.
Bryan, G. L. 2000 *Astrophys. J.* **544L**, 1B.
Buote, D. A., Canizares, C. R. & Fabian, A. C. 1999 *Mon. Not. R. Astr. Soc.* **310**, 483.
Davé, R. *et al.* 2001 *Astrophys. J.* **552**, 47.
Davis, D. S., Mulchaey, J. S. & Mushotzky, R. F. 1998 *Astrophys. J.* **511**, 34.
DeGrandi, S. & Molendi, S. 2001 *Astrophys. J.* **551**, 153D.
Donahue, M. *et al.* 2002 *Astrophys. J.* **569**, 689.
Ezawa, H. *et al.* 1997 *Astrophys. J.* **490**, L33.
Fabian, A. C., Crawford, C. S., Edge, A. C. & Mushotzky, R. F. 1994 *Mon. Not. R. Astr. Soc.* **267**, 779F.
Finoguenov, A., David, L. P. & Ponman, T. J. 2000 *Astrophys. J.* **544**, 188F.
Finoguenov, A., Matsushita, K., Böhringer, H., Ikebe, Y. & Arnaud, M. 2002 *Astron. Astrophys.* **381**, 21F.
Fukazawa, Y. *et al.* 2000 *Mon. Not. R. Astr. Soc.* **313**, 21F.
Gibson, B. K., Loewenstein, M. & Mushotzky, R. F. 1997 *Mon. Not. R. Astr. Soc.* **290**, 623G.
Hashimoto, Y., Hasinger, G., Arnaud, M., Rosati, P. & Miyaji, T. 2002 *Astron. Astrophys.* **381**, 841.
Heckman, T. M. 2002 In *Extragalactic Gas at Low Redshift* (ed. J. Mulchaey & J. Stocke). ASP Conference Series, vol. 254. San Francisco, CA: Astronomical Society of the Pacific.

Henry, J. P. 2000 *Astrophys. J.* **534**, 565.
Henry, R. B. C. & Worthey, G. 1999 *Publ. Astr. Soc. Pac.* **111**, 919H.
Horner, D. 2001 Ph.D. thesis, University of Maryland, USA.
Horner, D. *et al.* 2002 *Astrophys. J. Suppl.* (Submitted.)
Ikebe, Y., Reiprich, T. H., Böhringer, H., Tanaka, Y. & Kitayama, T. 2002 *Astron. Astrophys.* **383**, 773I.
Irwin, J. A. & Bregman, J. N. 2001 *Astrophys. J.* **546**, 150I.
Kaastra, J. S., Ferrigno, C., Tamura, T., Paerels, F. B. S., Peterson, J. R. & Mittaz, J. P. D. 2001 *Astron. Astrophys.* **365L**, 99K.
Kahn, S., Behar, E., Kinkhabwala, A. & Savin, D. W. 2002 *Phil. Trans. R. Soc. Lond.* A **360**, 1923–1933.
Kobulnicky, H. A. & Skillman, E. D. 1998 *Astrophys. J.* **497**, 601K.
Loewenstein, M. & Mushotzky, R. F. 1996 *Astrophys. J.* **466**, 69.
Loewenstein, M. 2001 *Astrophys. J.* **557**, 573L.
Makishima, K. *et al.* 2001 *Publ. Astr. Soc. Jpn* **53**, 401M.
Martin, C. L., Kobulnicky, H. A. & Heckman, T. M. 2000 In *Proc. American Astronomy Society Conf. no. 197*, paper no. 7911M. San Francisco, CA:ASP.
Matsushita, K., Makishima, K., Etsuko, R., Noriko, Y. & Ohashi, T. 1997 *Astrophys. J.* **488L**, 125M.
McWilliam, A. 1998 *Astrophys. J.* **115**, 1640.
Mohr, J. J., Mathieson, B. & Errard, A. E. 1999 *Astrophys. J.* **517**, 62.
Moscardini, L., Matarrese, S. & Mo, H. J. 2001 *Mon. Not. R. Astr. Soc.* **327**, 422.
Mushotzky, R. *et al.* 1996 *Astrophys. J.* **466**, 686M.
Nakata, F. *et al.* 2001 *Publ. Astr. Soc. Jpn* **53**, 1139N.
Peimbert, M., Carigi, L. & Peimbert, A. 2001 *Astrophys. Space Sci.* **277**, 147P.
Peterson, J. R. *et al.* 2002 In *Proc. Conf. New Visions of the X-ray Universe in the XMM-Newton and Chandra Era, 26–30 November 2001, ESTEC, The Netherlands.* (Preprint astro-ph/0202108.)
Ponman, T. J., Cannon, D. B. & Navarro, J. F. 1999 *Nature* **397**, 135.
Prochaska, J. X. & Wolfe, A. M. 2002 *Astrophys. J.* **566**, 6.
Sakelliou, I. *et al.* 2002 *Astron. Astrophys.* **391**, 903.
Scharf, C. A. & Mushotzky, R. F. 1997 *Astrophys. J.* **485L**, 65.
Schuecker, P. *et al.* 2001 *Astron. Astrophys.* **368**, 86S.
Sembach, K. R., Howk, J. C., Ryans, R. S. I. & Keenan, F. 2000 *Astrophys. J.* **528**, 310.
Stanford, S. A. *et al.* 2001 *Astrophys. J.* **552**, 504.
Tamura, T., Bleeker, J. A. M., Kaastra, J. S., Ferrigno, C. & Molendi, S. 2001 *Astron. Astrophys.* **379**, 107.
Thielemann, F.-K. *et al.* 2002 In *Proc. 27th Int. Cosmic Ray Conf., Hamburg, 7–15 August 2001.* (Preprint astro-ph/0202453.)
Timmes, F. X., Woosley, S. E. & Weaver, T. A. 1995 *Astrophys. J. Suppl.* **98**, 617T.
Trager, S. C., Faber, S. M., Worthey, G. & González, J. J. 2000 *Astrophys. J.* **119**, 1645T.
Wheeler, J. C., Sneden, C. & Truran Jr, J. W., 1989 *A. Rev. Astron. Astrophys.* **27**, 279W.
White, D. A. 2000 *Mon. Not. R. Astr. Soc.* **312**, 663.
Woosley, S. E. & Weaver, T. A. 1995 *Astrophys. J. Suppl.* **101**, 181W.
Wyse, R. F. G. 1997 *Astrophys. J.* **490L**, 69W.
Xu, H. *et al.* 2002 *Astrophys. J.* **579**, 600.

10

X-rays from active galactic nuclei: relativistically broadened emission lines

BY A. C. FABIAN

University of Cambridge

Introduction

Active galactic nuclei (AGN) are powered by accretion onto a massive black hole. The accreting matter forms into a disc within which the magnetorotational instability causes the angular momentum to be transported outward and the matter to flow inward. Most of the accretion power is liberated as heat within the innermost few tens of gravitational radii. In luminous AGN the disc is optically thick and much of the heat appears as quasi-thermal ultraviolet (UV) and far-UV black-body radiation from the disc surface. About one-third of the heat is released as a power law of Comptonized hard X-rays from a magnetic corona above the disc. Half of this radiation is absorbed and backscattered, or 'reflected' from the dense disc to produce what is called the 'reflection component to the X-ray emission.

A key aspect of the reflection spectrum is that it may contain fluorescent and/or recombination lines. When hard X-rays are photoelectrically absorbed by the disc matter, the ions, especially iron, often emit line radiation. The observed profile of such an emission line is then shaped by the high orbital velocities of the disc and the deep potential well close to the black hole (i.e. the Doppler effect and gravitational redshift) to form a broadened, skewed line (Fabian *et al.* 1989; Laor 1991). The line profile is a diagnostic of the accretion flow, and the strong gravitational field, of the black hole.

The observations and properties of the reflection component and especially the broad iron line form the basis of this chapter. The first clear example was discovered with the ASCA (Tanaka *et al.* 1995; Fabian *et al.* 1995) in the Seyfert 1 galaxy MCG–6-30-15, and now others have been found. With excellent new data from Chandra and XMM-Newton, the situtaion is found to be complex, but strongly indicates that the emission originates from very close to the black hole.

Frontiers of X-Ray Astronomy, ed. A.C. Fabian, K.A. Pounds and R.D. Blandford. Published by Cambridge University Press.

Broad iron lines from the ASCA

Following on from the broad line in MCG–6-30-15, Nandra *et al.* (1997a; see also Yaqoob *et al.* 2002) showed that the average spectrum of a number of Seyfert 1 galaxies has a pronounced red wing. The reason that an average of many objects was required was to match the signal-to-noise ratio of the data from MCG–6-30-15 which has both a high equivalent width and X-ray flux, and exceptionally long exposures. Clear broad lines were also found in IRAS18325 (Iwasawa *et al.* 1996b) and NGC3516 (Nandra *et al.* 1999). Iron-K emission lines appear to be weak or non-existent in the X-ray spectra of quasars (Iwasawa & Taniguchi 1993; Nandra *et al.* 1997b). Strong broad lines are generally only found in objects with a 2–10 keV luminosity less than 10^{44} erg s^{-1}.

Of great interest has been the variability of the iron line. In simple models it should follow flux variations of the continuum, with a time lag which can potentially be used to determine the mass of the black hole (Fabian *et al.* 1989; Stella 1990; Reynolds *et al.* 1999). The expected time-scales of such reverberation lags are too short to have yet been detected. Variations in the line from several long observations of MCG–6-30-15 have been found by Iwasawa *et al.* (1996a,b, 1999) including evidence that the red wing is so extensive as to require that the black hole be spinning rapidly.

Curiously, the line variations seen during a long joint ASCA/RXTE observation of MCG–6-30-15 during 1997 show no correlation with the continuum flux (Vaughan & Edelson 2001; see also Lee *et al.* (2000) and Reynolds (2000)). This result extends to an even longer observation made in 1999 (Matsumoto *et al.* 2002; Shih *et al.* 2002). Here there is little evidence for the line varying at all. Why this should be has led to a number of explanations. One is that the disc is so highly ionized by intense flares as to give no line (Lee *et al.* 2000; Reynolds 2000). A second is that the corona is not static but undergoes mildly relativistic motions during flaring. Then if the emitting gas is moving towards us we see an increase in continuum, but the disc is less irradiated than usual (Matsumoto *et al.* 2003; Lu & Yu 2001). The problem with this explanation is that it would be expected that the continuum would become harder when approaching us (since the Comptonizing emission region would see fewer soft disc photons), whereas the opposite effect is observed. Obscuration of the emission region by a marginally optically thick corona is another possibility (the ‘thundercloud’ model of Merloni & Fabian (2001)).

A final simple empirical model is discussed by Shih *et al.* (2002). They suggest that the spectrum consists of a slowly varying, flat-spectrum continuum which produces the line, together with a steeper, rapidly varying continuum which produces no line. This explains the observed behaviour including that of the continuum, and is supported by the apparent saturation of the spectral index at high fluxes (Shih *et al.* 2002). The two components are tentatively identified with the main disc (the line-bearing component) and the magnetized plunge region from just within the innermost stable orbit of the disc (see, for example, Krolik (1999), Gammie (1999), Hawley & Krolik (2002), Armitage *et al.* (2001)).

Chandra

The high spatial resolution of Chandra means that bright point sources, such as the AGN discussed here, tend to suffer from pile up unless viewed through the gratings. The high-energy transmission gratings (HETG) enable any narrow components to iron-K lines to be detected and their properties measured. Such components have been reported from NGC 5548 (Yaqoob *et al.* 2001) and from NGC 3783 (Kaspi *et al.* 2001). The throughput of the HETG is relatively small and not ideal for finding any broad components. However, by binning up the data, the broad components found with the ASCA are being confirmed. For example, the HETG data on MCG–6-30-15, when heavily binned, closely resemble the ASCA result (Lee *et al.* 2002). They show that any narrow component to the iron line is weak; the observed narrow core probably being the blue wing of the broad line.

The HETG is able to cope with the high flux of galactic black-hole candidates (BHCs) and has shown a nice broad line in Cyg X-1 (Miller *et al.* 2002a) when in an 'intermediate high' state. Another BHC with a very broad line is XTE 1650-500 (Miller *et al.* 2002b), this time detected with XMM-Newton.

Before we concentrate on the much brighter BHCs and ignore the AGN, it is worth noting the relative count rate *per light-crossing time* for the two classes. The relevant light-crossing time here is of the event horizon. Although well-studied BHC are about 100 times brighter than well-studied AGN, the light-crossing time of BHC is about 10^5 or more times shorter than that for AGN. Therefore, we detect about 1000 times more counts per crossing time from AGN than from BHCs. This will be relevant if we want to use timing to study the immediate environment of a black hole, by reverberation, for example. Of course, from the point of view of the time-averaged profile there is little difference, except that the disc temperature of a BHC is more than 10 times higher than that of an AGN so, even before irradiation of the disc by the corona is considered, the disc is more highly ionized and thus likely to be more complex to understand. Nevertheless, the discovery that they have broad lines is very important and opens up another path for studying accretion flows onto black holes.

XMM-Newton

Complex iron lines have been reported from several Seyfert 1 galaxies and even some quasars by Reeves *et al.* (2001) and O'Brien *et al.* (2001) using data from the

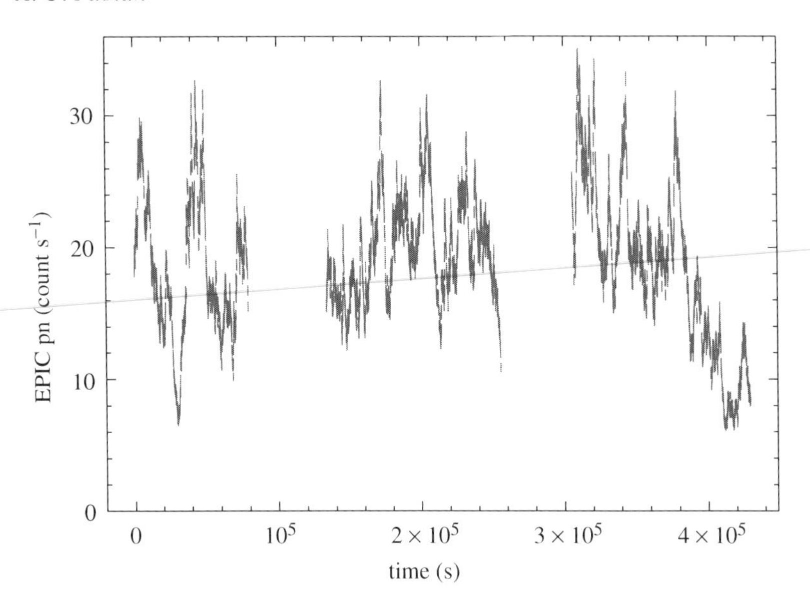

Figure 10.1. Light curve of the 325 ks XMM-Newton observation of MCG–6-30-15.

EPIC CCD cameras on XMM-Newton. Narrow components dominate in several examples, although broad, possibly highly ionized, iron lines are also seen. An exciting iron-K line result has been that of MCG–6-30-15 reported by Wilms *et al.* (2001). The source was caught in a relatively faint state, like the deep minimum of 1994 (Iwasawa *et al.* 1996a,b), showing a highly extended red wing. The emissivity profile required on the disc is so steep into such small radii that the authors claim that the spin energy of a rapidly spinning black hole is being tapped (Wilms *et al.* 2001). Martocchia *et al.* (2002) have since argued that the steep emissivity profile could be a result of 'returning radiation' (Cunningham 1975) in the extreme space-time geometry close to a spinning black hole.

We have also observed MCG–6-30-15 with XMM-Newton (Fabian *et al.* 2002), this time for 325 ks, and with simultaneous coverage by *Beppo*SAX. The source shows typical behaviour with much rapid variability (Fig. 10.1). The preliminary analysis has concentrated on the EPIC MOS data until a public pn response matrix is available and reveals a beautiful broad skewed line (Figs. 10.2 and 10.3). Spectral fits to the data indicate an inner radius of only $2G\dot{M}/c^2$ and an emissivity index of about 3. The precise values of these quantities depends on exactly where the continuum lies. We hope to be doing a good job since we are jointly fitting the MOS and *Beppo*SAX data, which extend to 200 keV, and it seems that the warm absorber has little effect above *c*. 2.5 keV. However, when the line is so broad we do have to rely on an intrinsic power law shape to the continuum, which is difficult to test.

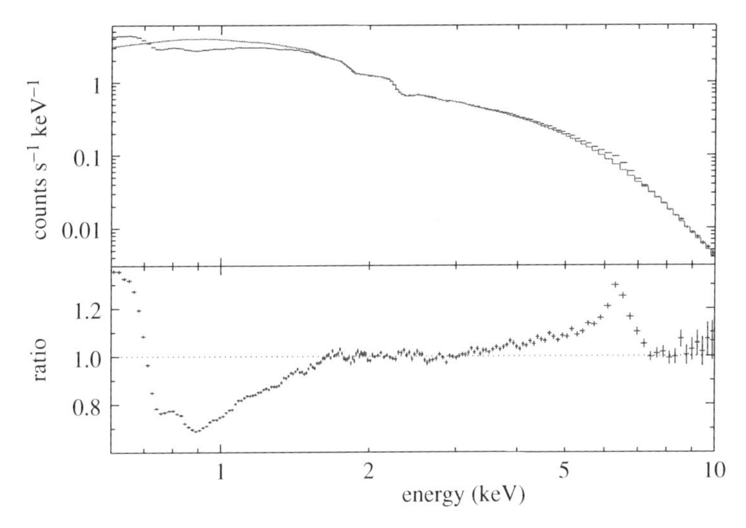

Figure 10.2. EPIC/MOS spectrum of MCG–6-30-15 (Fabian *et al.* 2002). The lower panel shows the ratio of the data and a power-law model.

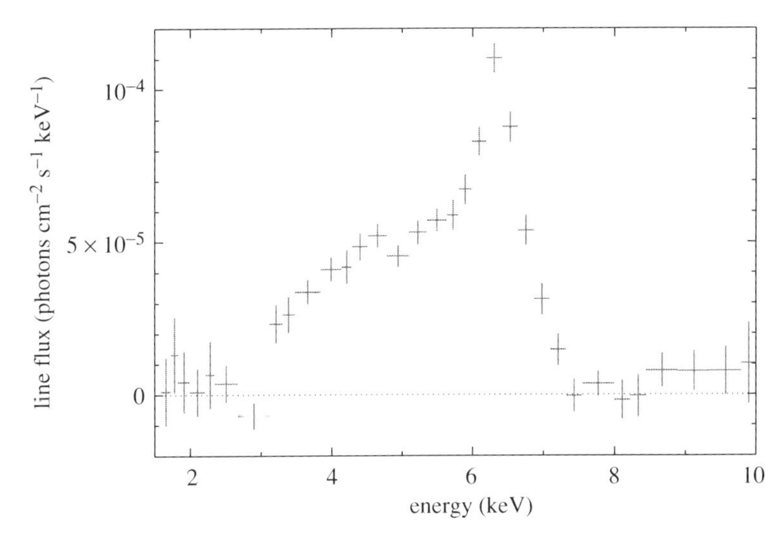

Figure 10.3. XMM-Newton broad iron line profile of MCG–6-30-15 (from Fabian *et al.* (2002)).

Interestingly, we confirm a lack of variability of the iron line. If we use the spectrum of a faint part of the light curve as background for a bright part, then the difference can be well fitted by a power law only, i.e. the broad iron line has a similar flux when the source is faint as when it is bright. Further work is being carried out to characterize any other variability, or lack of variability.

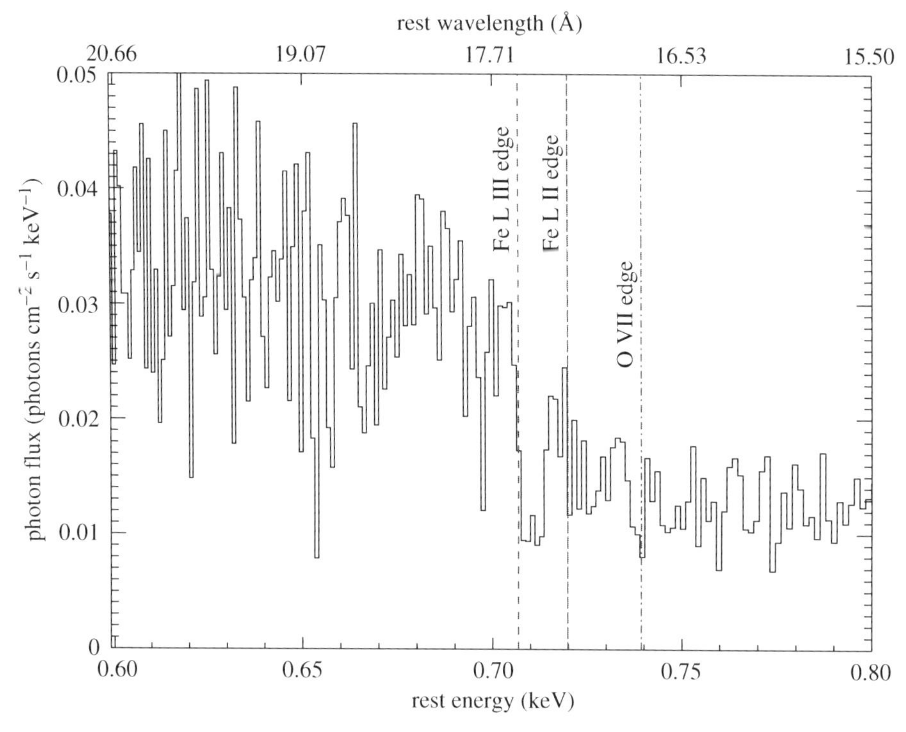

Figure 10.4. Chandra HETG spectrum of MCG–6-30-15 from Lee *et al.* (2001). The steep drop at 707 eV is interpreted by Branduardi-Raymont *et al.* (2001) as the blue wing of a broad OVIII line. Lee *et al.* argue that it is due to a neutral iron LIII edge. Note its sharpness.

Broad oxygen lines?

Branduardi-Raymont *et al.* (2001) argue, on the basis of XMM-Newton Reflection Grating Spectrometer (RGS) data of MCG–6-30-15 and Mrk766, that the abrupt spectral jump spectrum observed at 707 eV is due to the sharp blue wing of an OVIII emission line, rather than a warm absorber such as was assumed in the past (e.g. Fabian *et al.* 1994; Otani *et al.* 1996). A key feature of their argument is that the jump is redshifted (by 16 000 km s^{-1}) relative to the expected OVII absorption edge. Lee *et al.* (2001) counter using Chandra HETG spectra of MCG–6-30-15 (Fig. 10.4) that the apparent shift is due a neutral iron LIII absorption edge in a dusty warm absorber in the source. Sako *et al.* (2001) have revised the RGS interpretation to now include a weak warm absorber but appear to rule out the presence of any dust (despite optical and UV reddening being clearly seen in the object (Reynolds *et al.* 1997)).

It would of course be very exciting and important if clear relativistically broadened emission lines from O and other species were seen in AGN spectra. Weak lines are expected (see, for example, Ross & Fabian (1993), Nayakshin *et al.* (2000), Ballantyne *et al.* (2001), Różańska *et al.* (2002)) but not (Ballantyne *et al.* 2002) the

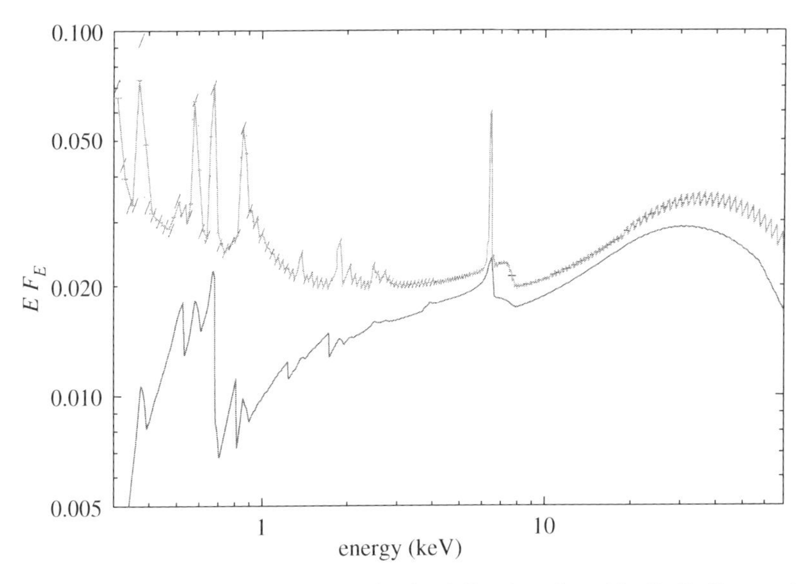

Figure 10.5. Example model of an ionized disc (produced by D. Ballantyne) with a warm absorber which fits the XMM-Newton data shown in Fig. 10.2 (lower line). The upper jagged line shows the intrinsic (unblurred and unabsorbed) spectrum of the ionized disc. Note the iron-K line at 6.5 keV, iron-L at 0.8 keV, OVIII at 0.65 keV and OVII at 0.57 keV. The jagged shape of this line is a binning effect.

strong 150 eV equivalent width line claimed by Branduardi-Raymont *et al.* (2001). A major problem for the emission-line interpretation of the 700 eV drop by the RGS team (Branduardi-Raymont *et al.* 2001; Sako *et al.* 2003) is its sharpness. It appears from the RGS spectra to be less than 10 eV wide and in the HETG spectra less than 3 eV. Now the bulk of the O lines produced by highly ionized, irradiated, discs suffer Compton scattering, since they are produced over a significant Thomson depth. Therefore, they are broadened considerably more than 10 eV (Ballantyne *et al.* 2002). Doppler and gravitational redshifts only broaden this further. Consequently, it is difficult to see how the sharp drop at 700 eV can be produced by the blue wing of a relativistically broadened line produced by reflection in a disc. It is much more plausible that it is an absorption edge.

It is of course likely that there are weak disc lines due to low-Z elements in the spectra of Seyfert 1 galaxies (Fig. 10.5), but disentangling them from each other in order to fully model them may prove difficult.

Discussion

The powerful combination of Chandra and XMM-Newton is demonstrating that iron emission lines are common in Seyfert galaxies and is enabling us to separate out narrow from broad components and lowly from highly ionized components.

A small but growing population has clear broad lines which are well explained as due to relativistic discs. Many more broad lines may have yet to be recognized due to a combination of low equivalent width, extreme broadness and high inclination. Complex or partial absorption may also mask or mimic broad lines.

The lack of broad lines in high luminosity objects needs to be understood. It could be due to a low disc density which leads to the gas being highly photoionized. A clear transition class has not, however, been recognized.

MCG -6-30-15 is continuing to display an excellent example of a broad line. The XMM-Newton data indicate that the disc in this object extends well within $6GM/c^2$ and probably that the black hole is rapidly spinning. Much work is now needed to understand exactly where the line is produced, how important the emissivity profile might be and whether returning radiation is crucial for its understanding. A key issue will surely be accounting for the lack of rapid variability shown by the iron line.

I thank Simon Vaughan, Kazushi Iwasawa, David Ballantyne, Paul Nandra, Randy Ross, Julia Lee and my other collaborators for help and discussions, and The Royal Society for support.

References

Armitage, P. J., Reynolds, C. S. & Chiang, J. 2001 *Astrophys. J.* **548**, 868–875.
Ballantyne, D. R., Ross, R. R. & Fabian, A. C. 2001 *Mon. Not. R. Astr. Soc.* **327**, 10–22.
Ballantyne, D. R., Ross, R. R. & Fabian, A. C. 2002 *Mon. Not. R. Astr. Soc.* **336**, 867–872.
Branduardi-Raymont, G., Sako, M., Kahn, S. M., Brinkman, A. C., Kaastra, J. S. & Page, M. J. 2001 *Astron. Astrophys.* **365**, L140–145.
Cunningham, C. T. 1975 *Astrophys. J.* **202**, 788–802.
Fabian, A. C., Rees, M. J., Stella, L. & White, N. E. 1989 *Mon. Not. R. Astr. Soc.* **238**, 729–736.
Fabian, A. C. *et al.* 1994 *Publ. Astr. Soc. Jpn* **46**, L59–L63.
Fabian, A. C., Nandra, K., Reynolds, C. S., Brandt, W. N., Otani, C., Tanaka, Y. & Inoue, H. 1995 *Mon. Not. R. Astr. Soc.* **277**, L11–L15.
Fabian, A. C., Iwasawa, K., Reynolds, C. S. & Young, A. J. 2000 *Publ. Astr. Soc. Pac.* **112**, 1145–1161.
Fabian, A. C. *et al.* 2002 *Mon. Not. R. Astr. Soc.* **335**, L1.
Gammie, C. F. 1999 *Astrophys. J.* **522**, L57–L60.
Hawley, J. F. & Krolik, J. H. 2002 *Astrophys. J.* **566**, 164–180.
Iwasawa, K. & Taniguchi, Y. 1993 *Astrophys. J.* **413**, L15–L18.
Iwasawa, K. *et al.* 1996a *Mon. Not. R. Astr. Soc.* **282**, 1038–1048.
Iwasawa, K., Fabian, A. C., Mushotzky, R. F., Brandt, W. N., Awaki, H. & Kunieda, H. 1996b *Mon. Not. R. Astr. Soc.* **279**, 837–846.
Iwasawa, K., Fabian, A. C., Young, A. J., Inoue, H. & Matsumoto, C. 1999 *Mon. Not. R. Astr. Soc.* **306**, L19–L24.
Kaspi, S. *et al.* 2001 *Astrophys. J.* **554**, 216–232.
Krolik, J. H. 1999 *Astrophys. J.* **515**, L73–L76.
Laor, A. 1991 *Astrophys. J.* **376**, 90–94.

Lee, J. C., Fabian, A. C., Reynolds, C. S., Brandt, W. N. & Iwasawa, K. 2000 *Mon. Not. R. Astr. Soc.* **318**, 857–874.
Lee, J. C. *et al.* 2001 *Astrophys. J.* **554**, L13–L17.
Lee, J. C., Iwasawa, K., Houck, J. C., Fabian, A. C., Marshall, H. L. & Canizares, C. R. 2002 *Astrophys. J.* **570**, L47–L50.
Lu, Y. & Yu, Q. 2001 *Astrophys. J.* **561**, 660–675.
Lubinski, P. & Zdziarski, A. A. 2001 *Mon. Not. R. Astr. Soc.* **323**, L37–L42.
Martocchia, A., Matt, G. & Karas, V. 2002 *Astron. Astrophys.* **383**, L23–L26.
Matsumoto, C., Inoue, H., Iwasawa, K. & Fabian, A. C. 2003 *Publ. Astr. Soc. Jpn.* **55**, 615–623.
Merloni, A. & Fabian, A. C. 2001 *Mon. Not. R. Astr. Soc.* **328**, 958–968.
Miller, J. M. *et al.* 2002a *Astrophys. J.* **578**, 348.
Miller, J. M. *et al.* 2002b *Astrophys. J.* **570**, L69–L72.
Nandra, K., George, I. M., Mushotzky, R. F., Turner, T. J. & Yaqoob, T. 1997a *Astrophys. J.* **477**, 602–622.
Nandra, K., George, I. M., Mushotzky, R. F., Turner, T. J. & Yaqoob, T. 1997b *Astrophys. J.* **488**, L91–94.
Nandra, K., George, I. M., Mushotzky, R. F., Turner, T. J. & Yaqoob, T. 1999 *Astrophys. J.* **523**, L17–L20.
Nayakshin, S., Kazanas, D. & Kallman, T. R. 2000 *Astrophys. J.* **537**, 833–852.
O'Brien, P. T., Page, K., Reeves, J. N., Pounds, K., Tunrer, M. J. L. & Puchnarewicz, E. M. 2001 *Mon. Not. R. Astr. Soc.* **327**, L37–L41.
Otani, C. *et al.* 1996 *Publ. Astr. Soc. Jpn* **48**, 211–218.
Reeves, J. N. *et al.* 2001 *Astron. Astrophys.* **365**, L134–L139.
Reynolds, C. S. 2000 *Astrophys. J.* **533**, 811–820.
Reynolds, C. S., Ward, M. J., Fabian, A. C. & Celotti, A. 1997 *Mon. Not. R. Astr. Soc.* **291**, 403–415.
Reynolds, C. S., Young, A. J., Begelman, M. C. & Fabian, A. C. 1999 *Astrophys. J.* **514**, 164–179.
Ross, R. R. & Fabian, A. C. 1993 *Mon. Not. R. Astr. Soc.* **261**, 74–82.
Rózańska, A., Dumont, A.-M., Czerny, B. & Collin, S. 2002 *Mon. Not. R. Astr. Soc.* **332**, 799.
Sako, M. *et al.* 2003 *Astrophys. J.* **596**, 114–128.
Shih, D., Iwasawa, K. & Fabian, A. C. 2002 *Mon. Not. R. Astr. Soc.* **333**, 687–696.
Stella, L. 1990 *Nature* **344**, 747–749.
Tanaka, Y. *et al.* 1995 *Nature* **375**, 659–661.
Vaughan, S. & Edelson, R. 2001 *Astrophys. J.* **548**, 694–702.
Wilms, J. *et al.* 2001 *Mon. Not. R. Astr. Soc.* **328**, L27–L31.
Yaqoob, T., George, I. M., Nandra, K., Turner, T. J., Serlemitsos, P. J. & Mushotzky, R. F. 2001 *Astrophys. J.* **546**, 759–768.
Yaqoob, T., Padmanabhan, U., Dotani, T. & Nandra, K. 2002 *Astrophys. J.* **569**, 487–492.

11

Obscured active galactic nuclei: the hidden side of the X-ray Universe

BY GIORGIO MATT

University 'Roma Tre'

Introduction

Both direct (surveys) and indirect (synthesis models of the cosmic-X-ray background (CXRB)) methods clearly indicate that most active galactic nuclei (AGN) are 'obscured' in X-rays, i.e. that their nucleus is hidden behind an absorbing material which prevents the nuclear emission from being directly observed up to the energy (if any, see below) at which the material becomes transparent (this energy depending, of course, on the column density of the absorber; see Figs. 11.1 and 11.2). Indeed, the three closest AGN (Circinus Galaxy, NCG 4945 and Centaurus A) are all heavily obscured. The absorbing matter is often very thick, its column density exceeding the value, $\sigma_T^{-1} = 1.5 \times 10^{24}$ cm^{-2}, for which the Compton-scattering optical depth equals unity (in these cases, the sources are called 'Compton-thick'; if the column density is smaller than σ_T^{-1} but still in excess of the galactic one, the source is called 'Compton-thin'). Compton-thick sources provide the most favourable case for studying the circumnuclear matter, because the emission from this matter (arising from reflection and reprocessing of the nuclear radiation) is not significantly diluted by the primary emission not only in soft X-rays but up to at least 10 keV (see, for example, Matt *et al.* (1999a)).

The unification model for Seyfert galaxies (see Antonucci (1993) for a review) assumes that Seyfert 1s (in which both broad and narrow lines are visible in the optical spectrum) and Seyfert 2s (where only narrow lines are observed) are intrinsically the same. The nucleus (i.e. the black hole with the accretion disc and the broad line regions) is surrounded by optically and geometrically thick matter, probably axisymmetric (usually called the 'torus'; see Fig. 11.3). If the line-of-sight does not intercept the torus, the nucleus is visible and the source is classified as type 1. Otherwise, only the narrow line regions are visible and the source is classified as

Frontiers of X-Ray Astronomy, ed. A.C. Fabian, K.A. Pounds and R.D. Blandford. Published by Cambridge University Press.

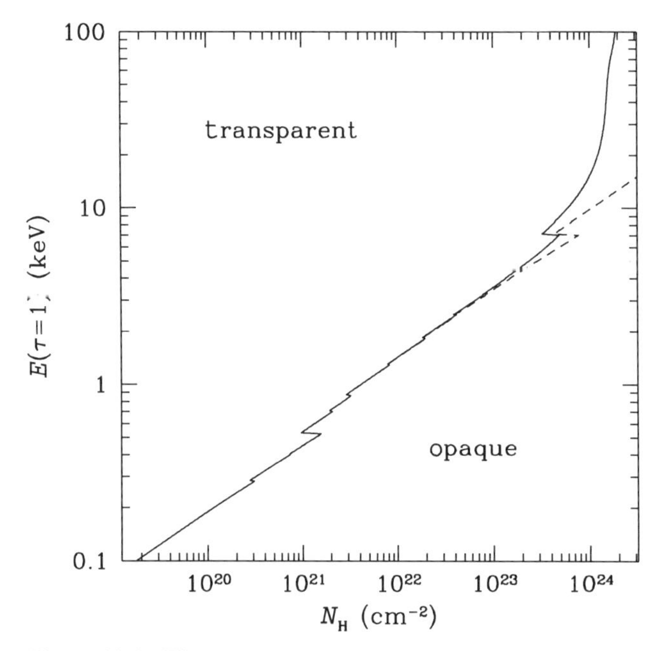

Figure 11.1. The energy corresponding to $\tau = 1$ as a function of the column density of the absorbing matter. The dashed line refers to photoabsorption only, the solid line includes Compton scattering.

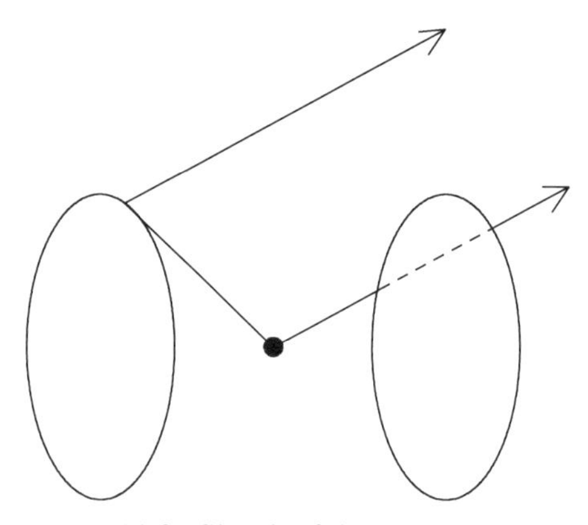

Figure 11.2. Sketch of the geometry adopted for the simulations described in the text.

type 2. (See also Elvis (2000) for a different version of the unification model.) In this scenario, a one-to-one relation between optical type 1 and X-ray unobscured sources, and between type 2 and X-ray obscured sources, is expected. I have used the term 'obscured' instead of the most common term 'type 2' intentionally. The reason is that there is increasing evidence that optical classifications (from which the terms type 1 and 2 derive) and X-ray classifications sometimes disagree with what expected from the Unification Model; in fact a number of sources clearly

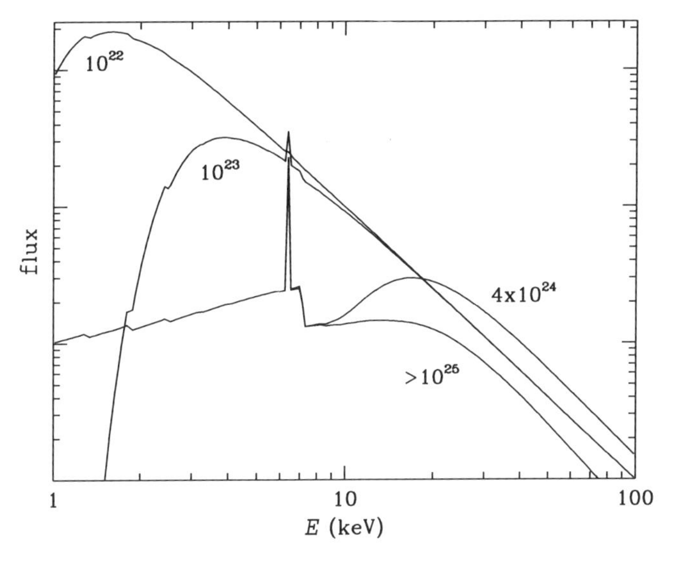

Figure 11.3. The X–ray spectrum of an obscured AGN is shown for different column densities of the absorber. The absorbing matter is assumed to form a geometrically thick torus (see previous figure), according to the unification model. In the Compton–thin case, the nuclear spectrum (assumed for simplicity to be a simple power law with photon spectral index 2) can be directly observed above a few keV, and an iron line with EW$\sim$10 (for $N_{\rm H} = 10^{22}$ cm^{-2}) or 100 eV (for $N_{\rm H} = 10^{23}$ cm^{-2}) is produced. In the moderately Compton-thick case ($N_{\rm H} = 4 \times 10^{22}$ cm^{-2}) the spectrum below 10 keV is dominated by the reflection component (see text) produced by the visible part of the inner wall of the torus itself, while the nuclear radiation can still be partly visible at higher energies. For column densities exceeding $N_{\rm H} = 10^{25}$ cm^{-2}, no nuclear radiation is transmitted, and only the reflection component is visible. Other components, arising from reflection off ionized circumnuclear matter, can also be present, as discussed in the text.

obscured in X–rays are either of type 1, on one extreme, or apparently inactive, on the other extreme, when observed in the optical. I will discuss this point in greater detailed later. To avoid confusion, I will here confine the terms type 1 and type 2 to their original meaning (based on the presence or not of broad permitted lines in the optical spectrum). In any case, I will take for granted, in agreement with the unification model, that the basic X-ray properties of unobscured and obscured AGN are the same, the latter being simply seen through a screen of cold matter.

The properties of circumnuclear matter

Physical properties of reflecting matter

The advent of high-spectral-resolution and high-spatial-resolution X-ray instruments on board Chandra and XMM-Newton is now allowing the study of the circumnuclear matter in obscured sources in much greater detail than before (and

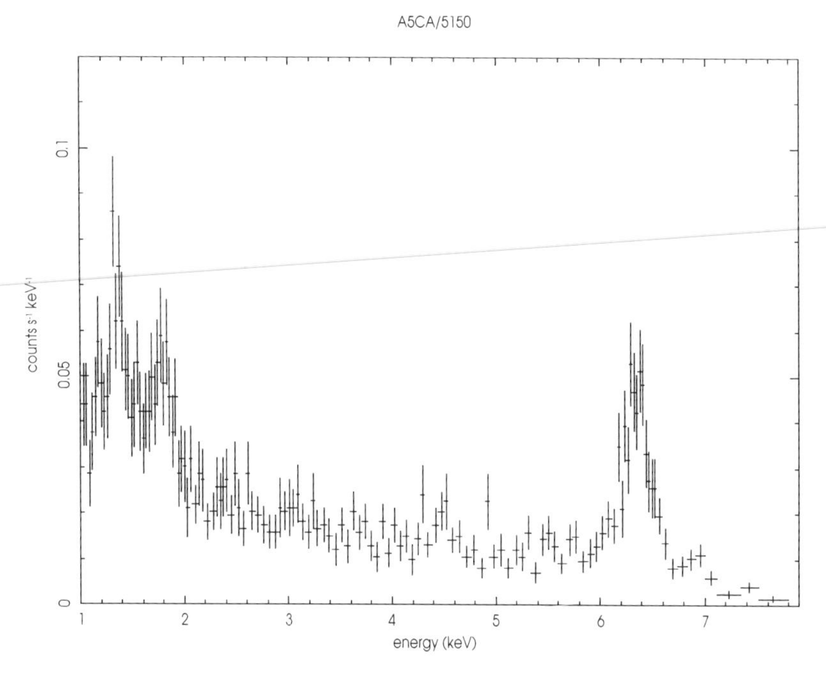

Figure 11.4. CCD (ASCA/SIS) spectrum of the Circinus Galaxy. More details in Matt *et al.* (1996a).

it is now becoming feasible to study this matter also in unobscured sources, despite the heavy dilution by the nuclear radiation). The results obtained by these instruments are improving and refining, but not revolutionizing, the previous scenario (to the relief and pleasure of those people who in the past struggled to find a coherent picture out of much poorer data). In Figs. 11.4 and 11.5 the differences in the line spectrum when observed with a charge-coupled device (CCD) and with a grating instrument are shown in the case of the Circinus Galaxy.

The spectrum of the reflected component depends on the ionization state of the matter. If the matter is highly ionized, Compton scattering is the most important process and the spectrum (at least up to a few tens of keV, where Compton recoil becomes important) is very similar to the primary one. If instead the matter is neutral (and optically thick), the resulting spectrum is the so-called 'Compton reflection' with a broad bump between 10 and 100 keV (e.g. George & Fabian (1991), Matt *et al.* (1991)). (If the matter is optically thin, the reflection spectrum is significantly different, see Fig. 11.6.) In both cases strong emission lines are also expected: iron lines from He- and H-like ions may be present in the former case; a 6.4 keV iron line with an equivalent width of *c.* 1–2 keV is present in the latter case (e.g. Matt *et al.* (1996b), see also Fig. 11.7). More complex continuum and line spectra are expected for mildly ionized material.

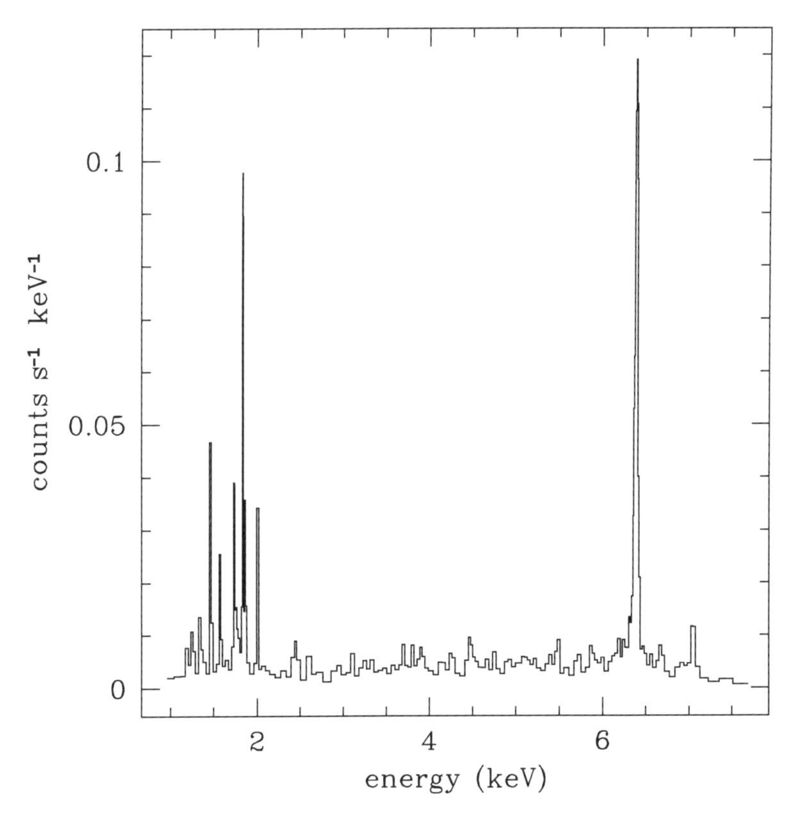

Figure 11.5. Grating (Chandra/HEG) spectra of the Circinus Galaxy. More details in Sambruna *et al.* (2001b).

The study of the circumnuclear matter is easier in Compton-thick sources, just because the nucleus is completely obscured up to at least 10 keV, i.e. in the band where imaging and high-resolution spectroscopic instruments work. Emission from off-nuclear regions is the only visible in these sources, and may be studied in great detail. The number and complexity of these regions differ greatly from source to source. In the Circinus Galaxy, the results obtained by Bianchi *et al.* (2001) from ASCA and *Beppo*SAX data have been basically confirmed by Chandra (Sambruna *et al.* 2001a,b; see also Fig. 11.5): the spectrum down to *c.* 2 keV is dominated by an optically thick reflecting region of low-ionization matter (unresolved even with Chandra), most likely the inner surface of the $N_{\rm H} \sim 4 \times 10^{24}$ cm^{-2} absorbing matter (Matt *et al.* 1999b) assuming, as customary, that the latter is (in the first approximation at least) axisymmetric. A second, ionized component extending over *c.* 50 pc becomes important below 2 keV, where it provides about half of the flux (Sambruna *et al.* 2001a,b).

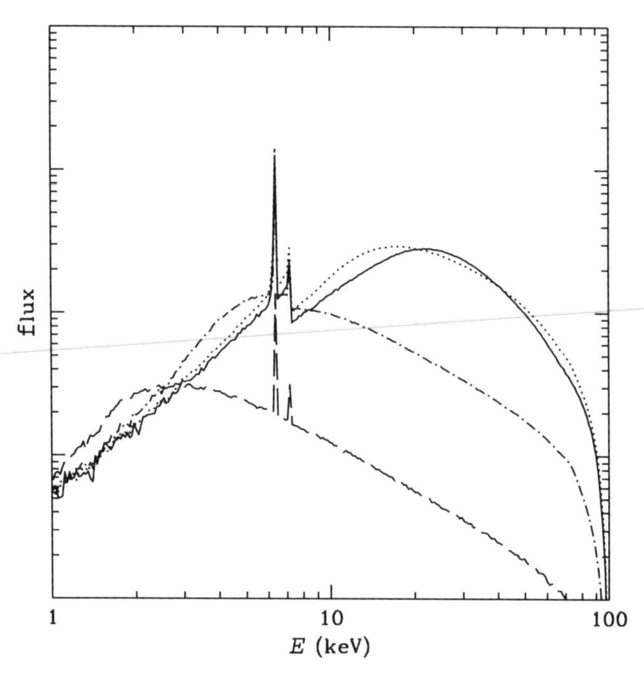

Figure 11.6. The reflection spectrum from the torus (seen face-on; see Fig. 11.2) for different column densities: 2×10^{22} cm^{-2} (dashed line), 2×10^{23} cm^{-2} (dotted–dashed line), 2×10^{24} cm^{-2} (dotted line), 2×10^{25} cm^{-2} (solid line). The illuminating spectrum is a power law with photon index 2 and exponential cut-off at 100 keV.

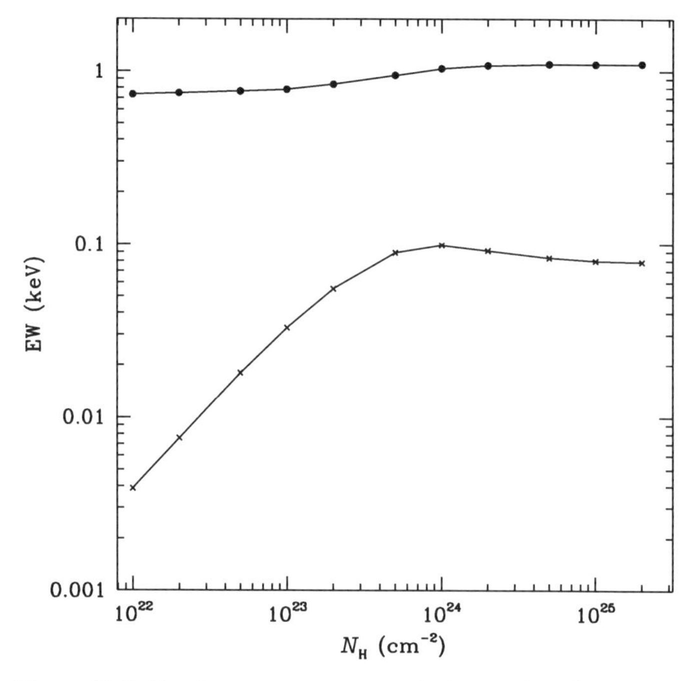

Figure 11.7. For the same geometry as in the previous figure, the equivalent width (EW) of the iron line against the pure reflection component (upper data) and against the total continuum (lower data).

The continuum and line spectra are instead much more complex in NGC 1068. Even if significant extended emission has been observed by Chandra, most of the emission is unresolved (Young *et al.* 2001). ASCA and *Beppo*SAX data already indicated the presence of more than one reflecting region. Bianchi *et al.* (2001; see also references therein) have indeed shown that at least three different regions are needed to explain the line spectrum: the first is optically thin and moderately ionized, and is responsible for the Ka lines of elements like Mg, Si and S; the second one is optically thin but highly ionized, and responsible for the He- and H-like Fe Kα lines; the third one is optically thick and of low ionization, and is responsible for the 6.4 keV Fe and the OVII Kα lines. The last region is again likely to be the inner surface of the absorber, which in this source has a column density of at least 10^{25} cm^{-2} (Matt *et al.* 1997). XMM-Newton grating spectra (see e.g. Behar *et al.* (2002), Kinkhabwala *et al.* (2002)) have definitely proved that the emitting gas responsible for the line spectrum is in photoionization equilibrium, the emission lines being due to both recombination and resonant scattering (as predicted by, for example, Band *et al.* (1990), Krolik & Kriss (1995) and Matt *et al.* (1996b)), and suggest that the scenario is even more complex than that deduced from low-resolution spectra. Photoionization equilibrium also seems able to explain the Chandra–HETG line spectrum of Mrk 3 (Sako *et al.* 2000). In the latter source the softest part (i.e. below *c.* 3 keV) of the spectrum is spatially extended along the [O III] ionization cone, while the high-energy spectrum is unresolved and consistent with reflection by cold and optically thick material, once more to be associated with the $N_{\rm H} \sim 10^{24}$ cm^{-2} absorber (Cappi *et al.* 1999).

Combining these results with those discussed by Matt *et al.* (2000), who studied a sample of bright Compton-thick sources observed by *Beppo*SAX, it is possible to conclude that reprocessing from optically thick, almost neutral matter is quite common. As said above, it seems natural to identify this matter with the inner surface of the absorber, as all these sources are Compton-thick. (We will see later that this matter is likely to also be present in most unobscured sources, according to the unification model, and even in Compton-thin obscured sources, which is less obvious.) In order not to exceed the dynamical mass, at least in Circinus and NGC 1068, the matter must be fairly close to the black hole, within a few tens of parsecs at most (Risaliti *et al.* 1999). (A completely independent estimate of the inner radius has been obtained by Bianchi *et al.* (2001) for these two sources, by modelling the X-ray line spectra. They found minimum distances of the torus of *c.* 0.2 and 4 pc, respectively.) Indeed, in the Circinus Galaxy the nuclear cold reflector is unresolved, which implies an upper limit to its size of *c.* 15 pc (Sambruna *et al.* 2001a).

Compton-thin or Compton-thick?

Unless the nucleus is directly observed above *c.* 10 keV (a task which is now possible with *Beppo*SAX and RXTE, but only for column densities not exceeding *c.* 10^{25} cm^{-2} (see Matt *et al.* 2000)), the signature of Compton-thick absorption is a reflection-dominated X-ray spectrum. Sometimes, however, a pure reflection spectrum may lead to an incorrect classification. The classical case is NGC 4051, which was observed by *Beppo*SAX in a low-flux state, in which the nucleus was switched off and only the reflection component was visible (Guainazzi *et al.* 1998). The source was a well-known Seyfert 1, otherwise it would have been classified as a Compton-thick absorbed AGN. A similar, even if somewhat less dramatic, change occurred in the Compton-thin ($N_H \sim 10^{22}$ cm^{-2}) source NGC 2992 (Gilli *et al.* 2000) which, during a *Beppo*SAX observation, was almost switched off, with the reflection component thus becoming very prominent. Subsequent *Beppo*SAX observations of NGC 4051 and NGC 2992 found that both sources had recovered their normal, bright state. In both cases it seems pretty obvious that what changed was the nuclear flux, rather than the properties of the absorption. It is also clear that the reflecting matter must be located at large distances from the black hole to echo the already switched-off primary emission; in fact, reflection from the accretion disc would disappear almost immediately. It is worth noting that, in the case of NGC 2992, the absorbing matter (which is Compton-thin) must be different from the reflecting matter (which is Compton-thick). Because the galaxy is edge-on, and given the rather small column density, it is possible that the thin absorber is the disc of the host galaxy.

Two more sources have undergone a similar transition. NGC 6300 and UGC 4203 were both Compton-thick when observed by RXTE (Leighly *et al.* 1999) and the ASCA (Awaki *et al.* 2000), respectively, but became Compton-thin when observed later by *Beppo*SAX (Guainazzi 2002) and XMM-Newton (Guainazzi *et al.* 2002). The ASCA and XMM-Newton spectra of UGC 4203 are presented in Fig. 11.8. While a change in the properties of the absorber cannot be completely ruled out, the explanation in terms of a 'switching-off' of the sources during their first observation seems the most natural. The 'thin' absorbers in these two sources have column densities of a few times 10^{23} cm^{-2}, and the host galaxies are seen at low-inclination angles (UGC 4203 is basically face-on), so the galactic disc cannot be the cause for the Compton-thin absorption. This therefore strongly suggests that in these two sources both Compton-thin (the absorber) and Compton-thick (the reflecting) cold, circumnuclear regions are present (even if for UGC 4203, for which the reflection component has been measured by ASCA only up to about 10 keV, it is still possible that the reflector and the absorber coincide, see Fig. 11.6).

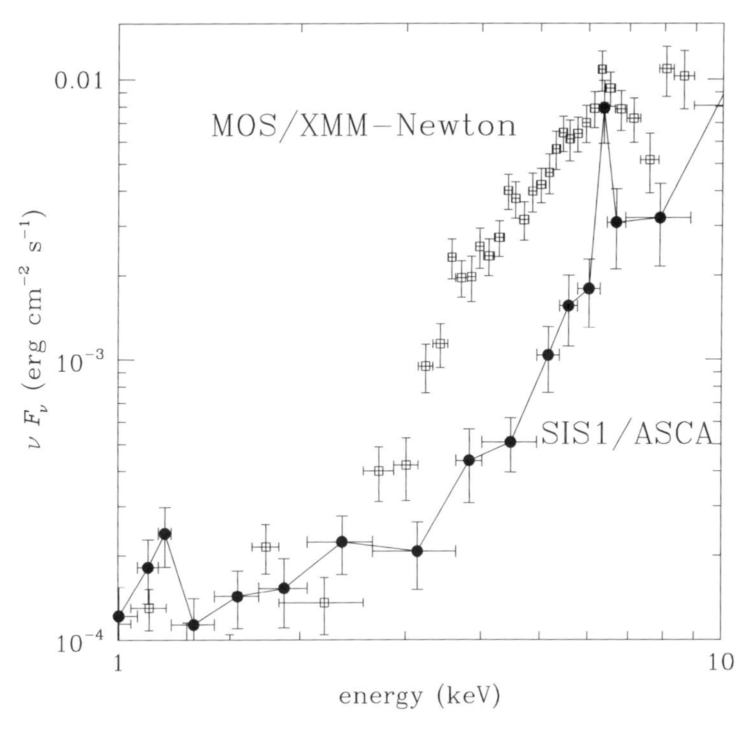

Figure 11.8. The ASCA and XMM-Newton spectra of UGC4203. From Guainazzi *et al.* (2002).

Two different regions?

Because X-ray absorption is very common in AGN, the covering factor of the absorbing matter must be large. Heavy absorption is also very common: about half of the optically selected Seyfert 2s in the local Universe are Compton-thick (see e.g. Maiolino *et al.* (1998)). Even allowing for a number of misclassifications due to the switching-off of the nucleus rather than a true Compton-thick absorption, as discussed in the previous subsection, it is clear that optically thick circumnuclear matter is common.

Risaliti *et al.* (1999) have shown that there is a relation between optical classification and column density, at least for optically selected AGN: intermediate (1.8–1.9) Seyferts are usually Compton-thin, while classical Seyfert 2s are Compton-thick. From the results discussed in the previous subsection, it seems that the two components may be present in the same source (another case of a Compton-thin absorber with Compton-thick reflection is NGC 5506 (Matt *et al.* 2001)).

All these pieces of evidence suggest that the Compton-thin and Compton-thick materials have different origins and locations, as originally proposed by Maiolino & Rieke (1995). Matt (2000) has suggested that the Compton-thin matter should be associated with the dust lanes at distances of hundred of parsecs which Malkan *et al.* (1998) found to be common in Seyfert galaxies, while the Compton-thick matter should be much closer to the nucleus, and associated with the 'torus' envisaged in the unification model (Antonucci 1993). A similar scenario has been proposed by Weaver (2002), who identifies the Compton-thin absorbers with starburst region clouds. In any case it is clear that the unification model should be somewhat revised to accommodate this further component. Other problems with the unification model have emerged from recent X-ray surveys and will be discussed in the next section.

Obscured AGN and the unification model

A first (actually not very serious) problem for the unification model has been already mentioned: the coexistence of two kinds of cold, circumnuclear matter, which suggests that the torus (i.e. the Compton-thick matter) is not the only possible absorbing/reflecting region. The inclusion of dust lanes and/or starburst regions in the basic ingredients should be all that is needed to explain the observations.

There are, however, more serious problems. First of all, type-1 sources may be obscured in X-rays. Maiolino *et al.* (2001a) found a number of broad-line quasi-stellar objects (QSOs) with significant X-ray absorption. Several sources in, for example, the *Beppo*SAX High Energy Large Area Survey (HELLAS) (Fiore *et al.* 2001), the ASCA HSS (Della Ceca *et al.* 2001) and the XMM-Newton HELLAS2XMM (Fiore *et al.* 2002) surveys appear to be absorbed in X-rays, but optically identified with type-1 AGN (see Fig.11.9). It has been shown (e.g. Gallagher *et al.* (2002)) that the X-ray weakness of BAL QSOs discovered by Röntgensatellit (ROSAT) is due to excess absorption (but it is still not clear whether the X-ray absorption is associated with the UV one, and therefore significantly ionized). For these absorbed blue QSOs, and in general for all the AGN for which optical extinction is lower than expected from X-ray absorption (there are several cases among Seyfert 2s), a possible explanation is that the dust/gas ratio of the absorber is much lower than in the galactic interstellar medium (ISM), probably due to dust sublimation. Another possibility is that the optical and X-ray absorbers cover different regions: the very nucleus for the X-ray absorber, a larger region for the optical one. Maiolino *et al.* (2001b) suggested an explanation in terms of different dust grain size with respect to the ISM. Whatever the solution, it is clear that optical and X-ray observations may sometimes lead to different classifications.

There are also X-ray loud sources (sometimes, but not always, with hard spectra, thus suggesting obscuration) that do not appear as AGN in the optical (or even do not have an optical counterpart, implying a very large X-to-optical ratio). Such a population is starting to emerge at fluxes of *c.* 10^{-14} erg cm^{-2} s^{-1} (see e.g. Mushotzky *et al.* (2000), Fiore *et al.* (2001), Barger *et al.* (2001), Alexander *et al.*

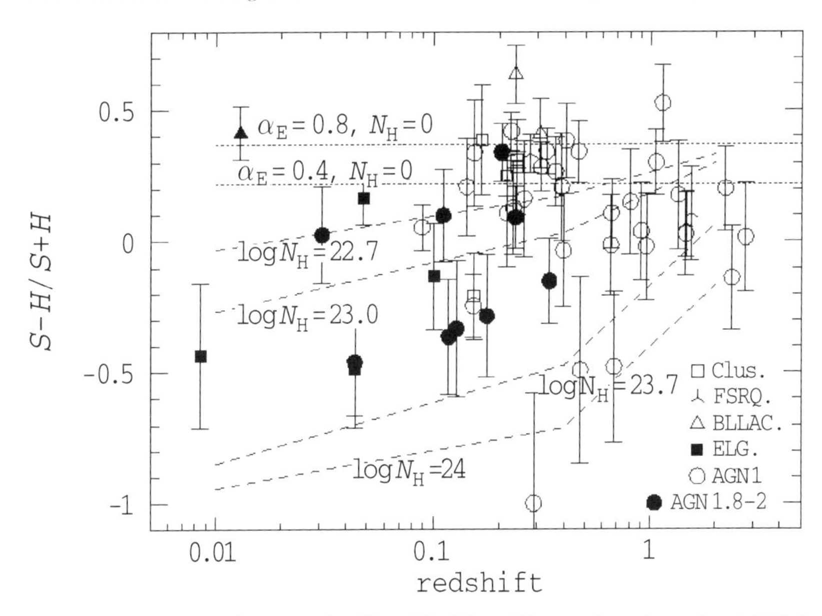

Figure 11.9. The softness ratio $(S - H)/(S + H)$ as a function of redshift for the identified *Beppo*SAX HELLAS sources. S and H are the counts in the 1.3–4.5 and 4.5–10 keV bands, respectively. Note that there are a number of absorbed type-1 AGN. From Comastri *et al.* (2001).

(2001)), now easily probed by Chandra and XMM-Newton. The nature of these X-ray bright, optically normal galaxies (XBONG, as they were christened by Comastri *et al.* (2002a), where a detailed discussion of the properties of ten bona fide sources of this class can be found) is still matter of debate. A large fraction of these sources have X-ray luminosities exceeding 10^{42} erg s^{-1}, well into the AGN regime. My personal feeling is that most, if not all, of them will turn out to be obscured AGN, in which the obscuration prevents the formation (or the visibility) of the narrow lines. Compton-thick obscuration may explain the not particularly hard spectrum of some of them, if reflection from ionized matter is not negligible with respect to reflection from cold matter (with an appropriate choice of the bands, even the archetypal Compton-thick source NGC 1068 would appear as a very soft source indeed!). It is worth recalling that cases of obscured AGN whose optical spectra, even if not really 'normal', are nevertheless different from that of an AGN are already known: NGC 6240 and NGC 4945 are the best examples; see Matt (2002) and references therein.

The most extensively studied among XBONG is probably 'P3', one of the sources in the sample of Fiore *et al.* (2000). This object was observed in radio (ATCA), near-infrared (IR) (ISAAC/VLT), optical (ESO 3.6 m) and X-rays (both Chandra and XMM-Newton), and is discussed in Comastri *et al.* (2002b). The X-ray luminosity (a few times 10^{42} erg s^{-1} if the source is Compton-thin; greater than *c*. 10^{44} erg s^{-1} if Compton-thick) and hard X-ray spectrum clearly indicate the presence of an AGN for which, however, there are no signatures at longer wavelengths. The analysis of the spectral energy distribution (SED) makes the ADAF solution rather unlikely. A heavy obscured AGN appears to be the most likely explanation, even if a rather extreme BL Lac cannot be ruled out.

Of course, what has just been said does not mean that there are no relations whatsoever between optical and X-ray appearances. More often than not, the optical (X-ray) appearance is just what one would predict from the unification model after observing the X-ray (optical) emission. Moreover, I am not aware of any certainly unobscured AGN which are not type 1, and of any type 2 which are definitely unobscured (Pappa *et al.* (2001) and Panessa & Bassani (2002) presented some possible cases, which deserve to be studied in more detail to search for short-term flux variability, i.e. the ultimate proof that they are really unabsorbed and not simply Compton-thick). So, at present the 'strict' (in the sense of no or very few exceptions found yet) relations between optical and X-ray classifications may be reduced to

$$\text{type 1} \leftarrow \text{unobscured},$$
$$\text{type 2} \rightarrow \text{obscured}.$$

It is worth noting that support for one of the predictions of the unification model, i.e. that Seyfert 1 galaxies have the 'torus' (and therefore would become Seyfert 2 if observed at different angles), is coming from Chandra and XMM-Newton observations, which are indicating that the presence of narrow iron Kα lines (therefore produced in distant matter; iron lines from the accretion disc being expected to be broad due to kinematic and relativistic effects (see Fabian 2002)) is rather common (e.g. Yaqoob *et al.* (2002), Weaver (2002), and references therein for Chandra results; Matt *et al.* (2001), Pounds & Reeves (2002), and references therein for XMM-Newton results).

Obscured AGN and the X-ray and IR cosmic backgrounds

Obscured AGN are a basic ingredient in synthesis models for the CXRB (e.g. Setti & Woltjer (1989), Comastri *et al.* (1995)). The mixture of unobscured and obscured (with a spread of column densities) AGN is able to reproduce well the spectral shape of the CXRB, so solving the long-standing problem known as the 'spectral paradox'. A large fraction of the CXRB below 10 keV have been now resolved in discrete sources, many of them optically identified as AGN (and part of them, as said above, *assumed* to be AGN due to their X-ray luminosity). Some problems are still to be solved (see Hasinger (2002), Brandt *et al.* (2002)) but, 40 years after its discovery, the main issue, i.e. what class of sources makes up the CXRB (or at least most of it), can now be considered settled (and this despite the bulk of the CXRB, which peaks at *c.* 30 keV, being still largely unresolved for lack of sensitive imaging instruments in that band).

Cosmic backgrounds have also been measured in other bands, notably the submillimetre and IR. When looking at the spectral energy distribution of the extragalactic backgrounds, the CXRB may appear almost negligible if compared with the cosmic IR background (CIRB). However, the luminosity we observe in the CXRB is only a fraction, probably of the order of 10–20%, of the energy actually emitted, the remaining flux having been absorbed by the circumnuclear matter, and re-emitted at longer wavelengths, mostly in the mid-IR. Granato *et al.* (1997) and Fabian & Iwasawa (1999) estimated that a by no means negligible fraction of the CIRB is actually due to X-ray photons absorbed and reprocessed from the obscuring medium (see also Risaliti *et al.* (2002)). Indeed, by cross-correlating IR and X-ray deep observations, Fadda *et al.* (2002) found an AGN contribution of *c.* 15% of the CIRB at 15 mm. From optical spectroscopic identifications of ISO/ELAIS sources, Matute *et al.* (2002) estimate a contribution at 15 mm of 10–15%. These two estimates are both likely to be lower limits, as in the former case some Compton-thick sources may have been missed, and in the latter case AGN which are bright in IR but do not show AGN-like optical lines, like NGC 6240 and NGC 4945 (Matt 2002), may have not been recognized as AGN.

Whatever the real number is, it is clear that accretion and star formation are processes of comparable importance in the history of the Universe.

It is a pleasure to thank all my collaborators in this research: S. Bianchi, A. C. Fabian, F. Fiore, M. Guainazzi, K. Iwasawa, G. C. Perola and all members of the HELLAS team. I acknowledge financial support from ASI and MURST (under grant COFIN-00-02-36).

References

Alexander, D. M. *et al.* 2001 *Astrophys. J.* **122**, 2156–2176.
Antonucci, R. R. J. 1993 *A. Rev. Astron. Astrophys.* **31**, 473–521.
Awaki, H., Ueno, S., Taniguchi, Y. & Weaver, K. A. 2000 *Astrophys. J.* **542**, 175–185.
Band, D. L., Klein, R. I., Castor, J. I. & Nash, J. K. 1990 *Astrophys. J.* **362**, 90–99.
Barger, A. *et al.* 2001 *Astrophys. J.* **560**, L23–L28.
Behar, E. *et al.* 2002. In *Mass Outflows in AGNs: New Perspectives* (ed. D. M Crenshaw, S. B Kraemer & I. M. George). (In the press.)
Bianchi, S., Matt, G. & Iwasawa, K. 2001 *Mon. Not. R. Astr. Soc.* **322**, 669–680.
Brandt, W. N., Alexander, D. M., Bauer, F. E. & Hornschemeier, A. E. 2002 *Phil. Trans. R. Soc. Lond.* A **360**, 2057–2075.
Cappi, M. *et al.* 1999 *Astron. Astrophys.* **344**, 857–867.
Comastri, A., Setti, G., Zamorani, G. & Hasinger, G. 1995 *Astron. Astrophys.* **296**, 1–12.
Comastri, A. *et al.* 2001 *Mon. Not. R. Astr. Soc.* **327**, 781–787.

Comastri, A. *et al.* 2002a. In *New Visions of the X-ray Universe in the XMM-Newton and Chandra Era*. (In the press.)
Comastri, A. *et al.* 2002b *Astrophys. J.* **571**, 771.
Della Ceca, R., Braito, V., Cagnoni, I. & Maccacaro, T. 2001. In *X-ray Astronomy* (ed. N. E. White, G. Malaguti & G. G. C. Palumbo), pp. 602–605. Maryland: AIP.
Elvis, M. 2000 *Astrophys. J.* **545**, 63–76.
Fabian, A. C. 2002 *Phil. Trans. R. Soc. Lond.* A **360**, 2035–2043.
Fabian, A. C. & Iwasawa, K. 1999 *Mon. Not. R. Astr. Soc.* **303**, L34–L36.
Fadda, D. *et al.* 2002 *Astron. Astrophys.* **303**, 838–853.
Fiore, F. *et al.* 2000 *New Astron.* **5**, 143–153.
Fiore, F. *et al.* 2001 *Mon. Not. R. Astr. Soc.* **327**, 771–780.
Fiore, F. *et al.* 2002. In *Issues in Unification of AGN* (ed. R. Maiolino, A. Marconi & N. Nagar). pp. 205–210. San Francisco, CA: ASP.
Gallagher, S. C., Brandt, W. N., Chartas, G. & Garmire, G. 2002 *Astrophys. J.* **567**, 37–41.
George, I. M. & Fabian, A. C. 1991 *Mon. Not. R. Astr. Soc.* **249**, 352–367.
Gilli, R. *et al.* 2000 *Astron. Astrophys.* **335**, 485–498.
Granato, G. L., Danese, L. & Franceschini A. 1997 *Astrophys. J.* **486**, 147–159.
Guainazzi, M. 2002. *Mon. Not. R. Astr. Soc.* **329**, L13–L17.
Guainazzi, M. *et al.* 1998 *Mon. Not. R. Astr. Soc.* **301**, L1–L5.
Guainazzi, M., Matt, G., Fiore, F. & Perola, G. C. 2002 *Astron. Astrophys.* **388**, 787.
Hasinger, G. 2002 *Phil. Trans. R. Soc. Lond.* A **360**, 2077–2090.
Kinkhabwala, A. *et al.* 2002. In *New Visions of the X-ray Universe in the XMM-Newton and Chandra Era*. (In the press.)
Krolik, J. H. & Kriss, G. A. 1995 *Astrophys. J.* **447**, 512–525.
Leighly, K. M. *et al.* 1999 *Astrophys. J.* **522**, 209–213.
Maiolino, R. & Rieke, G. H. 1995 *Astrophys. J.* **454**, 95–105.
Maiolino, R. *et al.* 1998 *Astron. Astrophys.* **338**, 781–794.
Maiolino, R. *et al.* 2001a *Astron. Astrophys.* **365**, 28–36.
Maiolino, R., Marconi, A. & Oliva, E. 2001b *Astron. Astrophys.* **365**, 37–48.
Malkan, M. A., Gorjian, V. & Tam, R. 1998 *Astrophys. J.* **117**, 25–88.
Matt, G. 2000 *Astron. Astrophys.* **335**, L31–L33.
Matt, G. 2002. In *Issues in Unification of AGN* (eds. R. Maiolino, A. Marconi & N. Nagar). pp. 3–14 San Francisco, CA: ASP.
Matt, G., Perola, G. C. & Piro, L. 1991 *Astron. Astrophys.* **247**, 25–34.
Matt, G. *et al.* 1996a *Mon. Not. R. Astr. Soc.* **281**, L69–L73.
Matt, G., Brandt, W. N. & Fabian, A. C. 1996b *Mon. Not. R. Astr. Soc.* **280**, 823–834.
Matt, G. *et al.* 1997 *Astron. Astrophys.* **325**, L13–L16.
Matt, G., Pompilio, F. & La Franca, F. 1999a *New Astron.* **4/3**, 191–195.
Matt, G. *et al.* 1999b *Astron. Astrophys.* **341**, L39–L42.
Matt, G. *et al.* 2000 *Mon. Not. R. Astr. Soc.* **318**, 173–179.
Matt, G. *et al.* 2001 *Astron. Astrophys.* **327**, L31–L34.
Matute, I. *et al.* 2002 *Mon. Not. R. Astr. Soc.* **332**, L11.
Mushotzky, R. F., Cowie, L. L., Barger, A. J. & Arnaud, K. A. 2000 *Nature* **404**, 459–464.
Panessa, F. & Bassani, L. 2002 *Astron. Astrophys.* **394**, 435.
Pappa, A., Georgantopoulos, I., Stewart, G. C. & Zezas, A. L. 2001 *Mon. Not. R. Astr. Soc.* **326**, 995–1006.
Pounds, K. A. & Reeves, J. N. 2002 In *New Visions of the X-ray Universe in the XMM-Newton and Chandra Era* (ed. F. Jansen). (In the press.)
Risaliti, G., Maiolino, R. & Salvati, M. 1999 *Astrophys. J.* **522**, 157–164.
Risaliti, G., Elvis, M. & Gilli, R. 2002 *Astrophys. J.* **566**, L67.

Sako, M., Kahn, S. M., Paerels, F. & Liedahl, D. A. 2000 *Astrophys. J.* **543**, L115–L118.
Sambruna, R. *et al.* 2001a *Astrophys. J.* **546**, L9–L12.
Sambruna, R. *et al.* 2001b *Astrophys. J.* **546**, L13–L17.
Setti, G. & Woltjer, L. 1989 *Astron. Astrophys.* **224**, L21–L23.
Weaver, K. A. 2002. In *The Central kpc of Starbursts and AGN: the La Palma Connection* (eds. J. H. Knapen, J. E. Beckman, I. Shlosman & T. J. Mahoney). (In the press.)
Yaqoob, T., George, I. M. & Turner, T. J. 2002. In *High Energy Universe at Sharp Focus: Chandra Science* (eds. E. M. Schlegel & S. Vrtilek). (In the press.)
Young, A. J., Wilson, A. S. & Shopbell, P. L. 2001 *Astrophys. J.* **556**, 6–23.

12

The Chandra Deep Field-North Survey and the cosmic X-ray background

BY W. NIELSEN BRANDT, DAVID M. ALEXANDER, FRANZ E. BAUER AND ANN E. HORNSCHEMEIER

The Pennsylvania State University

Introduction

The cosmic X-ray background

The cosmic X-ray background (XRB) was the first cosmic background radiation to be discovered (Giacconi *et al.* 1962) and has *c.* 10% of the energy density of the cosmic microwave background (CMB) (see, for example, Fabian & Barcons (1992) and Hasinger (2000) for reviews). In contrast to the CMB, the XRB comprises the integrated contributions from a large number of discrete sources. The XRB thus represents the summed emission from all X-ray sources since the Universe was less than a billion years old, and by surveying it we can learn about the nature and evolution of X-ray-emitting objects over most of the history of the Universe.

The XRB is detected over a broad energy band from *c.* 0.1 to 200 keV, peaking in energy density from 20 to 40 keV; the logarithmic frequency/energy coverage is comparable with that of the far-infrared to extreme ultraviolet bandpass. Current deep-imaging surveys of the XRB, however, have been primarily conducted in the narrower *c.* 0.1–12 keV band, due largely to technological limitations.

Observations with the new generation of X-ray observatories, Chandra (Weisskopf *et al.* 2000) and XMM-Newton (Jansen *et al.* 2001), have revolutionized studies of the XRB and the sources that comprise it. Early observations with Chandra resolved most of the 2–8 keV background into point sources (e.g. Brandt *et al.* (2000), Mushotzky *et al.* (2000), Giacconi *et al.* (2001)); most of the 0.5–2 keV background had already been resolved by the Röntgensatellit (ROSAT) (e.g. Hasinger *et al.* (1998)). The accurate positions from the new observatories, particularly Chandra, also allow X-ray sources to be matched unambiguously to (often faint) multi-wavelength counterparts. X-ray surveys have finally reached the depths needed to complement the most sensitive surveys in the radio, submillimetre,

Frontiers of X-Ray Astronomy, ed. A.C. Fabian, K.A. Pounds and R.D. Blandford. Published by Cambridge University Press.

infrared and optical bands. The focus has now shifted from simply resolving the XRB to understanding in detail the sources that comprise it at all X-ray fluxes. While it is clear that massive, accreting black holes produce the bulk of the XRB, the deepest X-ray surveys are now allowing the study of other source classes, including starburst and normal galaxies.

In this chapter, we will review some of the current results from the ongoing Chandra Deep Field-North (CDF-N) survey. Other ongoing deep X-ray surveys that have published results at present include the Chandra Deep Field-South survey (e.g. Giacconi *et al.* (2002), Rosati *et al.* (2002)), the Lockman Hole Survey (e.g. Hasinger *et al.* (2001), Lehmann *et al.* (2001)), and the SSA13 Survey (e.g. Mushotzky *et al.* (2000) Barger *et al.* (2001a)). Many other important X-ray surveys are being performed over larger solid angles to shallower depths.

Throughout this chapter we adopt $H_0 = 65$ km s^{-1} Mpc^{-1}, $\Omega_M = 1/3$ and $\Omega_\Lambda = 2/3$.

The Chandra Deep Field-North survey (CDF-N)

The CDF-N survey is currently the deepest X-ray survey ever performed. It comprises 1.4 Ms (16.2 days) of exposure with the Chandra Advanced Charge-Coupled Device (CCD) Imaging Spectrometer (ACIS), covering a *c.* $18' \times 22'$ field (see Fig. 12.1) centred on the Hubble Deep Field-North (HDF-N; see Ferguson *et al.* (2000)); these observations are publicly available. This is an excellent field for study, since it already has intensive coverage at optical, infrared, submillimetre and radio wavelengths (see Ferguson *et al.* (2000) for a review). More than 800 spectroscopic redshifts have been obtained in the field (e.g. Cohen *et al.* (2000)). In the soft (0.5–2 keV) and hard (2–8 keV) X-ray bands, the Chandra data reach fluxes *c.* 55 and *c.* 550 times fainter than surveys by previous X-ray missions (see Fig. 12.2). Near the centre of the field, sources with count rates as low as *c.* 1 count per 2 days can be detected. Source densities near the soft-band and hard-band flux limits are *c.* 7200 deg^{-2} and *c.* 4200 deg^{-2} (e.g. Brandt *et al.* (2001b), Cowie *et al.* (2002)), the fraction of the XRB resolved in the soft and hard bands is greater than 90% and 80%, respectively (the main uncertainty in the resolved fraction is the absolute normalization of the background itself). Source positions are typically accurate to better than $1''$ over the entire field, allowing multi-wavelength counterparts to be identified reliably.

The CDF-N team includes *c.* 20 researchers from the Pennsylvania State University, the University of Hawaii, the University of Wisconsin, the Massachusetts Institute of Technology, the California Institute of Technology and Carnegie Mellon University. To see all the contributors to the various CDF-N projects, please refer to the author lists of the cited CDF-N publications.

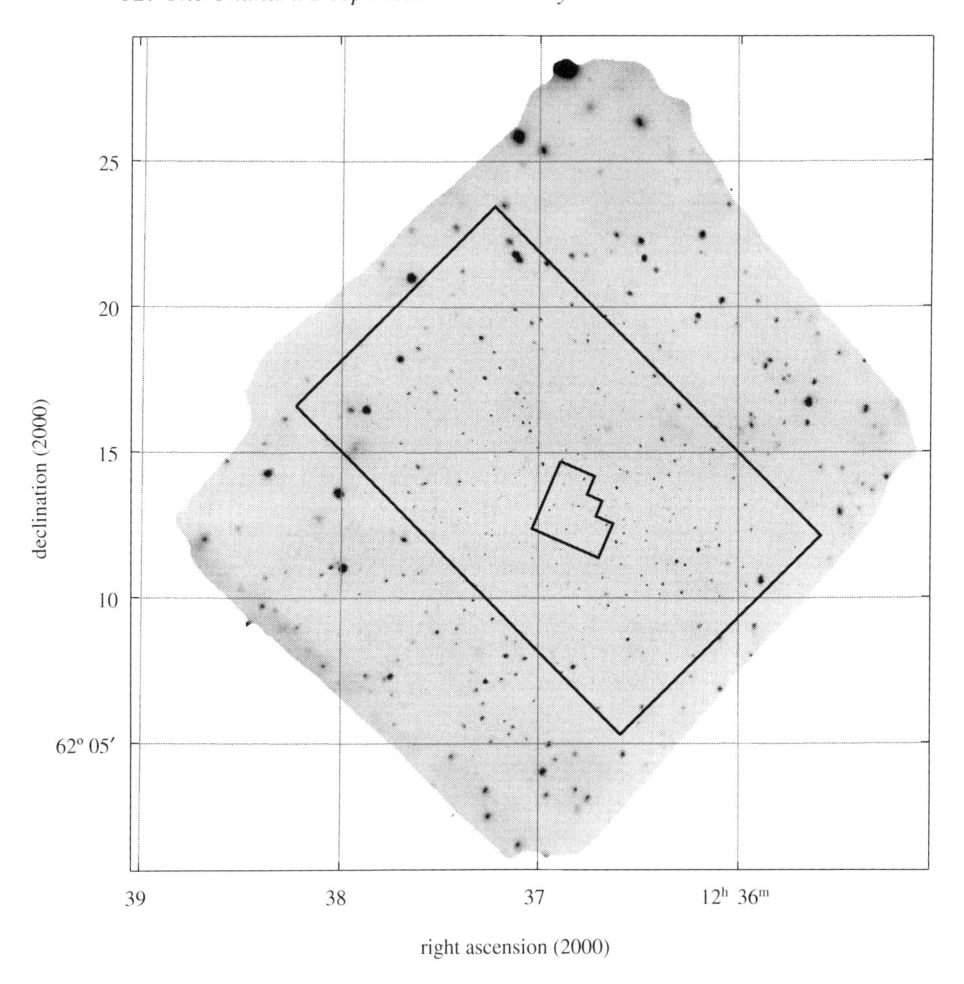

Figure 12.1. Adaptively smoothed and exposure-map corrected image of the CDF-N in the soft band (0.5–2 keV). The adaptive smoothing has been performed using the code of Ebeling *et al.* (2002) at the 2.5σ level, and the greyscale is linear. Source sizes appear to change across the field due to the spatial dependence of the instrumental point spread function. The small polygon indicates the HDF-N, and the large rectangle indicates the GOODS area .

Some key results from the survey

Detected point and extended sources

The CDF-N data have been searched intensively for point sources using the Chandra X-ray Center's WAVDETECT algorithm (Freeman *et al.* 2002). At a WAVDETECT false-positive probability threshold of 1×10^{-7}, *c.* 430 independent sources are detected in the 1.4 Ms exposure. In addition, a substantial number of sources have been found at lower significance levels that are spatially correlated with objects

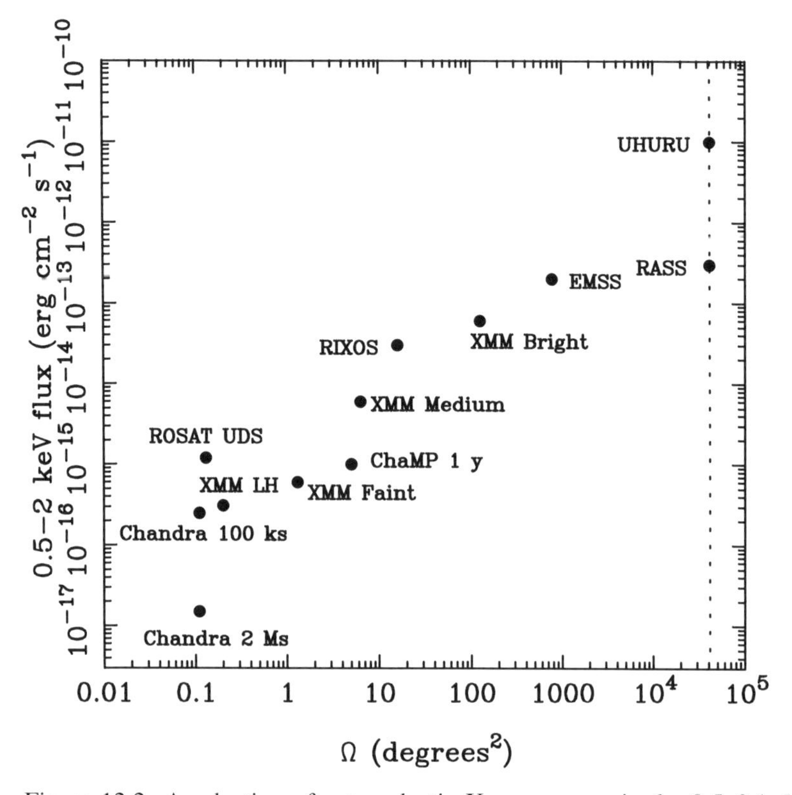

Figure 12.2. A selection of extragalactic X-ray surveys in the 0.5–2 keV flux-limit versus solid-angle, Ω, plane. Shown are the UHURU survey, the ROSAT All-Sky Survey (RASS), the Einstein Extended Medium-Sensitivity Survey (EMSS), the ROSAT International X-ray/Optical Survey (RIXOS), the XMM-Newton Serendipitous Surveys (XMM Bright, XMM Medium, XMM Faint), the Chandra Multi-wavelength Project (ChaMP), the ROSAT Ultra Deep Survey (ROSAT UDS), the deep XMM-Newton survey of the Lockman Hole (XMM LH), Chandra 100 ks surveys, and Chandra 1.4 Ms surveys (i.e. the CDF-N). Although each of the surveys shown clearly has a range of flux limits across its solid angle, we have generally shown the most sensitive flux limit. The vertical dashed line shows the solid angle of the whole sky. Adapted from Brandt *et al.* (2001b).

found at other wavelengths (e.g. optically bright galaxies). Many of these sources are real, but it is more difficult to define a complete X-ray flux-limited sample of these sources. In the full band (0.5–8 keV), the typical source has *c.* 90 counts, which is too small for detailed X-ray spectral analysis. Assessment of spectral hardness is possible, however, and X-ray spectral analysis can be performed for the brighter sources in the field. Catalogues of the detected sources and related analysis products have been made publicly available for the 1 Ms exposure (see Brandt *et al.* 2001b).†

† www.astro.psu.edu/users/niel/hdf/hdf-chandra.html

Six extended X-ray sources have also been detected in the CDF-N (Bauer *et al.* 2002a). Their X-ray spectral properties, angular sizes and likely luminosities (*c.* 10^{42} erg s^{-1}) are generally consistent with those found for nearby groups of galaxies. Two extended X-ray sources of note are: (1) a group in the HDF-N associated with the $z = 1.013$ FR I radio galaxy VLA J123644.3+621133, and (2) a likely poor-to-moderate cluster at $z \gtrsim 0.7$ that is coincident with an overdensity of very red objects (VROs; $I - K \geqslant 4$), optically faint ($I \geqslant 24$) radio sources, and optically faint X-ray sources. The surface density of extended X-ray sources is 167^{+97}_{-67} deg^{-2} at a limiting soft-band flux of *c.* 3×10^{-16} erg cm^{-2} s^{-1}. No evolution in the X-ray luminosity function of clusters is needed to explain this value.

Figure 12.3 shows the I-band magnitudes of the X-ray point sources near the centre of the CDF-N field. There is a wide spread of I-band magnitudes at faint X-ray fluxes, corresponding to a range of source types. The faint X-ray sources with $I \sim$ 15–22 counterparts are normal galaxies, starburst galaxies, low-luminosity active galactic nuclei (AGN) and stars. The faint X-ray sources with $I \gtrsim 22$ counterparts appear to be mostly luminous AGN; these often show evidence for X-ray absorption via large hardness ratios.

Optical spectra are presently available for *c.* 120 CDF-N objects. The majority of the sources with spectroscopic redshifts lie at $z \lesssim 1.3$ (see Fig. 12.4). This result is at least partly due to a selection effect: as implied by Fig. 12.4, the X-ray sources at $z \gtrsim 1.3$ will frequently be too optically faint for spectroscopy (with $I \gtrsim 24$). These sources are discussed further below. Even after making plausible corrections for selection effects, however, it appears likely that more X-ray power originates from $z \lesssim 1.3$ than is predicted by some XRB-synthesis models; these models will require revision.

Results for the Hubble Deep Field-North

Figure 12.5 shows the 23 HDF-N X-ray point sources detected thus far (Hornschemeier *et al.* 2000; Brandt *et al.* 2001a, 2002b); 16 are found with a false-source probability threshold of 1×10^{-7}, while the other 11 are found with lower significance but align spatially with optical galaxies (see above). As expected from Figs. 12.3 and 12.4, most sources are at $z < 1.5$, and their optical counterparts have a wide range of brightnesses. In at least one HDF-N object (CXOHDFN J123641.7+621131 at $z = 0.089$) the X-ray source is offset from the nucleus; the X-ray emission may arise from a starburst region in the host galaxy as it is coincident with a bright, blue 'knot'.

We have been able to find likely optical or near-infrared counterparts for all of the X-ray sources in the HDF-N; this contrasts with the case in the radio where some truly 'blank-field' sources have been found (e.g. Richards *et al.* (1999)). We find a

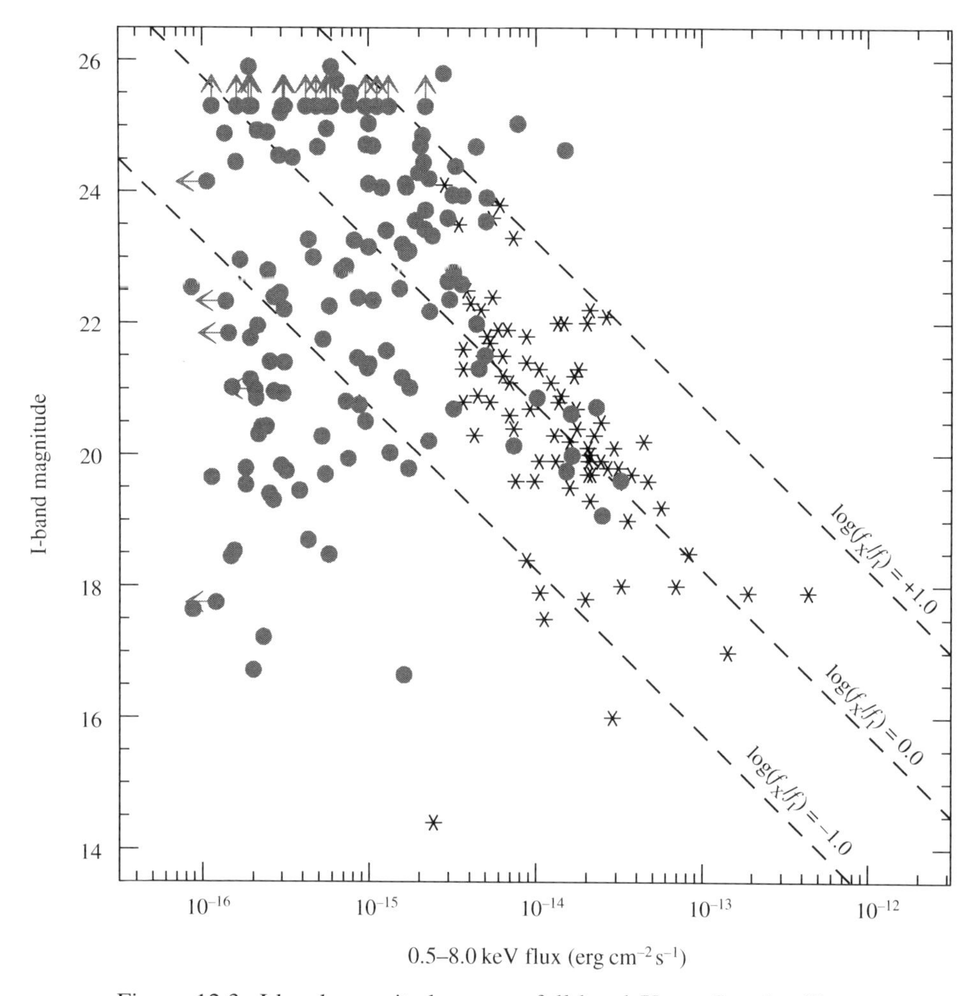

Figure 12.3. I-band magnitude versus full-band X-ray flux for X-ray sources from an $8.4' \times 8.4'$ region of the CDF-N field centred on the HDF-N (solid dots; Alexander *et al.* 2001) and the ROSAT UDS of the Lockman Hole (stars; converted to 0.5–8 keV; Lehmann *et al.* 2001). The dashed diagonal lines indicate constant X-ray-to-I-band flux ratios; luminous AGN typically have $\log(f_X/f_I)$ in the range from -1 to $+1$. Note the wide spread of I-band magnitudes for the faintest X-ray sources.

good correspondence between our brighter X-ray sources and μJy radio (1.4 GHz and 8.5 GHz) sources, but this trend does not continue for our faintest X-ray sources. For instance, 10 of the 16 brightest Chandra sources have radio matches, but only 12 of the 27 total Chandra sources have radio matches. The properties of the X-ray/radio sources suggest a broad range of emission mechanisms (e.g. Richards *et al.* (1998), Brandt *et al.* (2001a)). We also find a good correspondence between Chandra and ISO (6.7 μm and 15 μm (Aussel *et al.* 1999)) sources in the HDF-N. Ten of the 16 brightest Chandra sources have ISO matches, and 14% of the 27 total

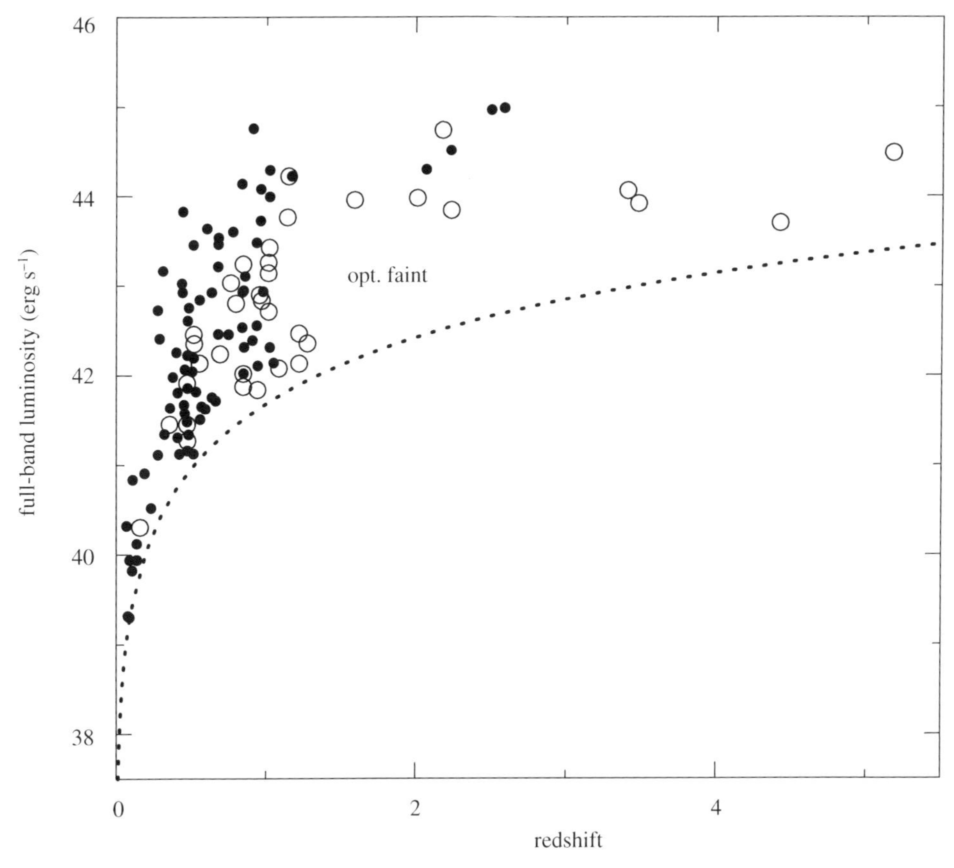

Figure 12.4. Full-band X-ray luminosity versus redshift for X-ray sources with optical spectra. Solid dots are sources with $I < 21.5$, and open circles are sources with $I > 21.5$. The dotted curve represents the X-ray detection limit. Note that most of the sources at $z \sim 1$–1.3 have $I > 21.5$. Similar sources at $z \gtrsim 1.3$ (in the region marked 'opt. faint') will frequently have $I \gtrsim 24$ and thus be too optically faint for spectroscopy (see also Fig. 12.8(*b*)).

Chandra sources have ISO matches. This good X-ray/IR correspondence bodes well for future IR follow-up of X-ray sources with the Space Infrared Telescope Facility (SIRTF) (see below). In this field with both exceptionally sensitive X-ray and IR coverage, we find a broad range of X-ray/IR source types including starburst galaxies, obscured AGN and normal elliptical galaxies. In the majority of the sources, the IR emission appears to be from dust re-emission of primary X-ray and ultraviolet radiation.

X-ray connections with infrared, radio and submillimetre sources

More detailed studies of the X-ray/IR sources in the CDF-N have been completed by Alexander *et al.* (2002b) and Fadda *et al.* (2002). These use the 21.5 arcmin^2 region

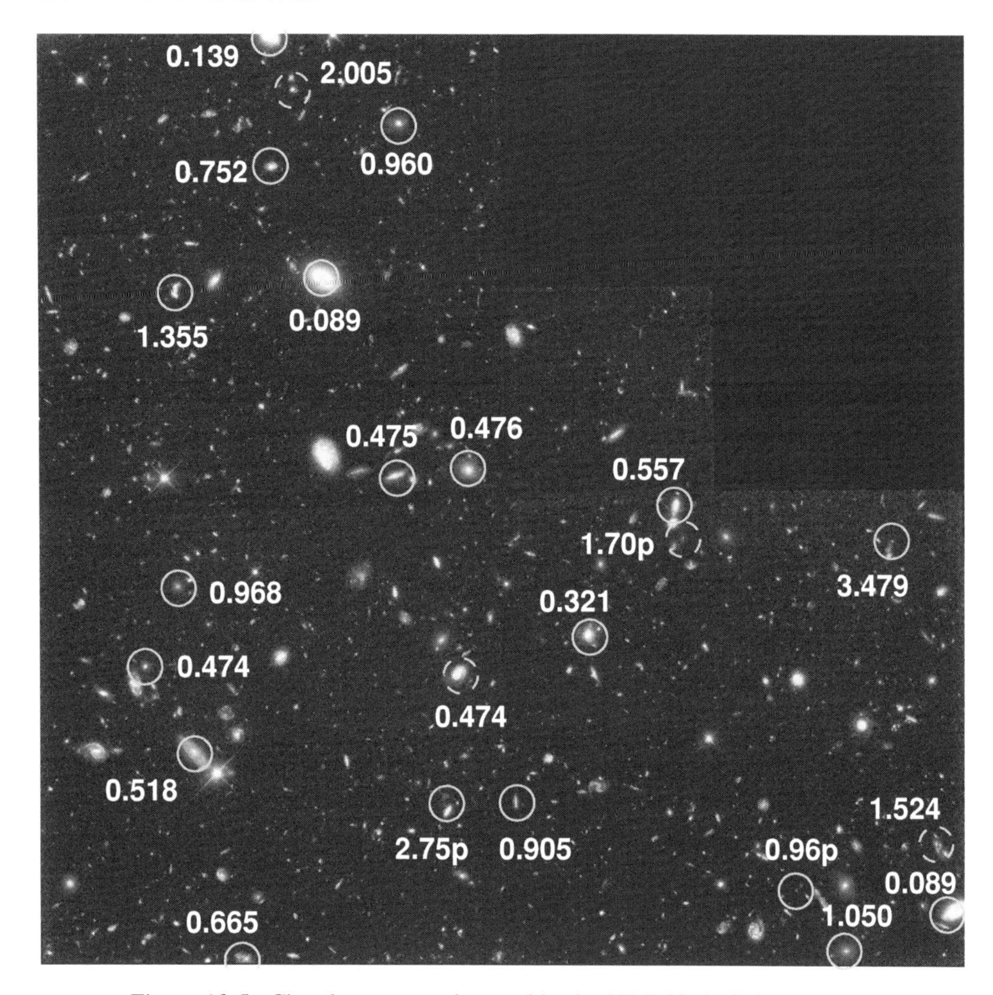

Figure 12.5. Chandra sources detected in the HDF-N circled on the HST optical image. Solid circles indicate sources detected with a false-positive probability threshold of 1×10^{-7}, and broken circles indicate sources found at lower significance levels that are spatially correlated with optical galaxies. The numbers are source redshifts; redshifts followed by a 'p' are photometric rather than spectroscopic. The extended source in the HDF-N is located near the $z = 1.013$ FR I radio galaxy at the bottom of the image.

with uniform ISO coverage at 15 μm rather than just the HDF-N (5.3 arcmin^2). Only *c*. 20% of the X-ray/IR sources are likely to be AGN. The majority rather appear to be $z \sim 0.4$–1.3 starburst galaxies and $z < 0.2$ normal galaxies (see Fig. 12.6). A notable finding is that up to 100% of the X-ray detected emission-line galaxies (see Cohen *et al.* (2000) for the optical spectral classification) have 15 μm counterparts (Alexander *et al.* 2002b); the majority of these are luminous IR starburst galaxies representative of the population making the bulk of the IR background. The

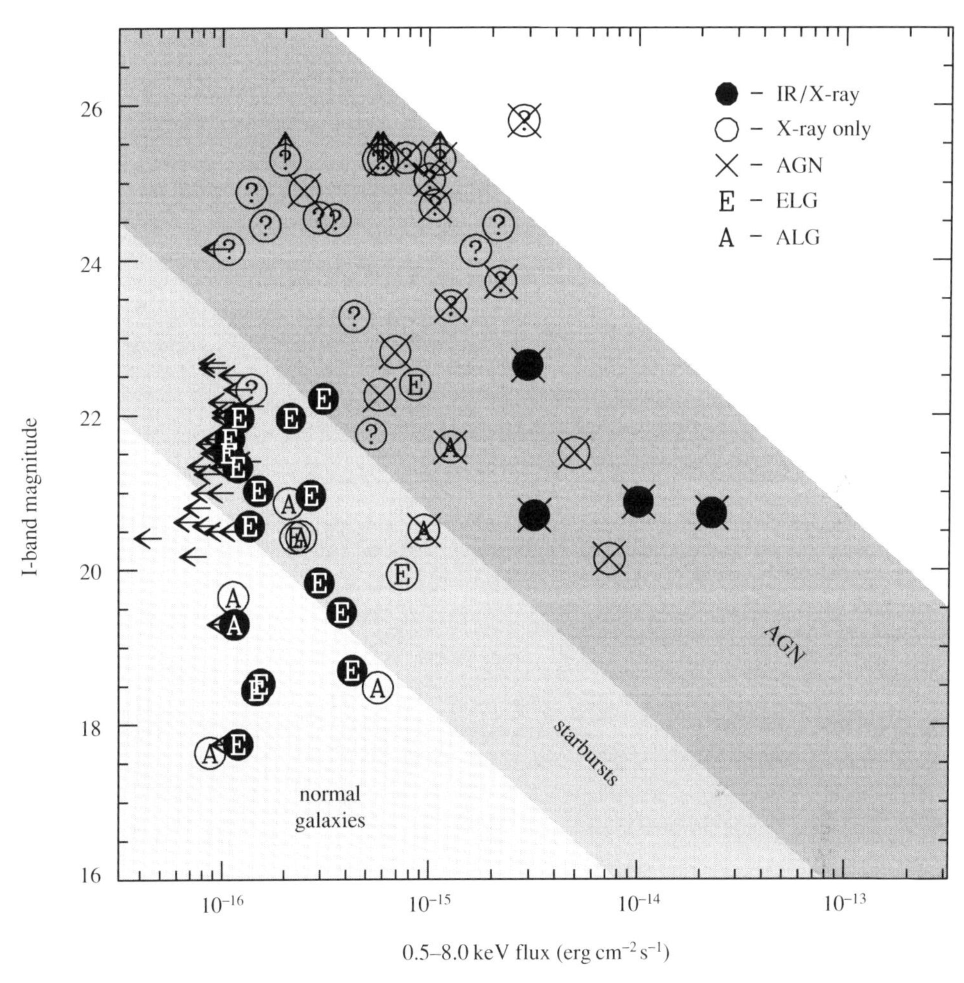

Figure 12.6. I-band magnitude versus full-band X-ray flux for sources with both deep Chandra and ISO coverage. The filled circles are the X-ray/IR matched sources, and the open circles are X-ray sources without IR counterparts. Characters inside the circles indicate different source types: 'E' indicates emission-line galaxies, 'A' indicates absorption-line galaxies, '?' indicates sources that do not have spectroscopic classifications, and overlaid crosses indicate AGN-dominated sources. IR sources without detected X-ray emission are plotted only as upper-limit arrows. The shaded regions delineate the approximate range of the X-ray-to-optical flux ratio for AGN-dominated sources, starburst galaxies and normal galaxies. Adapted from Alexander *et al.* (2002b).

X-ray/radio-matched galaxies trace the same star-forming population as the X-ray/IR sources, with nearly 100% of the matched galaxies in common (Bauer *et al.* 2002b). In addition, the X-ray/radio matches appear to trace a population of optically faint AGN.

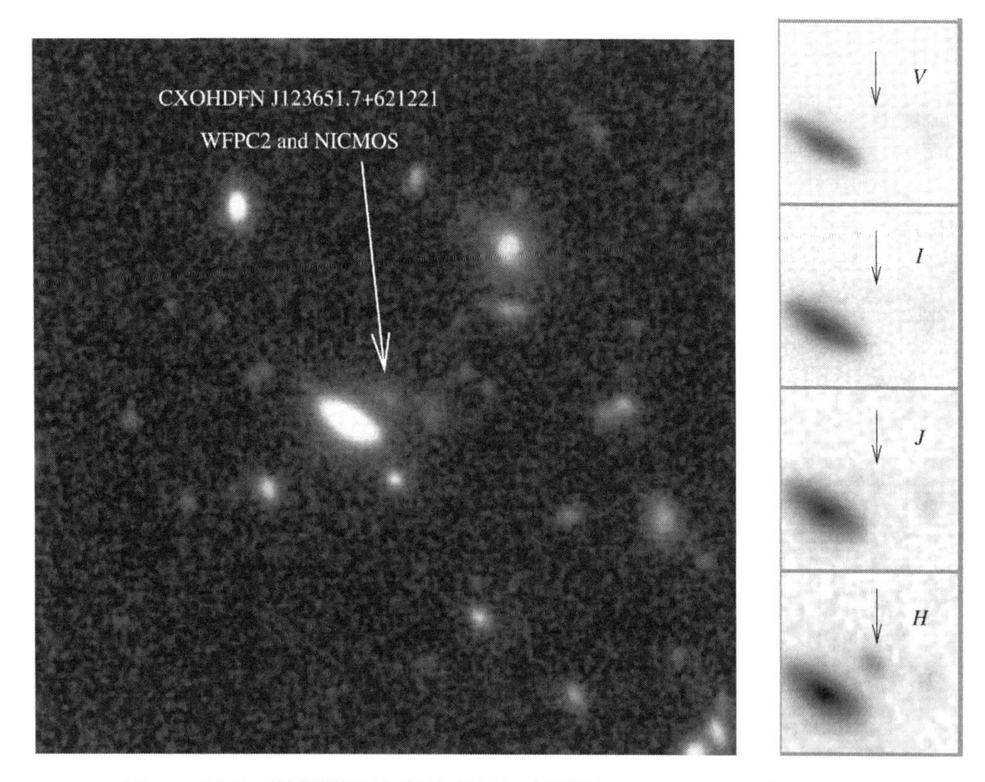

Figure 12.7. CXOHDFN J123651.7+621221, an example of an optically faint X-ray source lying in the HDF-N. This $I = 25.8$ source has a photometric redshift of $z = 2.75$ (e.g. Alexander *et al.* (2001)) and is remarkably red (compare the V-band and H-band images shown). It is the hardest as well as the second brightest X-ray source found in the HDF-N. The X-ray luminosity (*c.* 3×10^{44} erg s^{-1}) and hard spectral shape indicate this source is a luminous, obscured AGN. It was previously noted as an optically faint μJy radio source (e.g. Richards *et al.* (1999)) and thought to be a dusty starburst galaxy (e.g. Muxlow *et al.* (1999)). The Near-Infrared Camera and Multi-Object Spectrometer (NICMOS) images were kindly provided by M. Dickinson (see Dickinson *et al.* (2000)).

Barger *et al.* (2001b) have shown that the ensemble of CDF-N X-ray sources contributes about 15% of the extragalactic submillimetre background light at 850 μm, with the strongest submillimetre emission is seen from optically faint X-ray sources that are also detected at 1.4 GHz. At the current CDF-N flux limit, *c.* 20% of the submillimetere sources have X-ray couterparts.

Optically faint and high-redshift X-ray sources

As mentioned earlier, a significant fraction of the CDF-N X-ray sources are too optically faint for easy spectroscopic follow-up studies. One notable example from the HDF-N is shown in Fig. 12.7. Alexander *et al.* (2001) present a detailed study of the 47 optically faint X-ray sources (defined as having $I \geqslant 24$) in an $8.4' \times 8.4'$ region centred on the HDF-N. The number of optically faint X-ray sources increases at faint X-ray fluxes. However, the fraction of optically faint sources within the X-ray source population stays roughly constant at *c.* 35% for full-band fluxes from 10^{-16} to 10^{-14} erg cm^{-2} s^{-1} due to the emergence of a population of optically bright sources (see earlier and Fig. 12.3). Many of the optically faint X-ray sources have red-optical-to-near-IR colours, and a significant fraction are classified as VROs (see also Alexander *et al.* (2002a) for a detailed X-ray study of CDF-N VROs). They also have harder X-ray spectra on average than the sources with $I < 24$ (see Fig. 12.8(*a*)). Roughly half of the optically faint sources with enough X-ray counts to allow an effective search for X-ray variability show it, and the redshifts of the majority of the optically faint X-ray sources are estimated to be $z \sim 1$–3 (see Fig. 12.8(b)). All of these facts support an interpretation where most of the optically faint X-ray sources are luminous, obscured AGN (e.g. Seyfert-2 galaxies and type-2 quasars) at intermediate redshifts.

A minority of the optically faint X-ray sources, however, may be AGN at high-to-extreme redshifts ($z \sim 4$–10). One notable example is CXOHDFN J123642.0+621331 (Brandt *et al.* 2001a; $I = 25.3$), which lies just outside the HDF-N and is a μJy radio source at a likely redshift of $z = 4.424$ (Waddington *et al.* 1999). With an X-ray luminosity of *c.* 5×10^{43} erg s^{-1}, this is by far the lowest luminosity X-ray source known at $z > 4$. Its detection directly demonstrates that Chandra is achieving the sensitivity needed to study Seyfert-luminosity AGN at high redshift.

A large population of Seyfert-luminosity AGN at $z \sim 4$–10 has been postulated by Haiman & Loeb (1999). These objects would represent the first massive black holes to form in the Universe. At $z \gtrsim 6.5$ an AGN should appear not only optically faint but optically blank; the Lyman α forest and Gunn–Peterson trough will absorb essentially all flux through the I-band. Thus, an upper limit on the space density of extreme-redshift AGN can be set simply by counting the number of X-ray sources that lack any optical counterpart.† Unfortunately, however, confusion between truly optically blank sources at extreme redshift and very optically faint sources at moderate redshift (e.g. objects like that in Fig. 12.7) occurs without exceptionally deep optical imaging. At present, the CDF-N data suggest that there is less than *c.* 1 AGN detected at $z \gtrsim 6.5$ per *c.* 20 arcmin^2 (Alexander *et al.* 2001). This limit should soon be tightened substantially via the Great Observatories Origins Deep Survey (GOODS) project (see later).

† Of course, this method requires the plausible assumption that extreme redshift AGN be X-ray luminous. The best available data suggest that this should be the case (Vignali *et al.* 2001; Brandt *et al.* 2002a). Note also that at $z \approx 10$ Chandra provides rest-frame sensitivity up to *c.* 90 keV; such high-energy X-rays can penetrate a substantial amount of obscuration.

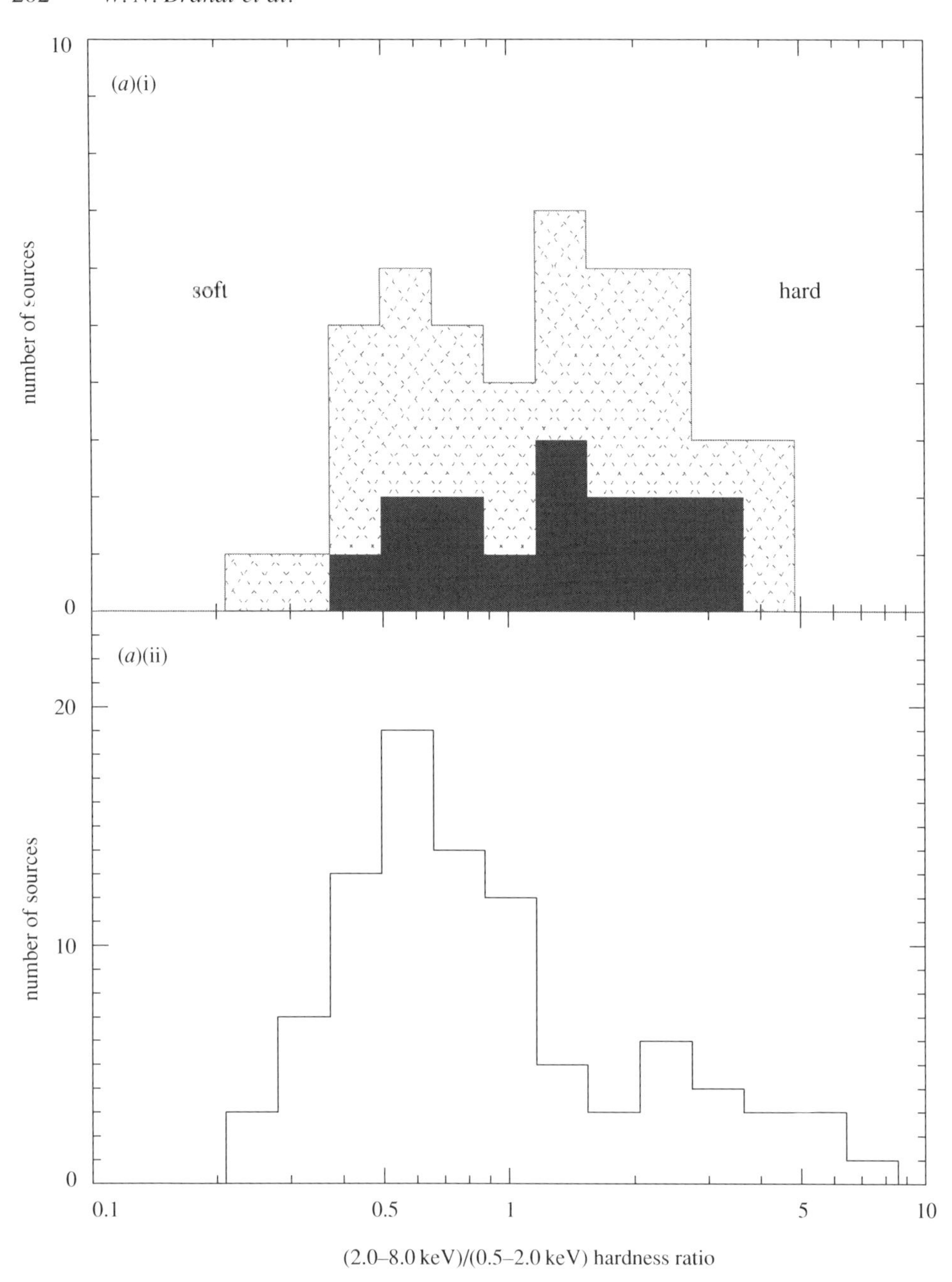

Figure 12.8. (*a*) X-ray hardness ratio distributions for (i) optically faint ($I \geqslant 24$) and (ii) optically bright source samples. While both samples have a wide spread in hardness ratio, the optically faint sources are harder on average. The solid shading in the upper panel shows the subset of optically faint sources without I-band detections (corresponding to $I > 25.3$).

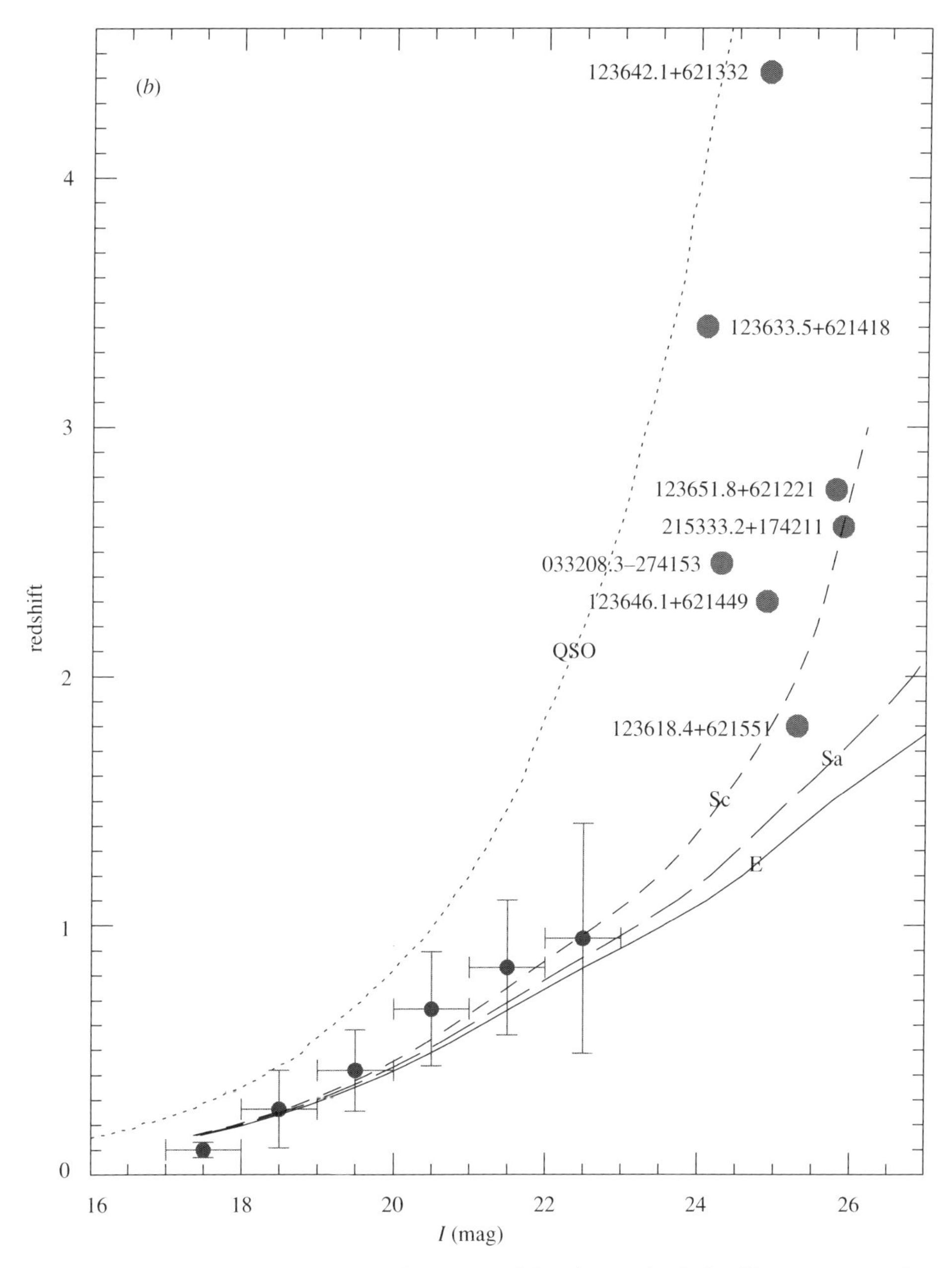

Figure 12.8. (*Cont.*) (*b*) Redshift versus I-band magnitude for X-ray sources. The small solid dots with error bars show the average spectroscopic redshifts of the $I < 23$ X-ray sources as a function of I-band magnitude. The large solid dots show individual optically faint X-ray sources with spectroscopic, photometric or millimetric redshifts. The labelled curves in the diagram show tracks for an $M_I = -23$ E galaxy, Sa galaxy, Sc galaxy and QSO. Note that the $I < 23$ X-ray sources follow the galaxy tracks fairly well; their optical emission appears to be dominated by that from the host galaxy. If the same holds for the optically faint X-ray sources, extrapolation along the curves indicates the majority should lie at $z \sim 1$–3. This is indeed consistent with that found for the few optically faint sources with redshift determinations. Adapted from Alexander *et al.* (2001).

The highest redshift AGN discovered in CDF-N follow-up studies has $z = 5.18$ (Barger *et al.* 2002). This is the highest redshift X-ray selected AGN, and it has relatively narrow emission lines. This source is not optically faint ($I = 22.7$) and has an X-ray luminosity of *c.* 3×10^{44} erg s^{-1}; it could have been detected by Chandra out to $z \approx 10$.

Stacking studies of galaxies at cosmologically interesting distances

At the fainter X-ray fluxes of the CDF-N, a significant number of 'normal' galaxies are detected at $z \lesssim 0.3$, where the observed emission appears to originate from X-ray binaries, ultraluminous X-ray sources, supernova remnants, and perhaps low-luminosity AGN (e.g. Hornschemeier *et al.* (2001, 2002b)). For example, in the HDF-N, Chandra has detected all optically luminous galaxies out to $z = 0.15$ (Brandt *et al.* 2001a).

X-ray studies of normal galaxies at cosmological distances are of importance, since normal galaxies are expected to be the most numerous extragalactic X-ray sources at the faintest X-ray fluxes; in Fig. 12.5 there are *c.* 3000 galaxies but only 27 X-ray sources, so 'there's plenty of room at the bottom' (cf. Feynman (1960)). Normal galaxies are expected to dominate the number counts at soft-band fluxes of *c.* 5×10^{-18} erg cm^{-2} s^{-1} (e.g. Ptak *et al.* (2001), Hornschemeier *et al.* (2002a), Miyaji & Griffiths (2002)), and they should be one of the main source types detected by missions such as XEUS and Generation-X. Furthermore, the X-ray properties of galaxies might well have changed in response to the substantial change in the cosmic star-formation rate over the history of the Universe; changes in the star-formation rate should affect the production of X-ray binaries and supernovae (e.g. Ghosh & White 2001).

Most normal galaxies at $z \gtrsim 0.3$ cannot be individually detected in the current CDF-N data, but their average properties can be probed using stacking techniques, where the X-ray emissions from many individually undetected galaxies are added together. Using stacking techniques, Hornschemeier *et al.* (2002a) have measured the average X-ray luminosities of *c.* L_B^* spiral galaxies out to $z = 1.2$, corresponding to a look-back time of *c.* 9.0 Gy. These measurements allow the first reliable predictions of the number of normal galaxies that should be seen in deeper X-ray exposures; typical observed soft-band fluxes are *c.* $(3\text{–}7)\times 10^{-18}$ erg cm^{-2} s^{-1}. There is evidence for a factor of 2–3 increase in X-ray luminosity (per unit B-band luminosity) with redshift (see Fig. 12.9), although this is weaker than some theoretical models have predicted.

At higher redshifts ($z = 2$–4), X-ray stacking analyses have allowed average detections of the Lyman break galaxies in the CDF-N (see Fig. 12.10 (Brandt *et al.* 2001c)). In the rest-frame ultraviolet and optical bands these galaxies share many of the properties of local starburst galaxies and have estimated star-formation rates

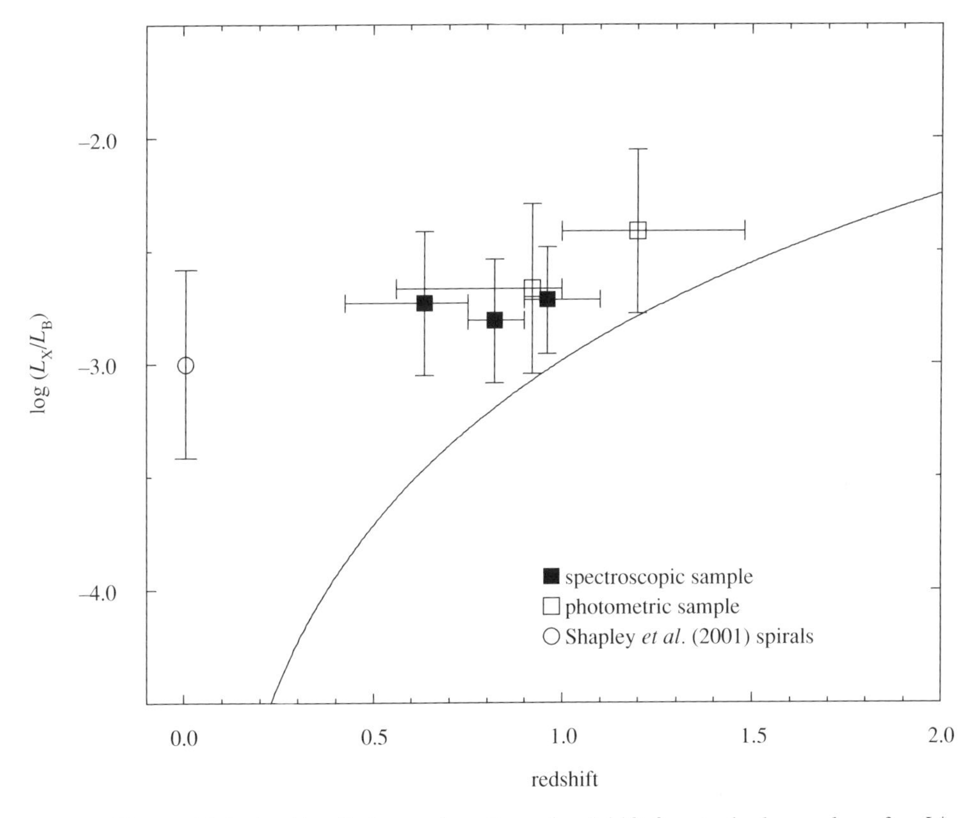

Figure 12.9. $\log(L_X/L_B)$ as a function of redshift for stacked samples of *c.* L_B^* spiral galaxies (L_X is for 0.5–2 keV); samples with both spectroscopically and photometrically derived redshifts are shown. The solid curve indicates the 2σ X-ray sensitivity limit normalized by L_B^*. The data point at $z \approx 0$ is derived from Shapley *et al.* (2001). Adapted from Hornschemeier *et al.* (2002a).

of *c.* 20–50 $M_\odot$ y^{-1}. Their average X-ray luminosity (*c.* 3×10^{41} erg s^{-1} in the rest-frame 2–8 keV band) is similar to those of the most luminous local starburst galaxies (these have star-formation rates comparable with those estimated for the Lyman break galaxies). For Lyman break galaxies, the observed ratio of X-ray to B-band luminosity is somewhat, but not greatly, higher than that seen from local starburst galaxies.

The future

Additional X-ray coverage

The current CDF-N survey is far from the limit of Chandra's capability. The full-band detector background is so low that, with appropriate grade screening, Chandra will not fully enter the background-limited regime near the aim point for exposure times of less than *c.* 5 Ms. In the soft band the situation is even better, due to

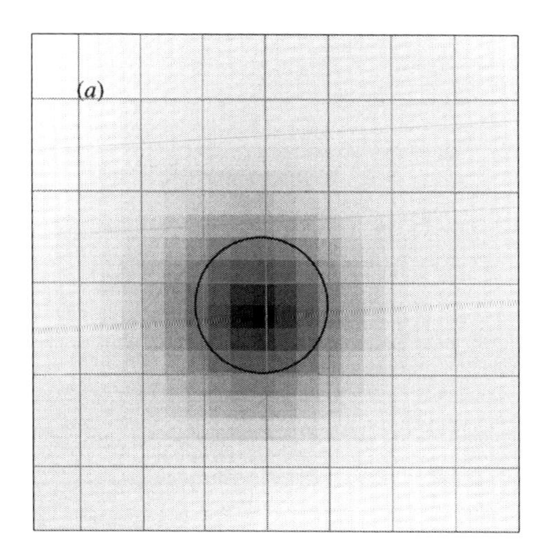

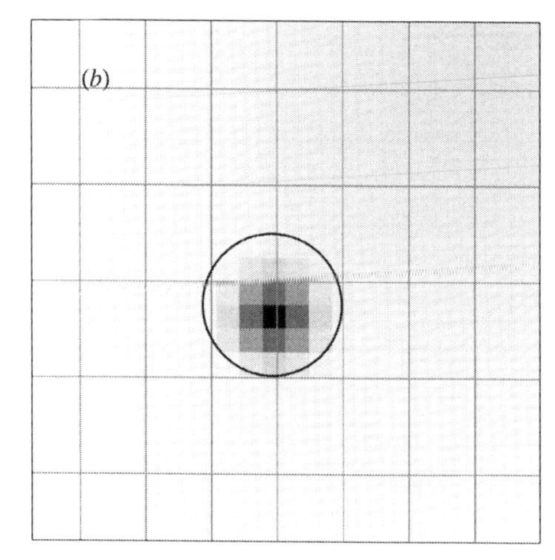

Figure 12.10. Stacked (*a*) soft-band and (*b*) full-band Chandra images of 24 HDF-N Lyman break galaxies at $z = 2.0$–4.0. The black circles are centred on the stacking position and have radii of 1.5″. The effective exposure time for the average Lyman break galaxy is 22.4 Ms (260 days). Both bands give highly significant detections ((*a*) 99% confidence, (*b*) 99.5% confidence), as assessed with a Monte Carlo technique. Adapted from Brandt *et al.* (2001c).

the lower background, and Chandra should remain photon-limited to *c*. 10 Ms. At present, Chandra observation time is allocated to extend the survey to 2 Ms, and additional observations will be proposed.

Figure 12.11 shows that, with sufficient exposure (5–10 Ms), Chandra surveys can reach depths comparable with those discussed for missions such as XEUS. They would thereby bolster the science cases for missions such as XEUS and Generation-X, providing key information on the existence and nature of the sources to be targeted by these missions. Important targets would include the first massive black holes at high redshift as well as normal and starburst galaxies at intermediate redshift. Detailed spectral, temporal and spatial constraints would be obtained for all the sources currently detected. Furthermore, Fig. 12.11 shows that Chandra positions are likely to be the best available for at least another 15–20 years; these will be essential for the reliable identification of optically faint sources at high redshift.

Observations at other wavelengths

Multi-wavelength follow-up studies of the CDF-N sources continue, and a catalogue presenting the current optical photometric and spectroscopic results will be completed shortly (Barger *et al.* 2002). The GOODS project will soon obtain deep, public Hubble Space Telescope Advanced Camera for Surveys and SIRTF coverage

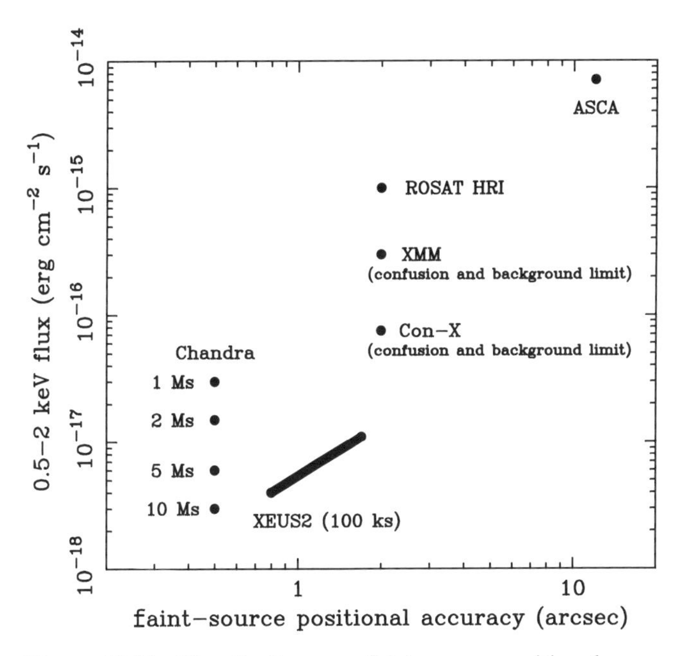

Figure 12.11. Flux limit versus faint-source positional accuracy for some past, present and future X-ray missions. Note that the locations in the diagram for future missions should be taken as approximate, and that Constellation-X is focused on high-throughput spectroscopy rather than deep surveys. Both XMM-Newton and Constellation-X are background-limited and suffer from source confusion at approximately the positions shown. With sufficient exposure, Chandra can achieve sensitivities comparable with those discussed for future missions such as XEUS. Furthermore, Chandra positions are likely to be the best available for at least another 15–20 years.

over the deepest third of the CDF-N (see Fig. 12.1).† These projects, along with others already completed and in progress, will provide a panchromatic dataset with the sensitivity and angular resolution needed to complement fully the CDF-N.

We thank all the members of the CDF-N team. This work would not have been possible without the efforts of the entire Chandra and ACIS teams. We gratefully acknowledge the financial support of NASA grant NAS 8-38252 (Gordon P. Garmire, PI), NSF CAREER award AST-9983783 (W.N.B., D.M.A., F.E.B.), and NASA GSRP grant NGT 5-50247 and the Pennsylvania Space Grant Consortium (A.E.H.).

References

Alexander, D. M. *et al.* 2001 *Astron. J.* **122**, 2156–2176.
Alexander, D. M. *et al.* 2002a *Astron. J.* **123**, 1149–1162.
Alexander, D. M. 2002b *Astrophys. J.* **568**, L85–L88.

† www.stsci.edu/science/goods/.

Aussel, H., Cesarsky, C. J., Elbaz, D. & Starck, J. L. 1999 *Astron. Astrophys.* **342**, 313–336.
Barger, A. J., Cowie, L. L., Mushotzky, R. F. & Richards, E. A. 2001a *Astron. J.* **121**, 662–682.
Barger, A. J., Cowie, L. L., Steffen, A. T., Hornschemeier, A. E., Brandt, W. N. & Garmire, G. P. 2001b *Astrophys. J.* **560**, L23–L28.
Barger, A. J. *et al.* 2002 *Astron. J.* **124**, 1839.
Bauer, F. E. *et al.* 2002a *Astron. J.* **123**, 1163–1178.
Bauer, F. E. 2002b *Astrophys. J.* **124**, 2351.
Brandt, W. N. *et al.* 2000 *Astron. J.* **119**, 2349–2359.
Brandt, W. N. *et al.* 2001a *Astron. J.* **122**, 1–20.
Brandt, W. N. *et al.* 2001b *Astron. J.* **122**, 2810–2832.
Brandt, W. N. *et al.* 2001c *Astrophys. J.* **558**, L5–L9.
Brandt, W. N. *et al.* 2002a *Astrophys. J.* **569**, L5–L9.
Brandt, W. N. *et al.* 2002b (In preparation.)
Cohen, J. G. *et al.* 2000 *Astrophys. J.* **538**, 29–52.
Cowie, L. L., Garmire, G. P., Bautz, M. W., Barger, A. J., Brandt, W. N. & Hornschemeier, A. E. 2002 *Astrophys. J.* **566**, L5–L8.
Dickinson, M. *et al.* 2000 *Astrophys. J.* **531**, 624–634.
Ebeling, H., White, D. A. & Rangarajan, F. V. N. 2002 *Mon. Not. R. Astron. Soc.* (Submitted.)
Fabian, A. C. & Barcons, X. 1992 *A. Rev. Astron. Astrophys.* **30**, 429–456.
Fadda, D. *et al.* 2002 *Astron. Astrophys.* **383**, 838–853.
Ferguson, H. C., Dickinson, M. & Williams, R. 2000 *A. Rev. Astron. Astrophys.* **38**, 667–715.
Feynman, R. P. 1960 In *Engineering and Sci.*, Feb. issue, California Institute of Technology, USA.
Freeman, P. E., Kashyap, V., Rosner, R. & Lamb, D. Q. 2002 *Astrophys. J. Suppl.* **138**, 185–218.
Ghosh, P. & White, N. E. 2001 *Astrophys. J.* **559**, L97–L100.
Giacconi, R., Gursky, H., Paolini, F. R. & Rossi, B. B. 1962 *Phys. Rev. Lett.* **9**, 439–443.
Giacconi, R. *et al.* 2001 *Astrophys. J.* **551**, 624–634.
Giacconi, R. *et al.* 2002 *Astrophys. J. Suppl.* **139**, 369–410.
Haiman, Z. & Loeb, A. 1999 *Astrophys. J.* **521**, L9–L12.
Hasinger, G. 2000 In *ISO Surveys of a Dusty Universe* (eds. D. Lemke, M. Stickel & K. Wilke). Springer Lecture Notes in Physics, vol. 548, pp. 423–431. Heidelberg: Springer.
Hasinger, G., Burg, R., Giacconi, R., Schmidt, M., Trümper, J. & Zamorani, G. 1998 *Astron. Astrophys.* **329**, 482–494.
Hasinger, G. *et al.* 2001 *Astron. Astrophys.* **365**, L45–L50.
Hornschemeier, A. E. *et al.* 2000 *Astrophys. J.* **541**, 49–53.
Hornschemeier, A. E. *et al.* 2001 *Astrophys. J.* **554**, 742–777.
Hornschemeier, A. E. *et al.* 2002a *Astrophys. J.* **568**, 82–87.
Hornschemeier, A. E. *et al.* 2002b (Preprint astro-ph/0305086.)
Jansen, F. *et al.* 2001 *Astron. Astrophys.* **365**, L1–L6.
Lehmann, I. *et al.* 2001 *Astron. Astrophys.* **371**, 833–857.
Miyaji, T. & Griffiths, R. E. 2002 *Astrophys. J.* **564**, L5–L8.
Mushotzky, R. F., Cowie, L. L., Barger, A. J. & Arnaud, K. A. 2000 *Nature*, **404**, 459–464.
Muxlow, T. W. B., Wilkinson, P. N., Richards, A. M. S., Kellermann, K. I., Richards, E. A. & Garrett, M. A. 1999 *New Astron. Rev.* **43**, 623–627.

Ptak, A., Griffiths, R., White, N. & Ghosh, P. 2001 *Astrophys. J.* **559**, L91–L95.
Richards, E. A., Kellermann, K. I., Fomalont, E. B., Windhorst, R. A. & Partridge, R. B. 1998 *Astron. Astrophys.* **116**, 1039–1054.
Richards, E. A. *et al.* 1999 *Astrophys. J.* **526**, L73–L76.
Rosati, P. *et al.* 2002 *Astrophys. J.* **566**, 667–674.
Shapley, A., Fabbiano, G. & Eskridge, P. B. 2001 *Astrophys. J. Suppl.* **137**, 139–199.
Vignali, C. *et al.* 2001 *Astron. J.* **122**, 2143–2155.
Waddington, I., Windhorst, R. A., Cohen, S. H., Partridge, R. B., Spinrad, H. & Stern, D. 1999 *Astrophys. J.* **526**, L77–L80.
Weisskopf, M. C., Tananbaum, H. D., Van Speybroeck, L. P. & O'Dell, S. L. 2000 *Proc. SPIE* **4012**, 2–16.

13

Hunting the first black holes

BY GÜNTHER HASINGER

Max-Planck Institut für extraterrestrische Physik, and Astrophysikalisches Institut Potsdam

Introduction

Deep X-ray surveys indicate that the cosmic X-ray background (XRB) is largely due to accretion onto supermassive black holes, integrated over cosmic time. In the soft (0.5–2 keV) band more than 90% of the XRB flux has been resolved using 1.4 Ms observations with Röntgensatellit (ROSAT) (Hasinger *et al.* 1998) and more recently 1 Ms Chandra observations (Brandt *et al.* 2001b; Rosati *et al.* 2002) and 100 ks observations with XMM-Newton (Hasinger *et al.* 2001) (see Fig. 13.1). In the harder (2–10 keV) band, a similar fraction of the background has been resolved with the above Chandra and XMM-Newton surveys, reaching source densities of *c.* 4000 $\deg^{-2}$. Surveys in the very hard (5–10 keV) band have been pioneered using *Beppo*SAX, which resolved *c.* 30% of the XRB (Fiore *et al.* 1999). XMM-Newton and Chandra have now also resolved the majority (60–70%) of the very hard X-ray background. The log N–log S distribution shows a significant cosmological flattening in the softer bands (see Fig. 13.2), while in the very hard band it is still relatively steep, indicating that those surveys have not yet sampled the redshifts where the strong cosmological evolution of the sources saturates.

Optical follow-up programmes with 8–10 m telescopes have been completed for the ROSAT deep surveys and found predominantly active galactic nuclei (AGN) as counterparts of the faint X-ray source population (Lehmann *et al.* 2001; Schmidt *et al.* 1998; Zamorani *et al.* 1999), mainly X-ray and optically unobscured AGN (type-1 Seyferts and quasi-stellar objects (QSOs)) and a smaller fraction of obscured AGN (type-2 Seyferts). Optical identifications for the deepest Chandra and XMM-Newton fields are still far from complete; however, a mixture of obscured and unobscured AGN with an increasing fraction of obscuration seems to be the dominant population in these samples too (Barger *et al.* 2001a; Fiore *et al.* 2000;

Frontiers of X-Ray Astronomy, ed. A.C. Fabian, K.A. Pounds and R.D. Blandford. Published by Cambridge University Press.

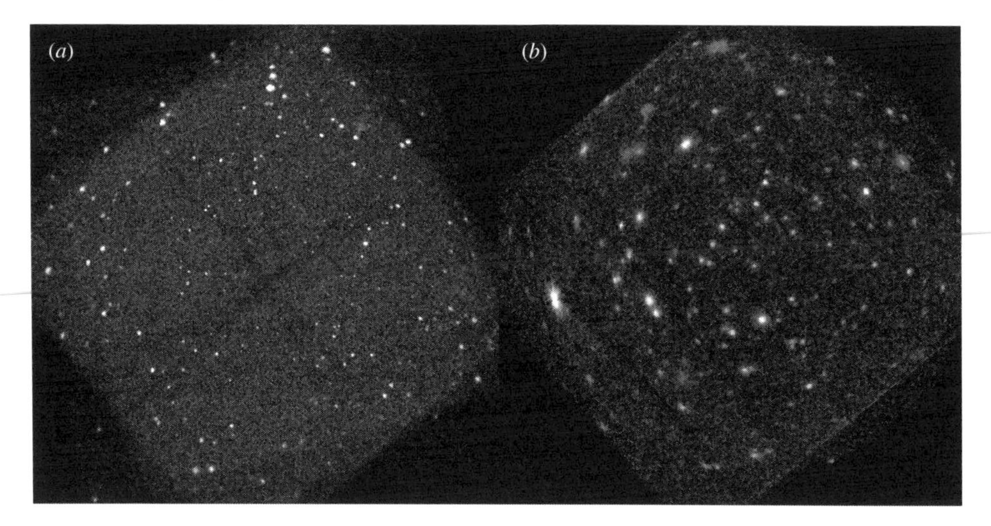

Figure 13.1. X-ray (*a*) XMM-Newton PN and MOS images of the Lockman Hole field (Hasinger *et al.* 2001) and (*b*) the Chandra ACIS-I image of the Chandra Deep Field South (Rosati *et al.* 2002). The field sizes are *c.* 30 × 30 arcmin^2 and 20 × 20 arcmin^2, respectively.

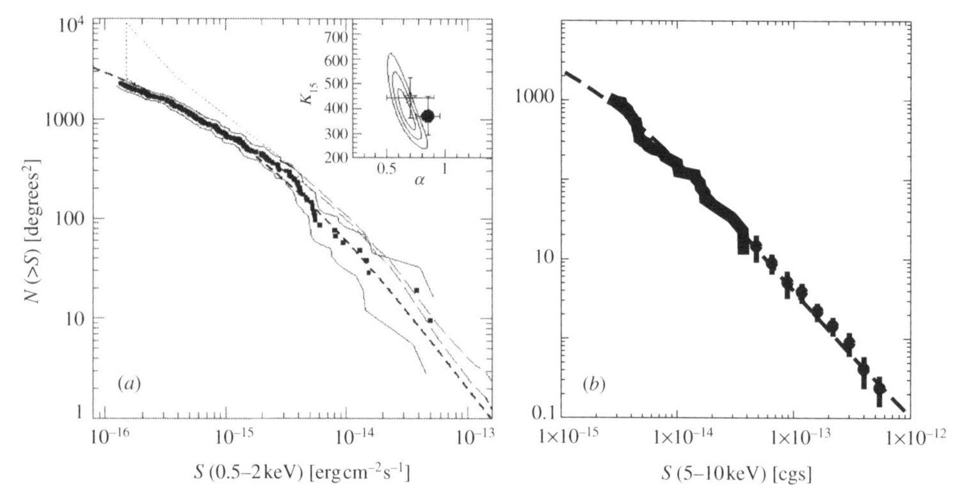

Figure 13.2. (*a*) log N–log S in the 0.5–2.0 keV band from the 300 ks Chandra Deep Field South observation (filled squares) (Tozzi *et al.* 2001). The long-dashed lines are source counts from the Lockman Hole field and the dotted contour is from the ROSAT fluctuation analysis (Hasinger *et al.* 1993). (*b*) log N–log S in the 5–10 keV band from the 100 ks XMM-Newton observation (thick line) in the Lockman Hole field (Hasinger *et al.* 2001) compared with the *Beppo*SAX source counts (data points with error bars) (Fiore *et al.* 1999). The short-dashed line in both figures gives the prediction based on the most recent background-synthesis models (Gilli *et al.* 2001).

Rosati *et al.* 2002; Tozzi *et al.* 2001) (see below). Interestingly, first examples of the long-sought class of high-redshift, high-luminosity, heavily obscured AGN (type-2 QSO) have been detected in deep Chandra fields (Norman *et al.* 2002; Stern *et al.* 2002) and in the XMM-Newton deep survey in the Lockman Hole field (Hasinger & Lehmann 2002).

After having understood the basic contributions to the X-ray background, the general interest is now focusing on understanding the physical nature of these sources, the cosmological evolution of their properties and their role in models of galaxy evolution. We know that basically every galaxy with a spheroidal component in the local Universe has a supermassive black hole in its centre (Gebhardt *et al.* 2000). The luminosity function of X-ray selected AGN shows strong cosmological density evolution at redshifts up to 2, which goes hand in hand with the cosmic star-formation history (Miyaji *et al.* 2000, 2001). At the redshift peak of optically selected QSOs of $c.\ z = 2.5$, the AGN space density is several hundred times higher than it is locally, which is in line with the assumption that most galaxies have been active in the past and that the feeding of their black holes is reflected in the X-ray background. While the comoving space density of optically and radio-selected QSOs declines significantly beyond a redshift of 3 (Fan *et al.* 2001; Schmidt *et al.* 1995; Shaver *et al.* 1996), a similar decline has not yet been observed in the X-ray selected AGN population (Miyaji *et al.* 2000), although the statistical quality of the high-redshift AGN samples needs to be improved. The new Chandra and XMM-Newton surveys are bound to give additional constraints here.

The X-ray observations have so far been about consistent with population-synthesis models based on unified AGN schemes (Comastri *et al.* 1995; Gilli *et al.* 2001), which explain the hard spectrum of the X-ray background by a mixture of absorbed and unabsorbed AGN, folded with the corresponding luminosity function and its cosmological evolution. According to these models, most AGN spectra are heavily absorbed and *c.* 80% of the light produced by accretion will be absorbed by gas and dust (Fabian *et al.* 1998). However, these models are far from unique and contain a number of hidden assumptions, so that their predictive power remains limited until complete samples of spectroscopically classified hard-X-ray sources are available. In particular, they require a substantial contribution of high-luminosity obscured X-ray sources (type-2 QSOs), which so far have only scarcely been detected. The cosmic history of obscuration and its potential dependence on intrinsic source luminosity remain completely unknown. Gilli *et al.* (2001), for example, assumed strong evolution of the obscuration fraction (ratio of type 2 : type-1 AGN) from 4 : 1 in the local Universe to much larger covering fractions (10 : 1) at high redshifts (see also Fabian *et al.* (1998)). The gas-to-dust ratio in high-redshift, high-luminosity AGN could be completely different from the usually assumed galactic value due to sputtering of the dust particles in the strong radiation field (Granato *et al.* 1997). This might provide objects that are heavily absorbed at X-rays and are unobscured at optical wavelengths.

In this chapter I discuss briefly the current status of the optical identification work in the ROSAT/XMM-Newton/Chandra deep survey in the Lockman Hole, which is largely based on optical work with the Keck telescope led by Maarten Schmidt (see Schmidt *et al.* (1998)) and Hasinger & Lehmann (2002) for more details). I also present preliminary results of optical identifications in the Chandra Deep Field South, obtained with the European Southern Observatory (ESO) Very Large Telescope (VLT), published in Szokoly *et al.* (2002) (see also Rosati *et al.* (2002), Tozzi *et al.* (2001)). I then discuss the results and come to some tentative conclusions about the evolution of X-ray sources at high redshifts.

Optical identifications of deep X-ray surveys

The Lockman Hole field

The Lockman Hole field has been observed with the XMM-Newton observatory during the performance verification phase (see Fig. 13.1(a)). About 100 ks of good data, centred on the same sky position as the ROSAT HRI image, have been accumulated with the European Photon Imaging Camera (EPIC), reaching minimum fluxes of 0.31, 1.4 and 2.4×10^{-15} erg cm^{-2} s^{-1} in the 0.5–2, 2–10 and 5–10 keV energy bands, respectively. Within an off-axis angle of 10 arcmin, 148, 112 and 61 sources, respectively, have been detected. In the 5–10 keV energy band a somewhat lower sensitivity compared with the 1 Ms Chandra Deep Field South observation has been reached (see Rosati *et al.* (2002)), resolving *c.* 60% of the very hard X-ray background (Hasinger *et al.* 2001). This is about a factor of 20 more sensitive than the previous *Beppo*SAX observations. A 300 ks observation of the Lockman Hole has been accumulated with the Chandra high-resolution camera (HRC) in the 0.5–7.0 keV band, reaching a similar flux limit compared with the XMM-Newton pointing. The Chandra HRC data provide very accurate source positions, whereas XMM-Newton allows spectrophotometry of very faint intrinsically absorbed X-ray sources due to its unprecedented sensitivity in the hard band.

The optical counterparts for *c.* 60 X-ray sources are already known from the spectroscopic identification of the ROSAT Ultradeep Survey sample (Lehmann *et al.* 2001). Among them are one of the most distant X-ray selected quasars at $z = 4.45$ (Schneider *et al.* 1998) and one of the highest redshift clusters of galaxies at $z = 1.26$ (Hashimoto *et al.* 2002; Hasinger *et al.* 1999; Thompson *et al.* 2001). We have identified 25 new XMM-Newton sources using low-resolution multi-slit mask spectra taken with the Low Resolution Imaging Spectrograph at the Keck II telescope in March 2001 (M. Schmidt 2001, personal communication; Lehmann *et al.* 2002). Among the new XMM sources we have found only a few new broad-emission-line AGN (type 1), while the optical spectra of most new sources show narrow emission lines and/or only galaxy-like continuum emission at redshifts $z < 1.0$. In several cases, high ionization emission lines like [Ne V] l 3426 are absent and thus we see no sign of AGN activity in the optical spectrum; however, their high X-ray luminosity ($L_X > 10^{43}$ erg s^{-1}) and/or the strong intrinsic absorption ($\log N_H > 22.0$ cm^{-2}) reveal a type-2 AGN in these sources. Three new sources showing typical galaxy spectra have been detected only in the 0.5–2.0 keV band. Due to their relatively low X-ray luminosities ($\log L_X < 42.0$) and their soft-X-ray spectra (no indication of intrinsic absorption) we classify them as normal galaxies. Several sources with X-ray luminosities in the range $42.0 < L_X < 43.0$, which show galaxy-like optical spectra, are hard to classify due to the small number of photons in their X-ray spectra. We preliminarily classify them as a type-2 AGN/galaxy. The completeness of the identification ranges from 61% in the soft sample to 79% in the ultra-hard sample (5–10 keV energy band) (Lehmann *et al.* 2002). The majority of the sources identified spectroscopically so far are type-1 and type-2 AGN. Although we have no complete identification so far, we find a strong indication for a larger fraction of type-2 AGN, especially in the ultra-hard sample, compared with that (*c.* 20%) of the UDS. Nearly all spectroscopically identified type-2 AGN are at moderate redshift ($z < 1$). One type-2 QSO candidate (X174A) at $z = 3.240$ has been identified in the Lockman Hole region so far (Hasinger & Lehmann 2002).

Most of the unidentified faint XMM-Newton sources have very faint optical counterparts (optical magnitude $R > 24.0$) and at least half of them are extremely red objects (EROs, $R - K' > 5.0$). The new XMM-Newton sources with EROs as optical counterparts are similar to those objects in the UDS with photometric redshifts, suggesting obscured AGN at redshifts of $1 < z < 3$. The photometric-redshift technique is probably the only tool for identifying such faint optical objects. The XMM-Newton source population at faint fluxes is therefore likely to be dominated by obscured AGN (type 2), as predicted by the AGN population-synthesis models for the XRB.

The Chandra Deep Field South (CDFS)

The Chandra X-ray Observatory has performed deep X-ray surveys in a number of fields with ever increasing exposure times (Brandt *et al.* 2001a; Giacconi *et al.* 2001; Hornschemeier *et al.* 2000; Mushotzky *et al.* 2000; Tozzi *et al.* 2001) and has completed two 1 Ms Chandra exposures, in the CDFS (Giacconi *et al.* 2002; Rosati *et al.* 2002) and in the Hubble Deep Field North (HDF-N (Brandt *et al.* 2001b)). The latter exposure is currently being increased to 2 Ms.

Here I discuss results from the 940 ks CDFS observation. The source counts (see Fig. 13.2) have been extended to 5.5×10^{-17} erg cm^{-2} s^{-1} in the soft 0.5–2 keV band and 4.5×10^{-16} erg cm^{-2} s^{-1} in the hard 2–10 keV band, reaching a space density of almost 4000 deg^{-2}, resolving more than 80% of the background in both bands. A total of *c.* 360 sources has been detected.

Optical imaging and spectroscopy were performed on *c.* 10 nights with the ESO VLT in the time frame April 2000 to December 2001, using deep optical imaging and low-resolution multi-object spectroscopy with the FORS instruments for individual exposure times ranging from 1 to 5 h. Some preliminary results including the VLT optical spectroscopy have already been presented (Norman *et al.* 2002; Rosati *et al.* 2002; Tozzi *et al.* 2001). The complete optical spectroscopy appears in Szokoly *et al.* (2002). Figure 13.3 shows examples of four VLT spectra. The two spectra in Fig. 13.3(*a*) and (*b*) show high-redshift QSOs with rest frame UV-absorption features (broad absorption line (BAL) or mini-BAL QSOs), which both have some indication of intrinsic absorption in their X-ray spectra. The object in Fig. 13.3(*c*) is the famous highest-redshift type-2 QSO detected in the CDFS with heavy X-ray absorption in the QSO rest frame (Norman *et al.* 2002). The spectrum in Fig. 13.3(*d*) shows a Seyfert-2 galaxy with heavy X-ray absorption and an AGN-type luminosity. The last spectrum is characteristic for the bulk of the detected galaxies, which show either no or very faint high-excitation lines, indicating the AGN nature of the object, so that we have to resort to a combination of optical and X-ray diagnostics to classify them as AGN (see below). Up to now, redshifts could be obtained for 169 of the 360 sources in the CDFS, of which 123 are very reliable (high-quality spectra with two or more spectral features), while the remaining optical spectra contain only a single emission line, or are too noisy. For objects fainter than $R = 24$, reliable redshifts can be obtained (see also Fig. 13.5) if the spectra contain strong emission lines. For the remaining optically faint objects we have to resort to photometric redshift techniques. About 11% of the CDFS sources have no counterpart, even in deep VLT optical images ($R < 27.5$) or near-IR imaging (15% at $K < 22$) (Rosati *et al.* 2002). Nevertheless, for a subsection of the sample at off-axis angles smaller than 8 arcmin we obtain a spectroscopic completeness of *c.* 60%.

Optical/X-ray classification

Type-1 AGN (Seyfert-1 and QSOs) can often be readily identified by the broad permitted emission lines in their optical spectra. Luminous Seyfert-2 galaxies show strong forbidden-emission lines and high-excitation lines, indicating photoionization by a hard continuum source. However, in the spectroscopic identifications of the ROSAT Deep Surveys it became apparent that an increasing fraction of faint X-ray-selected AGN show a significant, sometimes dominant, contribution of stellar

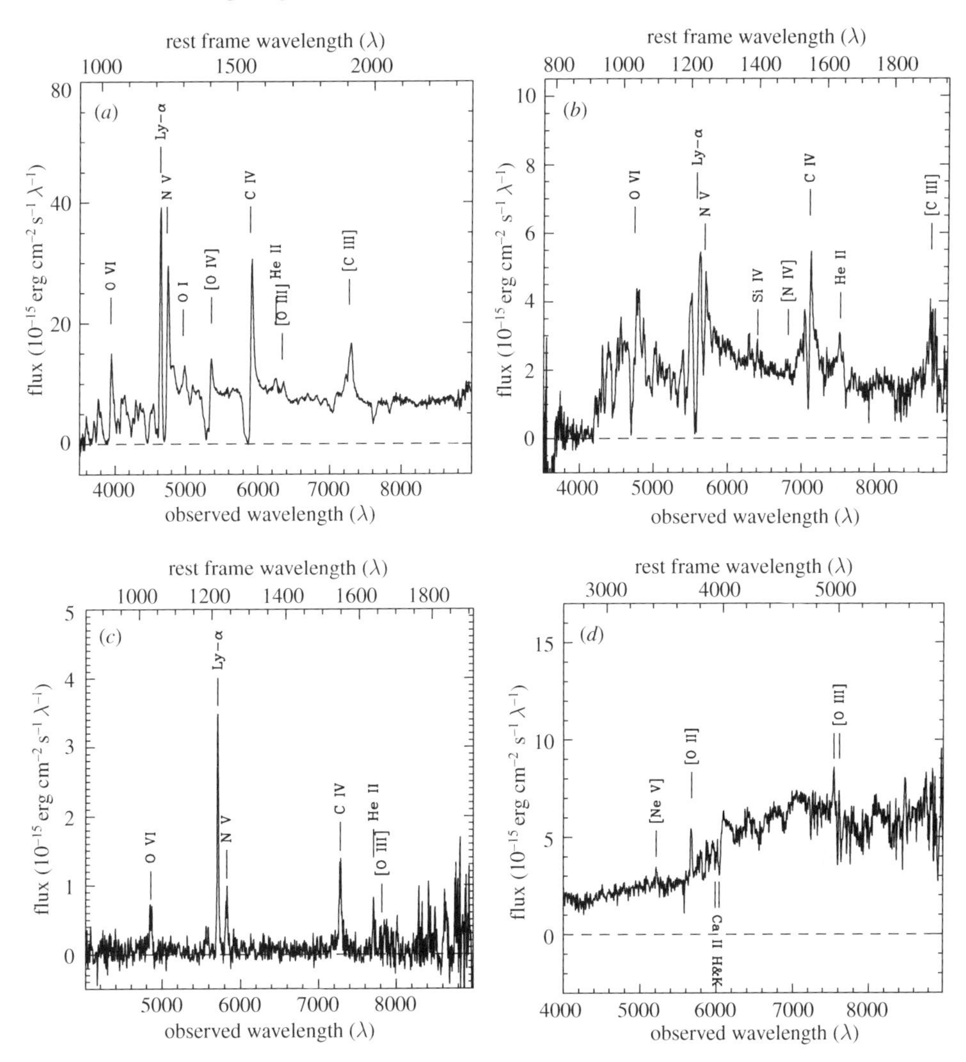

Figure 13.3. Optical spectra of some selected CDFS sources obtained using multi-object spectroscopy with FORS at the VLT (Szokoly *et al.* 2002): (*a*) BAL QSO CDFS-062 at $z = 2.810$ (see also Giacconi *et al.* (2001)); (*b*) high-redshift QSO CDFS-024 at $z = 3.605$, showing strong absorption lines; (*c*) QSO-2 CDFS-202 at $z = 3.705$ (see Norman *et al.* (2002)); (*d*) Seyfert-2 CDFS-175b at $z = 0.522$, showing a weak high excitation line of [Ne V].

light from the host galaxy in their optical spectra, depending on the ratio of optical luminosity between nuclear and galaxy light (Lehmann *et al.* 2000, 2001). If an AGN is outshone by its host galaxy, it is not possible to detect it optically. Many of the counterparts of the faint X-ray sources detected by Chandra and XMM-Newton show optical spectra which are dominated by their host galaxy and only a minority has clear indications of an AGN nature (see also Barger *et al.* (2001a,b)).

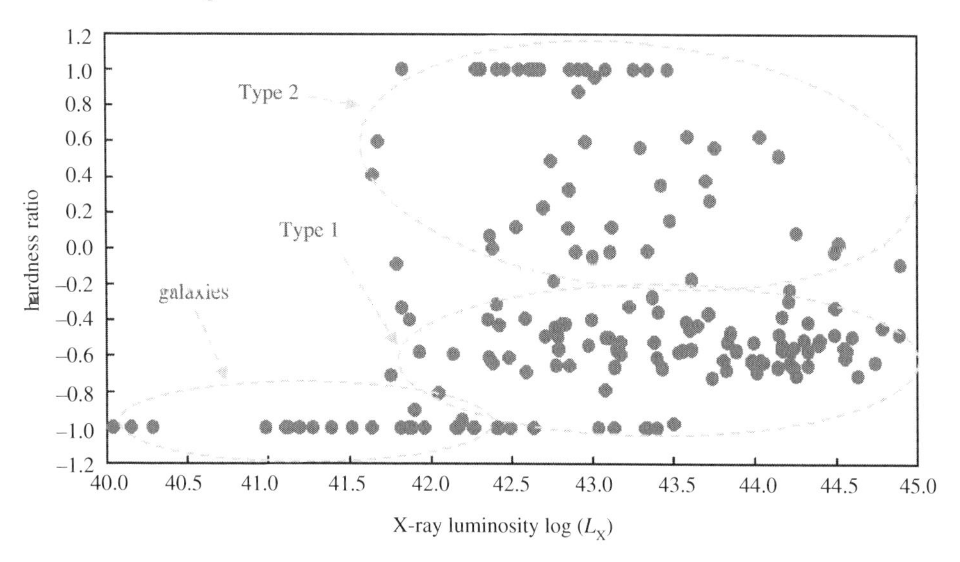

Figure 13.4. Hardness ratio of CDFS and Lockman Hole sources with redshifts as a function of X-ray luminosity. The dashed ellipses indicate the regions expected for type-1 and type-2 AGN as well as for galaxies whose X-ray emission is dominated by thermal processes.

In these cases, the X-ray emission could still be dominated by the AGN, while a contribution from stellar and thermal processes (hot gas from supernova remnants, starbursts and thermal haloes, or a population of X-ray binaries) can be important as well.

In these cases X-ray diagnostics in addition to the optical spectroscopy can be crucial to classify the source of the X-ray emission. AGN typically (but not always!) have X-ray luminosities above 10^{42} erg s^{-1} and power-law spectra, often with significant intrinsic absorption. Local, well-studied starburst galaxies have X-ray luminosities typically below 10^{42} erg s^{-1} and very soft-X-ray spectra. Thermal haloes of galaxies and the intergalactic gas in groups can have higher X-ray luminosities, but have soft spectra as well. The redshift effect in addition helps the X-ray diagnostic, because soft-X-ray spectra appear even softer at moderate redshift, while the typical AGN power-law spectra appear harder over a very wide range of redshifts.

In Fig. 13.4 the X-ray hardness ratio is shown as a function of the X-ray luminosity (in the 0.5–2 keV, 2–10 keV or 0.5–10 keV band, depending on the band in which the object was detected) for 170 sources for which we have optical spectra and reliable redshifts in the CDFS (Rosati *et al.* 2002; Szokoly *et al.* 2002) and the Lockman Hole (Lehmann *et al.* 2001, 2002) for X-ray sources detected by Chandra and XMM-Newton, respectively. The hardness ratio is defined as $R_H \equiv (H - S)/(H + S)$, where H and S are the net count rates in the hard (2–7 keV for Chandra and 2–4.5 keV for XMM-Newton) and the soft band (0.5–2 keV), respectively. The X-ray luminosities are not corrected for internal absorption and are computed assuming $H_0 = 50$ km s^{-1} Mpc^{-1} and $q_0 = 0.5$.

Although this diagram is for illustration purposes only and a correct treatment would have to properly take into account the different instrument characteristics and detection bands, it clearly shows a segregation of the different X-ray emitters (indicated by the dashed elliptical outlines). Type-1 AGN have luminosities above 10^{42} erg s^{-1} and hardness ratios scattered around $R_{\rm H} \simeq -0.6$, corresponding to a power-law photon index *c.* $\Gamma = 1.8$, which is typical for the intrinsic continuum of AGN. Type-2 AGN also have luminosities above 10^{42} erg s^{-1}, but are scattered to much higher hardness ratios ($R_{\rm H} > -0.2$). Direct spectral fits of the XMM-Newton and (some) Chandra spectra clearly indicate that these harder spectra are due to neutral gas absorption and not due to a flatter intrinsic slope (Hasinger & Lehmann 2002; Mainieri *et al.* 2002; Norman *et al.* 2002). It is interesting to note that no high-luminosity, very hard sources exist in this diagram. This is due to a selection effect of the pencil beam surveys: due to the small solid angle, the rare high-luminosity sources are only sampled at high redshifts, where the absorption cut-off of type-2 AGN is redshifted to softer X-ray energies. Indeed, the type-2 QSOs in this sample (Lehmann *et al.* 2002; Norman *et al.* 2002) are the objects at $L_{\rm X} > 10^{44}$ erg s^{-1} and $R_{\rm H} > 0$.

About 10% of the sources have optical spectra of normal galaxies, X-ray luminosities below 10^{42} erg s^{-1} and very soft spectra ($R_{\rm H} \sim -1$), typical for starburst galaxies or hot-gas haloes. The deep Chandra and XMM-Newton surveys therefore for the first time detect the population of normal starburst galaxies out to intermediate redshifts (Giacconi *et al.* 2001; Lehmann *et al.* 2002; Mushotzky *et al.* 2000), for which a significant contribution to the XRB had been claimed for a long time (McHardy *et al.* 1998), although at much lower fluxes and therefore almost negligible contribution to the background. These galaxies might become an important means to study the star-formation history in the Universe completely independently from optical/UV, submillimetre or radio observations. However, in the X-ray-luminosity range *c.* 10^{42} erg s^{-1}, where the emission from star-forming processes and the central AGN may be comparable, there will always remain ambiguities.

The joint optical/X-ray diagnostics scheme can also be applied to the spectroscopically identified X-ray sources in other deep Chandra fields in order to obtain as complete a sample of faint X-ray source classification as possible. Table 13.1 gives a summary of optical identifications and X-ray source types in the two deep fields discussed here, as well as in the Hawaii 13 h field, the Abell 370 cluster field and the HDF-N, which all have spectroscopically identified samples in the literature (Barger *et al.* 2001a,b).

Table 13.1. *Chandra and XMM-Newton survey identifications*

field	type 1	type 2	galaxies	total	reference
Lockman Hole	41	26	7	74	Lehmann *et al.* (2001, 2002)
CDFS	47	73	49	169	Szokoly *et al.* (2002)
Abell 370	9	5	6	20	Barger *et al.* (2001b)
13 h[a]	5	7	1	20	Barger *et al.* (2001a)
HDF-N[a]	10	10	0	20	Barger *et al.* (2001b)

[a] Only 2–7 keV band detections considered.

The Chandra/XMM-Newton redshift distribution

All the above samples have a spectroscopic completeness of *c*. 60%, which is mainly caused by the fact that *c*. 40% of the counterparts are optically too faint to obtain reliable spectra. This incompleteness is probably also reflecting some redshift bias, and it is most likely that higher-redshift objects are missing, as well as faint emission-line objects, where the strongest emission lines ([O II], Lya) fall outside the optical bands. On the other hand, the optically faintest identified sources ($R = 24$–25) are distributed throughout the whole redshift range $z = 0$–4 (see Fig. 13.5); therefore there is reason to believe that a substantial fraction of the sources so far unidentified follow the same redshift distribution as the identified sources. The completeness of 60% therefore allows us to compare the redshift distribution with predictions from XRB population-synthesis models (Gilli *et al.* 2001), based on the AGN X-ray luminosity function and its evolution as determined from the ROSAT surveys (Miyaji *et al.* 2000), which predict a maximum at redshifts of $z \sim 1.5$. Fig. 13.6 shows two predictions of the redshift distribution from the Gilli *et al.* (2001) model for a flux limit of 2.3×10^{-16} erg cm^{-2} s^{-1} in the 0.5–2 keV band with different assumptions for the high-redshift evolution of the QSO space density. The two models from Gilli *et al.* (2001) have been normalized at the peak of the distribution.

The redshift distribution actually observed does not vary significantly within the flux-limit range covered by the samples in Table 13.1, therefore the total observed redshift distribution is shown in Fig. 13.6 for the total number of sources, *c*. 300, in all samples. The observed redshift distribution, arbitrarily normalized to roughly fit the population-synthesis models in the redshift range 1.5–3 keV, is radically different from the prediction, with a peak at a redshift in the range 0.5–0.7. This is still the case when the objects belonging to the large-scale structures around $z = 0.7$ in the CDFS are removed. The total number of objects at a redshift of less than 1 is significantly higher than the model predictions, even ignoring the 40% spectroscopic incompleteness. The peak at redshifts below 1 is also significant, if the normal star-forming galaxies in the sample are removed. This clearly demonstrates

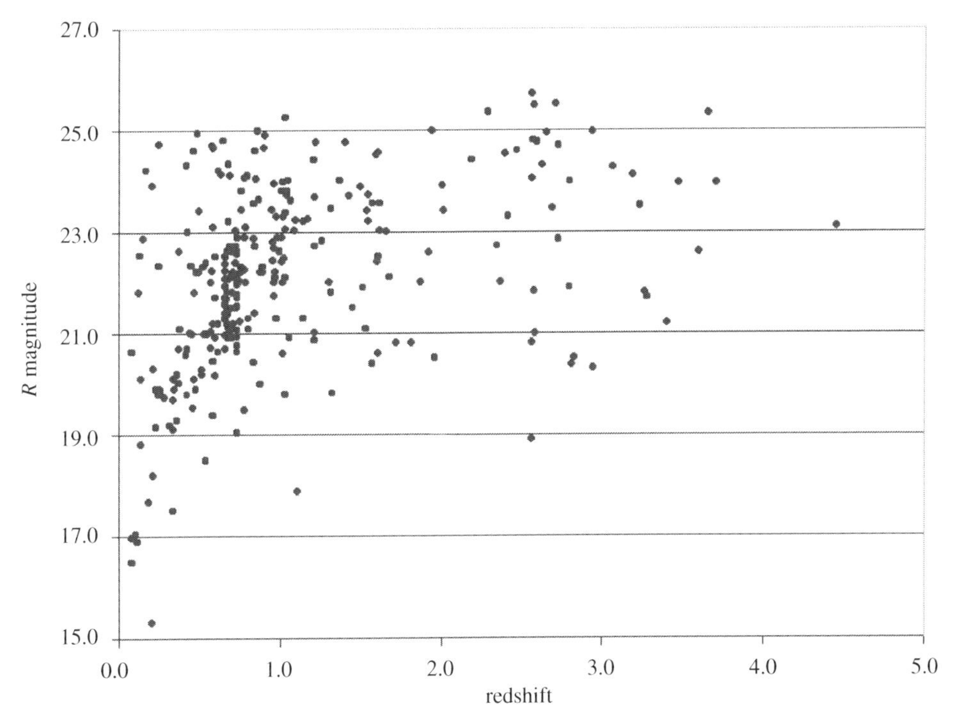

Figure 13.5. Optical magnitudes of AGN and galaxies from all the samples in Table 13.1 as a function of redshift. R magnitudes are taken from the Lockman Hole (Lehmann *et al.* 2001, 2002) and the CDFS samples (Giacconi *et al.* 2002). For the Hawaii 13 h (Barger *et al.* 2001a), Abell 370, and HDF-N samples (Barger *et al.* 2001b), where only I magnitudes are given, a colour of $R - I = 1$ has been assumed. An accumulation of objects in two redshift bins around $z = 0.7$ is due to a large-scale structure in the CDFS.

that the population-synthesis models will have to be modified to incorporate different luminosity functions and evolutionary scenarios for intermediate-redshift, low-luminosity AGN.

The AGN evolution at high redshift

The comparison between the observed and predicted $N(z)$ distributions at high redshifts is complicated by the possible existence of large-scale structure in the pencil-beam survey (there is, for example, a possibly significant excess of objects around $z = 2.5$ in the CDFS), but also by redshift-dependent selection effects and in general by the still relatively small volume sampled and therefore poor counting statistics in the number of objects. In addition, the overall normalization of the curves is uncertain because of the significant mismatch of the distribution at low z.

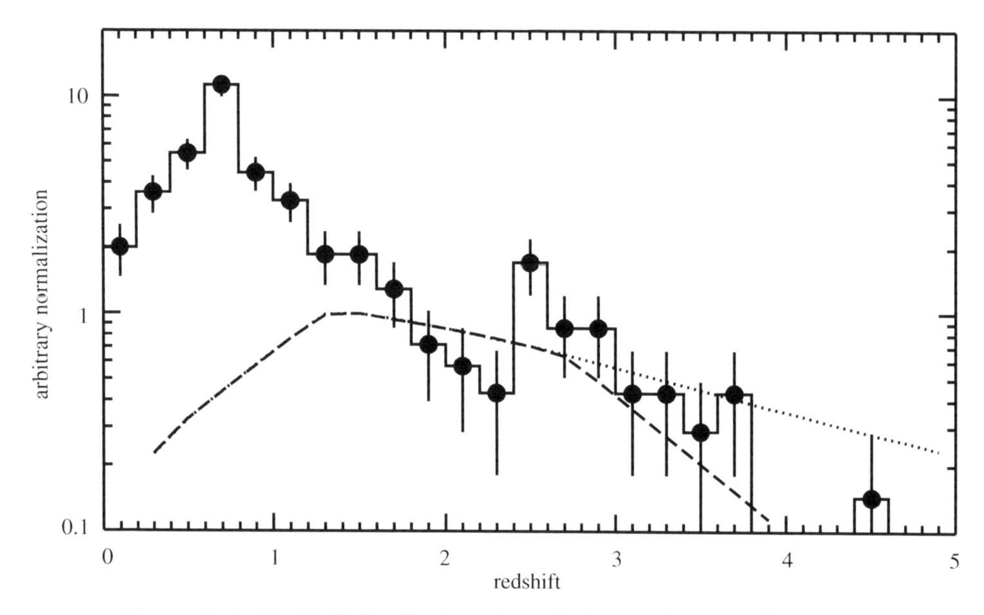

Figure 13.6. Redshift distribution of *c.* 300 X-ray selected AGN and galaxies in the deep Chandra and XMM-Newton survey samples given in Table 13.1 (solid circles and histogram), compared with model predictions from population-synthesis models (Gilli *et al.* 2001). The dashed line shows the prediction for a model, where the comoving space density of high-redshift QSO follows the decline above $z = 2.7$ observed in optical samples (Fan *et al.* 2001; Schmidt *et al.* 1995). The dotted line shows a prediction with a constant space density for $z > 1.5$. The two model curves have been normalized to their peak at $z = 1$, while the observed distribution has been normalized to roughly fit the models in the redshift range 1.7–3.

Nevertheless, the observed distribution is roughly consistent with both predictions in the redshift range $z = 1.6$–3.8. There is, however, a significant discrepancy between the observed distribution and the constant space density model (dotted line, Fig. 13.6) at redshifts above 4, where only one object was detected, while about eight objects would be predicted. From Fig. 13.5 it becomes apparent that the dearth of X-ray selected AGN is probably not due to optical spectroscopic selection effects. The one object detected at $z = 4.45$ already in the ROSAT data of the Lockman Hole (Schneider *et al.* 1998) has an optical magnitude of $R = 23$ and is therefore not at the spectroscopic limit of the samples. Also the $\mathrm{Ly}\alpha$ and CIV lines for QSOs in the redshift range 4–5 fall well into the optical range. The observed redshift distribution therefore gives a strong indication for a decline of the QSO space density beyond a redshift of 3.8.

A similar conclusion about a decline of the X-ray-selected AGN space density at high redshifts can be obtained from the absence of QSOs with $z > 5$ in all X-ray survey samples so far. (The QSO at $z = 5.2$ in the Chandra observation of

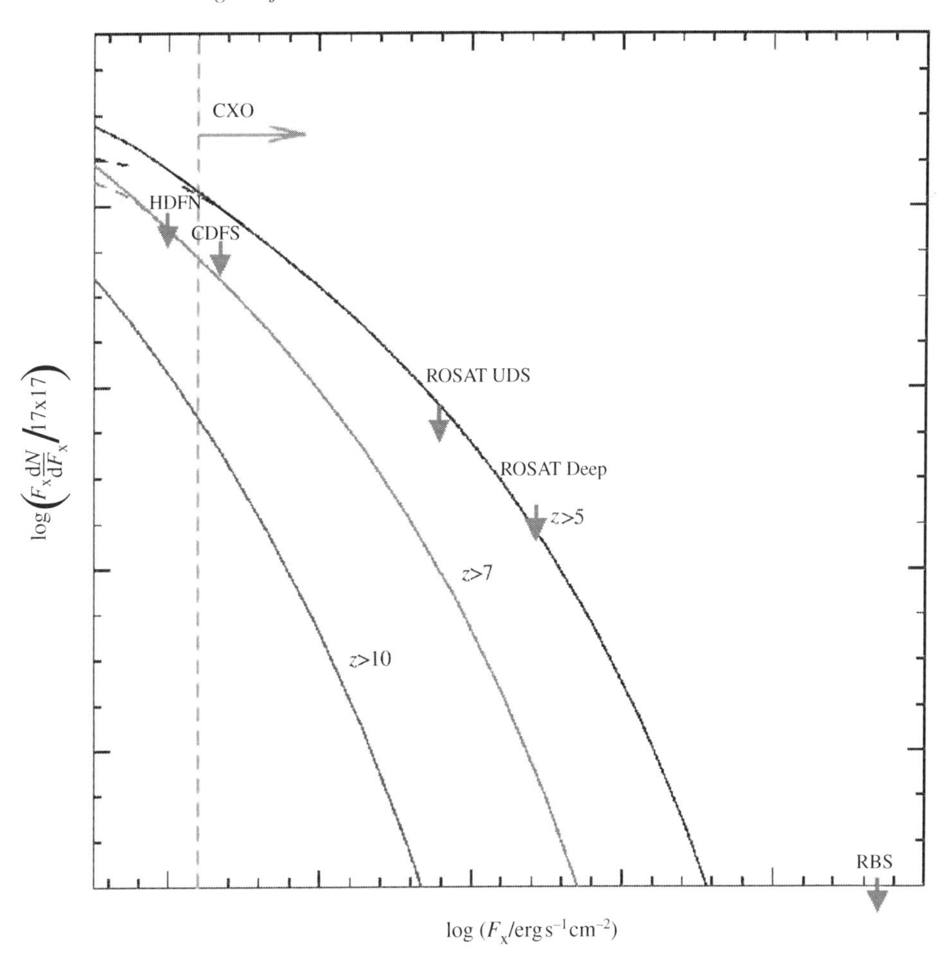

Figure 13.7. Prediction of the number density of AGN with redshifts larger than 5, 7 and 10, respectively, as a function of flux in a typical 17×17 arcmin^2 Chandra field of view from Haiman & Loeb (1999). Upper limits measured in X-ray surveys at various flux limits are indicated. Quasar counts 0.4 keV$< E <$ 6 keV.

the HDF-N does not change the conclusions discussed below.) Figure 13.7 shows a prediction of number counts for high-redshift QSO from Haiman & Loeb (1999), according to which a large number of $z > 5$ AGN should be detected with Chandra in any deep survey. This theoretical model assumes the X-ray luminosity function at $z = 3.5$ determined from the ROSAT surveys and extrapolates it backwards in time, assuming a simple hierarchical CDM model. The figure also shows limits for the number counts of $z > 5$ AGN from X-ray surveys at varying flux limits. The most distant QSO among *c.* 2000 objects in the ROSAT Bright Survey (RBS; Schwope *et al.* 2000) has a redshift of 2.8; the lack of higher-redshift objects is, however, not constraining given the high flux limit of this survey. The lack of $z > 5$ AGN in the ROSAT Deep and Ultradeep Surveys (Lehmann *et al.* 2000, 2001; Schmidt *et al.* 1998) is still just consistent with the Haiman & Loeb predictions: the highest-redshift object in the UDS is RX J105225.9+571905 at $z = 4.45$ (Schneider *et al.* 1998). The Chandra Deep survey, while only *c.* 60% spectroscopically identified, still provides an upper limit for the number counts of $z > 5$ AGN significantly lower than the prediction, using the conservative assumption that less than half of the unidentified objects are at redshifts larger than 5. Finally, the 400 ks Chandra observation in the Hubble Deep Field proper, providing 100% identifications for 12 sources in the field and their highest redshift object at $z = 4.42$ just outside the HDF-N, also gives an upper limit about a factor of 3 lower than the Haiman & Loeb prediction.

The information about the space density of X-ray-selected AGN is still limited by the small-number statistics in the deep X-ray surveys that cover too small a solid angle. More and wider fields have been surveyed by both Chandra and XMM-Newton. As soon as the tedious and time-consuming optical follow-up work in these fields is completed, we will be able to learn more about the decline of the X-ray AGN and therefore their formation at early redshifts. The possible discrepancy between a declining space density of optical and radio-selected QSOs above a redshift of 2.7 and an apparently constant space density of X-ray selected AGN with a decline beyond a redshift of *c.* 4 could still be understood in terms of the different luminosities, and therefore different black-hole masses, of the objects involved. The optical and radio surveys cover a large solid angle to a modest flux limit and therefore pick up only the most luminous and therefore most massive objects at high redshift. The deep pencil-beam surveys, on the other hand, sample a much smaller volume to much fainter flux limits and therefore select high-redshift AGN which are intrinsically more than a factor of 10 less luminous and therefore probably less massive than the objects selected in wide-angle surveys. In the hierarchical large-scale-structure formation, the smaller cold-dark-matter haloes collapse earlier than the larger ones. Given the correlation between black-hole mass and galaxy mass (and presumably dark-matter mass), it is expected that the lower-mass black holes are formed earlier than the most massive objects and thus that lower luminosity AGN appear earlier than the most luminous QSOs. This concept can be tested with more optical identifications of Chandra and XMM-Newton surveys and with future, even-more-sensitive X-ray telescopes, like the ESA/ISAS XEUS mission.

I thank the organizers of the Royal Society Discussion Meeting 'X-ray astronomy in the new millennium' for the invitation to present this review. I thank my co-workers in the Chandra Deep Field South and Lockman Hole identification teams for the fruitful collaboration and the permission to show data in advance of publication.

References

Barger, A. J., Cowie, L. L., Mushotzky, R. F. & Richards, E. A. 2001a *Astrophys. J.* **121**, 662.
Barger, A. J., Cowie, L. L., Bautz, M. W. *et al.* 2001b *Astrophys. J.* **122**, 2177.
Brandt, W. N., Hornschemeier, A. E., Alexander, D. M. *et al.* 2001a *Astrophys. J.* **122**, 1.
Brandt, W. N., Alexander, D. M., Hornschemeier, A. E. *et al.* 2001b *Astrophys. J.* **122**, 2810.
Comastri, A. *et al.* 1995 *Astron. Astrophys.* **296**, 1.
Fabian, A. C., Barcons, X., Almaini, O. & Iwasawa, K. 1998 *Mon. Not. R. Astron. Soc.* **297**, L11.
Fan, X. *et al.* 2001 *Astrophys. J.* **121**, 54.
Fiore, F. *et al.* 1999 *Mon. Not. R. Astron. Soc.* **306**, 55.
Fiore, F. *et al.* 2000 *New Astron.* **5**, 143.
Gebhardt, K. *et al.* 2000 *Astrophys. J.* **539**, 13.
Giacconi, R. *et al.* 2001 *Astrophys. J.* **551**, 624.
Giacconi, R. *et al.* 2002 *Astrophys. J. Suppl.* **139**, 369.
Gilli, R., Salvati, M. & Hasinger, G. 2001 *Astron. Astrophys.* **366**, 407.
Granato, G. L., Danese, L. & Francheschini, A. 1997 *Astrophys. J.* **486**, 147.
Haiman, Z. & Loeb, A. 1999 *Astrophys. J.* **519**, 479.
Hashimoto, Y. *et al.* 2002 *Astron. Astrophys.* **381**, 841.
Hasinger, G. & Lehmann, I. 2002 In *Proc. Where's the Matter?, Marseille, France, 25–29 June 2001* (eds. L. Tresse & M. Treyer). Paris: Editions Frontieres.
Hasinger, G. *et al.* 1993 *Astron. Astrophys.* **275**, 1.
Hasinger, G. *et al.* 1998 *Astron. Astrophys.* **329**, 482.
Hasinger, G. *et al.* 1999 *Astron. Astrophys.* **340**, 27.
Hasinger, G. *et al.* 2001 *Astron. Astrophys.* **365**, 45.
Hornschemeier, A. E. *et al.* 2000 *Astrophys. J.* **541**, 49.
Lehmann, I. *et al.* 2000 *Astron. Astrophys.* **354**, 35.
Lehmann, I. *et al.* 2001 *Astron. Astrophys.* **371**, 833.
Lehmann, I., Hasinger, G., Murray, S. S. & Schmidt, M. 2002 In *Proc. ASP Conf. Series* (eds. S. Vrtilek, E. M. Schlegel & L. Kuhi). San Francisco, CA:ASP.
McHardy, I. *et al.* 1998 *Mon. Not. R. Astron. Soc.* **295**, 641.
Mainieri, V. *et al.* 2002 *Astron. Astrophys. Astron. Astrophys*, **393**, 625.
Miyaji, T., Hasinger, G. & Schmidt, M. 2000 *Astron. Astrophys.* **353**, 25.
Miyaji, T., Hasinger, G. & Schmidt, M. 2001 *Astron. Astrophys.* **369**, 49.
Mushotzky, R. F., Cowie, L. L., Barger, A. J. & Arnaud, K. A. 2000 *Nature* **404**, 459.
Norman, C. *et al.* 2002 *Astrophys. J.* **571**, 218.
Rosati, P. *et al.* 2002 *Astrophys. J.* **566**, 667.
Schmidt, M., Schneider, D. P. & Gunn, J. E. 1995 *Astrophys. J.* **114**, 36.
Schmidt, M. *et al.* 1998 *Astron. Astrophys.* **329**, 495.
Schneider, D. P. *et al.* 1998 *Astrophys. J.* **115**, 1230.
Schwope, A. *et al.* 2000 *Astron. Nachr.* **321**, 1.
Shaver, P. A. *et al.* 1996 *Nature* **384**, 439.
Stern, D. *et al.* 2002 *Astrophys. J.* **568**, 71.
Szokoly, G. *et al.* 2002 (In preparation.)
Thompson, D. *et al.* 2001 *Astron. Astrophys.* **377**, 778.
Tozzi, P. *et al.* 2001 *Astrophys. J.* **562**, 42.
Zamorani, G. *et al.* 1999 *Astron. Astrophys.* **346**, 731.

14

X-ray astronomy in the new millennium: a summary

BY ROGER D. BLANDFORD

Caltech

Introduction

Out of the nearly 70 octaves of the electromagnetic spectrum that have been opened up to astronomical observation, X-ray astronomers can lay claim to roughly 10 (as opposed to the single octave explored by optical astronomers!). Although no one would pretend that all octaves are equally interesting in terms of physics, there are some special reasons why the X-ray band is particularly informative. It includes the K- and L-shell transitions of all the post-Big-Bang elements. It is where to find thermal emission from gas with sound speed greater than or approximately equal to 300 km s^{-1}, typical of the intergalactic medium, galaxies and stars. It lies right below 1 MeV, which is a characteristic energy scale for many non-thermal processes.

X-ray astronomy began in 1962 with the discovery of Sco X-1 (Giaconni *et al.* 1962; Pounds 2002) and was enthusiastically developed surprisingly soon after the dawn of the space age, perhaps because radio astronomy had, by this time, revealed a universe of sources with strengths and properties that were completely unexpected on the basis of optical observations. By the time Sco X-1 was identified, quasars and the microwave background had been discovered and pulsars were soon to follow. These four discoveries ushered in modern astronomy.

As this book celebrates, X-ray astronomy has come a long way. It is, arguably, ceasing to exist as a separate observational subfield of astronomy, so central have X-ray observations become to the study of essentially all classes of cosmic sources. The most improbable objects – brown dwarfs, all types of protostar and the Moon, for example – have been detected in X-rays (Güdel 2002). However, in spite of the fascination of this history, it is to the present and the future that we must turn, and earlier chapters in this book have taken stock of where we are after a couple of years of full operation of Chandra and XMM-Newton and what the prospects are

Frontiers of X-Ray Astronomy, ed. A.C. Fabian, K.A. Pounds and R.D. Blandford. Published by Cambridge University Press.

for the future. Even this turned out to be too ambitious to cover and the chapters presented here (and *a fortiori* in this brief summary) have had to be very selective and I shall defer to the other contributors for more representative bibliographies.

As the only non-observer among the chapter authors, I have organized this chapter around three themes that are somewhat 'orthogonal' to most of the preceding, source-centred ones. These are the physical processes that are ultimately responsible for X-ray emission, the peculiar importance of X-ray observations in rounding out our view of the structure and evolution of stars and galaxies and the under-acknowledged role of X-ray astronomy in defining the cosmological world model to which we have been led and which is now starting to raise some very important questions concerning what actually happened in the first 10^6 years of the life of the Universe. I conclude with a brief listing of the proposed next generation of X-ray observatories.

Physical processes

The description of many high-temperature and non-thermal sources is dependent upon some poorly understood physical processes. It is striking how often the same questions are asked of quite different physical environments. How important is thermal conduction? How effective is magnetic reconnection in heating plasma? And so on. The optimistic view, which links several of the chapters, is that we really only have to solve these problems, e.g. the structure of Mach 30 shocks, once and we should be prepared to combine astrophysical, space physical observational experiments with laboratory experiments and computational work to develop some confident answers.

Spectroscopy and its interpretation

The gratings on XMM-Newton and Chandra have unprecedented spectroscopic capability, and the ability to use them so effectively has benefitted from over 30 years of hard (and largely unacknowledged) work measuring and computing wavelengths, oscillator strengths, etc., for transitions of little terrestrial interest. Of particular importance are the helium-like triplets that combine observations of permitted, intercombination and forbidden transitions (Gabriel & Jordan 1969; Kahn *et al.* 2002) so as to provide density and temperature diagnostics. The low densities that allow forbidden transitions to be so important in cosmic sources are hard to work with experimentally. Conversely, the high radiation densities that allow radiative-ionization equilibria to be established are only just being achieved using powerful lasers.

By now, most of the important wavelengths, oscillator strengths, collision integrals, etc., for the strongest lines have been measured, although these measurements are still lacking for the majority of the weaker lines that can also be observed in the brightest sources. Observations of H-like and He-like transitions of the more common ions provide useful diagnostics of the density, temperature, abundance, ionization equilibrium and velocity in the emission regions, and we have seen in this book many examples of what can be done in a wide variety of sources, including accretion discs (stellar (Done 2002) and active galactic nuclei (AGN) (Fabian 2002)), stars and protostars (Güdel 2002), clusters of galaxies (Mushotzky 2002) and polars (Cropper 2002). The power of X-ray spectroscopy is most clearly brought out by the detailed observations of bright supernova remnants (Canizares 2002). Here, it is possible to use the *c*. 100 km s^{-1} velocity resolution to make three-dimensional (3D) abundance maps of the expanding debris and forensic analyses of the initial explosions.

Unfortunately, most other sources are unresolved and most direct analyses of the data are often limited to one-zone models and some very primitive radiative transfer. Now, it is possible, at least in principle, to include anisotropy, inhomogeneity and peculiar geometric effects in theoretical models of these sources. The problem is that we really have no clear idea of the disposition and structure of the emission region and the medium through which the radiation is propagating. A particularly important example is provided by the ongoing debate concerning the soft-X-ray spectra of Seyfert galaxies. Does the power-law, continuum source that is reflected by the disc arise in a local corona or at high altitude so that it can illuminate the whole disc? Are the carbon and oxygen lines produced by reflection (Sako *et al.* 2003) or in a dusty, warm absorber (Lee *et al.* 2001), again at some distance? Why does there appear to be no sign of a reverberative response in Seyfert galaxies with variable X-ray continua? Answering these questions using a more detailed analysis of line formation in the two cases is tantamount to understanding the source geometry. However, as broad emission lines are now being reported from binary X-ray sources like Cyg X-1 (Fabian 2002) and J1650-500 (Miller *et al.* 2002), it is reasonable to suppose a model that works for Seyferts should also work for black-hole binaries.

A second example is provided by stellar coronae, where hot, coronal gas is excited by twisted, magnetic loops. (Temperatures as high as 40 million degrees are now reported to be associated with the Galactic Cepheid, YY Mon (Güdel 2002).) Although we can image similar activity in the sun, we still do not understand it at all well. Again, we do not have an agreed story as to the sequence of events that leads to coronal lines being emitted and, consequently, how to convert raw line strengths from distant stars into physical conditions in their coronae. The observation of coronal activity from late-M stars that are thought to be cool enough not to possess surface convection zones suggests that other, non-magnetic processes could be at work. Again, there is probably a general theory that can be inferred by combining solar and stellar observations.

A third case, where we probably do understand the geometry and can compute the emissivity and opacity, is provided by cyclotron line formation in accreting white dwarfs (Cropper 2002). Although eclipse observations have confirmed the expected strongly inverted temperature gradient, the radiative transfer is quite subtle and a far more detailed theoretical treatment is likely to be necessary for us to reproduce the observed spectra as well as their polarization and time dependence. (Monte Carlo techniques will surely continue to play a major role here and some of the experience gained from working with Tokamak plasmas may be relevant.)

The plasma impasse

X-ray sources are, inevitably, fully ionized gases. It is therefore unfortunate that X-ray astronomers have been resistant to learning the principles of plasma physics and incorporating them into their science, preferring instead to limit their purview to gas dynamics and atomic physics. This evasion can no longer be excused. There are now several sources where progress awaits the answers to fundamental plasma physics questions. Again, I only have space for a few examples.

The first question is 'how fast do ions with some temperature T heat cooler electrons?' There is a minimal and standard heating rate resulting from two-body Coulomb scattering. However, there is an abundance of wave–particle interactions that might, for example, be excited by streams of fast particles. The empirical evidence comes from the observations of high Mach number, heliospheric and supernova remnants' shock fronts, which appear to transmit thermal electrons with a temperature well below the equipartition value so that the ions do not quickly attain collisional ionization equilibrium (Canizares 2002). This suggests that collective effects are not that important. This view is supported by observations of slowly accreting black holes that are also best interpreted in terms of a minimal, Coulomb, heating rate. The answer to this question is of direct relevance to the debate about the efficacy of electron heat conduction. There are really two issues. What is the mean free path of the electrons along the magnetic field and how quickly do the field lines wander in response to an imposed turbulence spectrum? As discussed below, our best laboratories are clusters of galaxies.

A related plasma question concerns the efficiency of strong shocks for accelerating cosmic rays. There is a linear theory which appears to have some validity, again based upon heliospheric measurements and X-ray observations of supernova remnants (Canizares 2002). However, in order to model the observations, we have to understand how the back reaction associated with the cosmic-ray pressure moderates the acceleration and what controls the rate of both ion and electron injection. Theoretically, it may soon be possible to perform $(3 + 3)$-dimensional kinetic simulations with sufficient resolution to address these questions. Observationally, the

Gamma-Ray Large Area Space Telescope (GLAST) should provide measurements of the energetically dominant GeV ions. As relativistic electrons are scattered by the same Alfvénic turbulence that operates on the ions, the combination of γ-ray and X-ray observations should enable us to infer the injection rate and perhaps determine the scaling with Mach number.

Another, quite controversial, feature of shock fronts is their role in amplifying magnetic fields. Theoretically, there is no very good reason why simple gas dynamical shocks should do any more than compress the pre-shock magnetic field. In addition, radio observations of many (though not all) supernova remnants seem to show that the emissivity and, presumably, the magnetic field strength only increase in the interaction region between the shocked ejecta and the swept-up interstellar medium. Conversely, it is possible that the hydromagnetic turbulence that is invoked to scatter the cosmic rays leads to an overall increase in the root-mean-squared magnetic field strength. This is highly relevant to the late-time evolution of γ-ray burst afterglows.

Even greater uncertainty surrounds our understanding of relativistic shocks, which are thought to be the primary acceleration site for X-ray-emitting relativistic electrons and magnetic-field amplification in γ-ray bursts, pulsar-wind nebulae and extragalactic jets. However, the diffusive mechanism for particle acceleration that operates non-relativistically is kinematically precluded. There is a promising relativistic variant (Achterberg *et al.* 2001). However, this assumes that the cosmic rays can move far enough upstream from the shock to scatter off the background flow and it is not clear how this can happen if there is an oblique magnetic field. Neither is it clear how the scatterers can be generated. The magnetic field itself is also believed to be strongly magnified at the shock front, though no good explanation of how this happens has been found. Indeed, the very existence of sudden, collisionless discontinuities, as opposed to a slow sharing of momentum between two fluids, has been questioned. X-ray observations should be especially instructive because they permit us to resolve these putative shocks, say in the Crab Nebula and jets like M87, at energies where the emitting, relativistic electrons quickly cool. The observation of what is presumably X-ray synchrotron radiation well away from the supposed strong shocks implies that relativistic electrons have to be accelerated *in situ*, rather than at strong shocks. These observations further raise the possibility, discussed elsewhere (Blandford 2002), that the observed sources, including their 'shocks', are actually relativistic, electromagnetic structures and are not described well by gas dynamics.

Another general process is magnetic reconnection, which has been invoked, for example, in explaining the energization of accretion-disc coronae. However, the manner in which it operates remains quite controversial. Most existing discussions (e.g. Priest & Forbes (2000)) have been essentially hydromagnetic, except within a small region where the magnitude of the field gradient becomes very large and where a scalar (and usually 'anomalous') resistivity is invoked. A development, which has serious implications for topological behaviour, is that the resistivity might be dominated by non-dissipative Hall terms (Bhattarcharjee *et al.* 2001). These embellishments of magnetohydrodynamics (MHD) are now finding their way into numerical simulations and it will be interesting to see what their implications are for X-ray sources.

In addition to numerical simulation and *in situ* observation of space plasmas, it is becoming possible to address some of these questions using the growing field of laboratory experimentation. It is now possible to create both ionic and pair relativistic plasmas, using powerful lasers, electron beams and magnetic pinches. Temperatures as high as 100 MeV, energy fluxes of *c.* 100 ZW m^{-2} and *c.* 1 MT magnetic field strengths are all attainable. High-energy-density investigations are likely to become much more versatile in the coming years (Takabe 2001).

Black-hole accretion

The problem of accretion onto a compact object, specifically a black hole, is generally well posed but has also not had a confident solution. However, through a combination of theoretical arguments and direct observation of accreting sources, it has been possible to make a lot of progress. The greatest excitement has probably centred around the occasional observation of broad iron lines from selected low-luminosity AGN and, as reported here, a couple of galactic binary X-ray sources (Fabian 2002; Done 2002). (We now know that broad lines are not seen as commonly as was once thought and that their formation must be more complicated than envisaged in early models. The prospects for performing useful 'reverberation mapping' do not look good.) In those sources where these features are undoubtedly seen, we can say that there is evidence that the second parameter that characterizes a classical black hole (the spin) is responsible for the line width. Indeed, it has even been argued that the role of the black hole is not just the passive one of allowing stable orbits from which highly redshifted photons can be observed, but is an active one, in which a magnetic connection of the gas to the spinning hole leads to an enhanced emissivity from the innermost and most redshifted orbits (Wilms *et al.* 2001).

Another very promising line of investigation is epitomized by the observations of Sgr A*, which show that the black hole is a strikingly underluminous X-ray source with an apparent luminosity of *c.* 10^{-8} $L_{\rm edd}$ and a radiative efficiency of *c.* $10^{-7}c^2$ relative to the inferred mass-accretion rate of *c.* 10^{22} g s^{-1}. There have been several explanations put forward, but most of these require that the rate of electron–ion equilibration be slow, as discussed above. It no longer seems possible that all of the mass supplied can accrete onto the hole and either most of the mass is lost (Blandford & Begelman 1999) or the accretion backs up to the Bondi radius at *c.* 10^7 gravitational radii. The whole matter has been made more interesting through the discovery of surprisingly rapid X-ray variability in Sgr A* (Baganoff *et al.* 2001) and the even more remarkable suggestion that the radio variation may be periodic (Zhao *et al.* 2001). These observations open up many more possibilities and will undoubtedly be quite constraining once the observational situation is clarified.

Galactic black holes provide more immediate gratification for observers than massive black holes, both on account of their larger fluxes and also because of their much more rapid variability time-scales (Done 2002). There is now a fairly convincing, qualitative explanation of the low and high states. The former arise when the luminosity is greater than or approximately equal to $0.03L_{\rm edd}$ and a thin (or slender) disc extends down to a radius $r_{\rm ms}$, the latter when there is a central hole filled by gas that cannot cool and radiate efficiently and where a non-thermal spectrum is created by Comptonization (Esin *et al.* 2001). It is not clear that all of this gas accretes onto the black hole.

Many of the questions raised by these observations are issues of theoretical principle that are still being debated. The approaches that will be necessary to address these questions are both observational and theoretical. For the former, the angular resolution of Chandra can be put to great advantage resolving the accretion radii in nearby dormant galactic nuclei. These observations are helping us to define the physical conditions and perhaps to deduce the rate of gas supply to the central black hole. Theoretically, there are opportunities for carrying out 3D numerical fluid-dynamical and MHD/electromagnetic (including general relativistic) simulations of discs and outflows.

Non-thermal emission

The capability to perform arc-second imaging at X-ray wavelengths is revolutionizing our view of non-thermal emission. Surely, the most famous instance of this is the discovery of a pair of axial jets in the Crab Nebula (Weisskopf *et al.* 2000) as well as other pulsar wind nebulae. This was relatively unexpected and shows that accretion discs are not necessary for 'jet' formation. However, it may suggest something even more fundamental, and to explain this I should return to one of the first models of a pulsar: the Goldreich–Julian axisymmetric model. Here, it was proposed that a spinning, magnetized neutron star acts as a unipolar inductor and generates an electromotive force (EMF), $\mathcal{E} \sim 30$ PV in the case of the Crab pulsar and that this drives a current of *c.* 300 TA around a quadrupolar circuit. (We now know, thanks to Ulysses, that the heliospheric electrical circuit is of this form although the EMF and current are only *c.* 100 MV and *c.* 1 GA, respectively.)

Now real pulsars are, by definition, non-axisymmetric and the electromagnetic field just beyond the light cylinder will contain both an AC and a DC component. The interaction between these two components is unclear, but it has generally been assumed that essentially all of the electromagnetic Poynting flux is quickly converted into the kinetic energy of a plasma-dominated, outflowing wind. In other words, the electrical circuit is completed quite close to the pulsar. In this case, a hypersonic wind is created which, it has been supposed, passes through a strong shock front with a Lorentz factor *c*. 10^6 close to the famous 'wisps', where particle acceleration and non-thermal emission can occur. However, the X-ray image really shows no evidence for this shock front except perhaps along the poles and the equator. The moving features that are seen optically also appear to be confined to the equatorial plane.

These observations suggest a different interpretation (Blandford 2002), specifically that the AC electromagnetic component dies away very quickly, perhaps non-dissipatively, while the DC component persists all the way into the nebula. In this case, the X-ray emission that is observed largely delineates the quadrupolar current flow. More specifically, there is no strong reverse shock and relatively little of the current circuit completes near the pulsar. The emission that is seen may well reflect MHD instability in the magnetic configuration – pinches and current sheets are notoriously unstable – and be a manifestation of ohmic dissipation. In this case, the observed Crab Nebula would be magnetically dominated rather than particle-dominated, except, perhaps, in the emission region, where the relativistic electron energy density might build up to an equipartition value. These ideas should be testable by examining the spectral index gradients and the polarization map.

This viewpoint has implications for ultrarelativistic jets and γ-ray bursts (GRBs), which are now also widely acknowledged to have an electromagnetic origin. However, where it is also supposed that the Poynting flux is quickly converted to fluid form (most commonly as an optically thin pair plasma in the former case and as a radiation-dominated fireball in the latter) and that the ultimate emitters are strong, relativistic shock fronts. By contrast, under the electromagnetic hypothesis, it is supposed that the energy released remains in an electromagnetic form all the way to the emission region and that the particle acceleration is a direct result of wave turbulence. There should be ample potential difference available for particle acceleration to take place. In the case of a quasar jet, the EMF is *c*. 100 EV and the current is *c*. 1 EA. (For GRBs the estimates are now *c*. 10 ZV and *c*. 100 EA, rather lower than in the past.)

Astronomical questions

Nuclear power

High-angular-resolution X-ray observations have, in an almost literal sense, transformed our view of AGN. They have amply confirmed the finding from optical observations that our classification of these sources is strongly aspect-dependent.

The simple geometrical model of an AGN invokes a thick torus that will absorb ultraviolet continuum and emission lines and all but the hardest X-rays and γ-rays and then re-emit the absorbed energy in the thermal infrared. This description has received impressive confirmation with the Advanced Satellite for Cosmology and Astrophysics (ASCA), XMM and Chandra measurements of hard-X-ray spectra (Matt 2002) which clearly exhibit the effects of absorption with hydrogen-column densities that can approach *c.* 10^{24} cm^{-2}. This, in turn, implies that a significant fraction of the infrared background, as well as the bolometric luminosity density of the young Universe, be associated with AGN. In round numbers, the energy density associated with the observed X-ray background (mostly in the energy interval *c.* 20–40 keV) is 3×10^{-17} erg cm^{-3}, a fraction *c.* 0.003 of the energy density measured in both the optical and in the far-infrared backgrounds (and *c.* 10^{-4} of the microwave-background energy density). In other words, if the intrinsic, integrated, ultraviolet to X-ray power of an AGN is, on average, 30 times the hard X-ray power and this is a reasonable guess based upon observations of local AGN, then AGN must account for *c.* 10% of the infrared background, with the remainder being presumably associated with stars. Estimates of the AGN fraction in the literature, based upon more detailed assumptions about the mean AGN spectrum and the redshift distribution of the sources, range from *c.* 3 to 30%. However, this fraction may not be very well defined because much of the luminosity associated with galactic nuclei could be attributed to starbursts in addition to accretion onto massive black holes (Ward 2002).

The characterization of the obscuring material as a torus is problematic, at least from a theoretical viewpoint. The difficulty is that it is very hard to see how a thick, cold ring of molecular gas can be supported. Individual clouds should collide inelastically and a torus would quickly deflate. A more plausible alternative (Sanders *et al.* 1989) is that there is a locally thin, though strongly warped, disc, where the thickness is maintained through the marginal growth of gravitational instabilities. However, there is still confusion about the size of this disc. Combined X-ray and Space Infrared Telescope Facility (SIRTF) observations should greatly improve our understanding of AGN obscuration.

Obscuration is not the only way to produce beaming. The X-ray study of relativistic jets, especially in blazars and nearby sources like M87 (which would probably be classified as a blazar were it pointed at us) is becoming more sophisticated. However, we are still not yet able to answer the quantitative questions about beaming fraction, the angular and radial variation of jet Lorentz factors, etc. These questions will undoubtedly be a focus of future X-ray research, especially after GLAST is launched.

The luminosity of galactic nuclei is not exclusively radiative. There are 'superwinds', which are driven by nuclear activity (Ward 2002). In addition, *c.* 10%

of optically selected quasars exhibit (X-ray-quiet) broad absorption-line outflows. Even if these flows only represent a minor fraction of the nuclear-power budget, they can still have a large impact on the host galaxies because the momentum flux of a wind scales inversely with the outflow speed. Not only can these outflows drive away much of the interstellar medium, they may even influence the overall galaxy morphology by, for example, inhibiting or limiting disc formation.

Stellar populations

X-ray observations are also presenting us with a complementary perspective to that obtained using optical and infrared observations on how the stellar populations of galaxies evolve. This is because they allow us to witness the endpoints of stellar evolution as opposed to the beginnings. So far, most attention has been on nearby galaxies, both spirals and ellipticals (Ward 2002). The primary targets are young supernova remnants and binary X-ray sources. Comparison with radio and infrared observations is particularly important in the former case, as it allows us to derive some useful empirical correlations. As an example of what can be learned in this way, it has been reported that the star-formation rate declines less rapidly than the supernova rate (Ward 2002).

Probably the biggest new discovery is the ultraluminous X-ray sources, which are now showing up quite regularly in all types of galaxy (Ward 2002). They are defined by having X-ray luminosity in excess of the Eddington limit for a *c.* 30 $M_\odot$ black hole. They have been associated with intermediate-mass black holes, conceivably relics from the Population III era. In this case the fuel supply is a bit of a puzzle. They must either be short-lived binaries or single stars moving slowly through molecular clouds and, consequently, a major constituent of the Universe, overall. Alternatively, they could be the long-sought stellar blazars, although here the absence of a radio emission is surprising. A third possibility is that they are normal-mass black-hole binaries with super-Eddington luminosity, which may be physically possible in the highly-clumped, radiation-dominated fluid that is expected in the innermost regions of accretion discs orbiting *c.* 10 $M_\odot$ black holes. Again, this phase would have to be quite short-lived. The heterogeneity of the observed properties of ultra-luminous X-ray sources suggests that more than one of these explanations could be correct.

Cosmological issues

Clusters

X-ray observations continue to be of central importance to the study of rich clusters of galaxies (Allen 2002; Mushotzky 2002). The most basic information concerns the shape and depth of the gravitational potential well. The various methods that have been used to determine this now seem to be in fairly good agreement. Arguably the most reliable is gravitational lensing – strong in the core and weak in the outer parts – especially when there are reliable source redshifts. As demonstrated here, this technique works well for clusters that appear to be nearly circular and which are dynamically relaxed. The total mass density can then be derived from Poisson's equation. (The smoothness of the known arcs assures us that the potential is also quite smooth.) It is then possible to use imaging spectroscopy to measure the baryon mass distribution, as *c*. 85% of this is believed to be in the form of hot gas. (In practice this is carried out by model-fitting rather than an unbiased inversion.) This procedure may be problematic near the centre of the cluster but is far safer in the outer parts of the cluster, where most of the mass resides and where we are most likely to sample fairly the cosmological mass distribution. If we are prepared to trust the baryon density derived from the measured deuterium (plus other light elements) abundance and the theory of Big Bang nucleosynthesis, the fraction of the critical density in the form of baryons is $\Omega_b = 0.04$. We can then deduce the contemporary matter fraction of the Universe and a value of $\Omega_0 = 0.31 \pm 0.03$ has been quoted (assuming a Hubble constant $H_0 = 70\,\mathrm{km\,s^{-1}\,Mpc^{-1}}$) (Allen 2002). This argument, which preceded the more publicized type-Ia supernova determination and which is less subject to systematic error, appears to be holding up very well. Repeating these measurements with larger redshift clusters, probably in conjunction with Sunyaev–Zel'dovich measurements, should allow us to infer the expansion history of the Universe, at least in principle. To date, this translates into an unsurprising bound on Ω_Λ.

It is also possible to explore the thermal history of the intergalactic gas by comparing the measured mass–temperature relation with the expectations of adiabatic simulations. It should not be too surprising that the agreement here is less good (Mushotzky 2002; Allen 2002). The entropy history of the gas, which affects this determination, is likely to be quite complicated. The gas sound speed, immediately after re-ionization by the first stars and quasars, is only *c*. 10 km $\mathrm{s^{-1}}$. However, as structure grows, this gas will acquire speeds of several hundred km $\mathrm{s^{-1}}$. Strong, large-scale shock fronts are likely to form and increase the entropy of the gas. However, as noted above, the post-shock electron temperatures will be significantly lower than the ion temperatures. (There is no observational evidence for strong accretion shocks surrounding observed clusters; it appears, quite reasonably, that the gas is heated much earlier in the merger hierarchy.) The gas will also be mixed with the cooler gas swept out of galaxies. It is very hard to compute the influence of these and other easily imagined effects from first principles. The dark-matter mass–velocity relation, which is probably best determined by gravitational lensing, is more likely to furnish a robust measure of the growth of large-scale structure.

There is an inescapable implication of the cluster gas having been heated by strong shock fronts, whatever their provenance. This is that the hot gas will be accompanied by cosmic-ray ions. The speeds and Mach numbers of the shock fronts are quite similar to those in the interplanetary and interstellar media and we expect that the post-shock, GeV cosmic-ray partial pressure will lie somewhere in the range 0.2–0.5 of the total pressure. This fraction will decrease slightly as the gas is adiabatically compressed, but if the gas starts to cool appreciably then cosmic rays may dominate the pressure and inhibit further cooling. This may be part of the explanation of the surprising results from X-ray spectroscopy by XMM of a few well-studied clusters (Mushotzky 2002). These seem to show that the gas starts to cool as it flows towards the central cD galaxy and then appears to vanish: the lines expected from gas with temperatures below *c.* 2 keV are absent. By resisting further compression, the cosmic rays will make it easier to keep the gas hot. Other factors that have been invoked to explain the failure to observe cooling flows include variable metallicity, thermal conduction and supernova heating. There is a further implication of having these cosmic rays present and this is that they may contribute to the heating and, in particular, may create γ-rays through pion production. The predicted γ-ray flux from nearby clusters should be detectable by GLAST and, under extreme assumptions, could contribute to the γ-ray background.

X-ray observations are also providing new information on the chemical history of the Universe as we try to reconstruct the history of clusters of galaxies (Mushotzky 2002). C, N, O, S, Fe and Ni have all been measured in a large sample, and abundance gradients in a smaller number. The iron abundance has now stabilized at $[\mathrm{Fe/H}] \sim 0.3$ (possibly increasing with the size of the cluster) and is consistent with a type-Ia origin. The supernovae may have occurred mostly in clusters with the processed gas being driven out in superwinds. The correlations of the relative abundances with the cluster properties as well as the radius are starting to become diagnostic of the evolutionary history of the cluster gas.

X-ray background

Another great success for Chandra and XMM-Newton has been the resolution of *c.* 80–90% of the X-ray background into *c.* 3×10^8 discrete sources (Brandt *et al.* 2002; Hasinger 2002). As discussed above, most of the energy density appears to derive from black-hole accretion in low-power AGN. The redshift distribution of these sources is controversial. On one hand, the *c.* 10^8 sources which contribute most of the background appear to have modest redshifts, $z < 1$ (Hasinger 2002); on the other, the faintest sources are seen out to $z \sim 3$ (Brandt *et al.* 2002). In addition, the surveys are so sensitive that nearby normal galaxies and luminous quasars with $z \gtrsim 5$ are also found to be minor contributors to the background. I suspect that these statements are approximately true and not in contradiction with each other.

The notion that most of these X-ray sources are obscured receives support from their association with Infrared Space Observatory sources and bodes well for the SIRTF observations. If, following the example above, we suppose that AGN account for *c.* 10% of the infrared background then the energy produced, allowing for the expansion of the Universe, corresponds to *c.* $10^7\ M_\odot c^2$ per *local* L^* galaxy. If we assume that black holes grow with *c.* 10% radiative efficiency, then we deduce a mean black-hole mass per L^* galaxy of *c.* $10^8\ M_\odot$, roughly compatible with what is measured. In addition, as there are *c.* 3×10^9 of these locally specified L^* galaxies out to $z \sim 2$, where most of the black-hole mass is grown, we conclude that a typical nucleus of one of these galaxies is active for *c.* 10^8 years, consistent with the Salpeter time.

A possible problem with this neat explanation is that if the spectrum below an observed energy *c.* 30 keV really does come from obscured sources, then they must have low redshifts, as it is hard to see how an absorption turnover could occur at a much higher energy than *c.* 40 keV when scattering dominates any plausible opacity. If this is true, then, when Swift or the proposed Energetic X-ray Imaging Survey Telescope (EXIST) identify the sources that contribute most of the hard-X-ray background, they should find that they have low redshifts. An alternative possibility is that the hard-X-ray sources are mostly at $z \sim 2$–3 and we are observing hard, Comptonized spectra in a corona with a temperature *c.* 100 keV. Ultimately, we should like to extend this analysis all the way up to γ-ray energies and make smooth contact with studies of the γ-ray background.

First light

The famous penetrating power of hard X-rays makes them excellent candidates for probing the very early Universe. Observations are encouraging. Quasars have been found with $z \sim 6.5$ and the X-ray powers suggest that the holes exceed a billion solar masses. This is a constraint on theories of the growth of black holes and a timely reminder that quasar activity must be intimately related to galaxy formation and evolution during the *c.* 0.5 Gy between re-ionization and $z \sim 6.5$, when the first quasars are seen.

In addition, GRBs have already been seen out to $z \sim 4.5$, where the redshift actually helps by allowing an observer with a fixed response time to observe an earlier and brighter part of the evolution, at greater emitted-photon energy. There is optimism that Swift will identify X-ray afterglows emitted earlier than the light from the first observed quasars.

Future missions

Although we look forward to many years of active service from Chandra and XMM-Newton as well as INTEGRAL, Swift (2004), ASTRO E-2 (2005) and GLAST (2006), there are also longer range plans to construct more powerful telescopes like Constellation-X, EXIST, LOBSTER, Generation-X and XEUS. The problems discussed in this book are already rewriting the scientific case for the longer-term missions and it is hoped that this will be reflected in further improvements in mission design and optimal use of over-subscribed international resources for space astronomy.

I thank the NSF and NASA for support under grants AST99-00866 and 5-2837.

References

Achterberg, A. A., Gallant, Y. A., Kirk, J. G. & Guthmann, A. W. 2001 *Mon. Not. R. Astron. Soc.* **328**, 393.
Allen, S. W. 2002 *Phil. Trans. R. Soc. Lond.* A **360**, 2005–2017.
Baganoff, F. K. *et al.* 2001 *Nature* **413**, 45.
Bhattarcharjee, A., Ma, Z. W. & Wang, X. 2001 *Phys. Plasmas* **8**, 1829.
Blandford, R. D. 2002 In *Lighthouses of the Universe* (ed. R. Sunyaev *et al.*). Heidelberg: Springer. (In the press.)
Blandford, R. D. & Begelman, M. C. 1999 *Mon. Not. R. Astron. Soc.* **303**, L1.
Brandt, W. N., Alexander, D. M., Bauer, F. E. & Hornschemeier, A. E. 2002 *Phil. Trans. R. Soc. Lond.* A **360**, 2057–2075.
Canizares, C. R. 2002 *Phil. Trans. R. Soc. Lond.* A **360**, 1981–1989.
Cropper, M. 2002 *Phil. Trans. R. Soc. Lond.* A **360**, 1951–1966.
Done, C. 2002 *Phil. Trans. R. Soc. Lond.* A **360**, 1967–1980.
Esin, A. *et al.* 2001 *Astrophys. J.* **555**, 483.
Fabian, A. C. 2002 *Phil. Trans. R. Soc. Lond.* A **360**, 2035–2043.
Gabriel, A. H. & Jordan, C. 1969 *Mon. Not. R. Astron. Soc.* **145**, 241.
Giaconni, R., Gursky, H., Paolini, F. & Rossi, B. 1962 *Phys. Rev. Lett.* **9**, 439.
Güdel, M. 2002 *Phil. Trans. R. Soc. Lond.* A **360**, 1935–1949.
Hasinger, G. 2002 *Phil. Trans. R. Soc. Lond.* A **360**, 2077–2090.
Kahn, S. M., Behar, E., Kinkhabwala, A. & Savin, D. W. 2002 *Phil. Trans. R. Soc. Lond.* A **360**, 1923–1933.
Lee, J. C. *et al.* 2001 *Astrophys. J.* **554**, L13.
Matt, G. 2002 *Phil. Trans. R. Soc. Lond.* A **360**, 2045–2056.
Miller, J. M. *et al.* 2002 *Astrophys. J.* **578**, 348.
Mushotzky, R. 2002 *Phil. Trans. R. Soc. Lond.* A **360**, 2019–2033.
Pounds, K. 2002 *Phil. Trans. R. Soc. Lond.* A **360**, 1905–1921.
Priest, E. R. & Forbes, T. G. 2000 *Magnetic Reconnection.* Cambridge: Cambridge University Press.

Sako, M. *et al.* 2003 *Astrophys. J.* **596**, 114.
Sanders, D. B., Phinney, E. S., Neugebauer, G., Soifer, B. T. & Matthews, K. 1989 *Astrophys. J.* **347**, 29.
Takabe, H. 2001 *Prog. Theor. Phys.* **143**, 202.
Ward, M. 2002 *Phil. Trans. R. Soc. Lond.* A **360**, 1991–2003.
Weisskopf, M. C. *et al.* 2000 *Astrophys. J.* **536**, L81.
Wilms, J. *et al.* 2002 *Mon. Not. R. Astron. Soc.* **328**, L27.
Zhao, J. H., Bower, G. C. & Goss, W. M. 2001 *Astrophys. J.* **547**, 29.

Index

Page numbers in *italic* refer to Figures and Tables.

2341402R00137

Made in the USA
San Bernardino, CA
09 April 2013